BOSTON STUDIES IN THE PHILOSOPHY AND HISTORY OF SCIENCE

Editors

ALISA BOKULICH, *Boston University*
ROBERT S. COHEN, *Boston University*
JÜRGEN RENN, *Max Planck Institute for the History of Science*
KOSTAS GAVROGLU, *University of Athens*

Managing Editor

LINDY DIVARCI, *Max Planck Institute for the History of Science*

Editorial Board

THEODORE ARABATZIS, *University of Athens*
HEATHER E. DOUGLAS, *University of Waterloo*
JEAN GAYON, *Université Paris 1*
THOMAS F. GLICK, *Boston University*
HUBERT GOENNER, *University of Goettingen*
JOHN HEILBRON, *University of California, Berkeley*
DIANA KORMOS-BUCHWALD, *California Institute of Technology*
CHRISTOPH LEHNER, *Max Planck Institute for the History of Science*
PETER McLAUGHLIN, *Universität Heidelberg*
AGUSTÍ NIETO-GALAN, *Universitat Autònoma de Barcelona*
NUCCIO ORDINE, *Universitá della Calabria*
ANA SIMÕES, *Universidade de Lisboa*
JOHN J. STACHEL, *Boston University*
SYLVAN S. SCHWEBER, *Harvard University*
BAICHUN ZHANG, *Chinese Academy of Science*

VOLUME 308

More information about this series at http://www.springer.com/series/5710

Gabriele Lolli • Marco Panza • Giorgio Venturi
Editors

From Logic to Practice

Italian Studies in the Philosophy
of Mathematics

Editors
Gabriele Lolli
Scuola Normale Superiore
Pisa, Italy

Marco Panza
IHPST, CNRS, University of Paris 1
Panthéon-Sorbonne
Paris, France

Giorgio Venturi
FAPESP and Centre for Logic, Epistemology
   and the History of Science
State University of Campinas
São Paulo, Brazil

ISSN 0068-0346   ISSN 2214-7942 (electronic)
ISBN 978-3-319-10433-1   ISBN 978-3-319-10434-8 (eBook)
DOI 10.1007/978-3-319-10434-8
Springer Cham Heidelberg New York Dordrecht London

Library of Congress Control Number: 2014956748

© Springer International Publishing Switzerland 2015
This work is subject to copyright. All rights are reserved by the Publisher, whether the whole or part of the material is concerned, specifically the rights of translation, reprinting, reuse of illustrations, recitation, broadcasting, reproduction on microfilms or in any other physical way, and transmission or information storage and retrieval, electronic adaptation, computer software, or by similar or dissimilar methodology now known or hereafter developed. Exempted from this legal reservation are brief excerpts in connection with reviews or scholarly analysis or material supplied specifically for the purpose of being entered and executed on a computer system, for exclusive use by the purchaser of the work. Duplication of this publication or parts thereof is permitted only under the provisions of the Copyright Law of the Publisher's location, in its current version, and permission for use must always be obtained from Springer. Permissions for use may be obtained through RightsLink at the Copyright Clearance Center. Violations are liable to prosecution under the respective Copyright Law.
The use of general descriptive names, registered names, trademarks, service marks, etc. in this publication does not imply, even in the absence of a specific statement, that such names are exempt from the relevant protective laws and regulations and therefore free for general use.
While the advice and information in this book are believed to be true and accurate at the date of publication, neither the authors nor the editors nor the publisher can accept any legal responsibility for any errors or omissions that may be made. The publisher makes no warranty, express or implied, with respect to the material contained herein.

Printed on acid-free paper

Springer is part of Springer Science+Business Media (www.springer.com)

Introduction

Gabriele Lolli, Marco Panza, and Giorgio Venturi

Integrating Logical, Historical and Philosophical Concerns

When, in the far away 70s of the twentieth century, Reuben Hersh urged for a revival of the philosophy of mathematics (Hersh 1979), two distinct but for many reasons converging lines of research predominated in the discipline. On the one side, attention focused on foundational issues, still connected to the logicism-formalism-intuitionism debate of the first part of the century. On the other side, Benacerraf's papers (Benacerraf 1965, 1973) had, among others, succeeded in grafting onto these issues a growing interest for a fresh version of traditional metaphysical questions concerning mathematical ontology and epistemology. In the following years, the latter line has largely prevailed on the former, which has progressively lost its centrality, or migrated toward logic, conceived as a related, but largely independent discipline. Still, Hersh's plea did not remain unheard.

A number of different studies, like Wilder's address to the 1950 International Congress of Mathematicians devoted to "The Cultural Basis of Mathematics" (Wilder 1950) or Polya's investigation of the heuristics of mathematics (Polya 1945, 1954, 1962), had preceded Lakatos's influential claims of mathematical fallibilism, quasi-empiricism, and dialectical development (Lakatos 1976, 1978). Though essentially polemical in nature, and addressed against a caricature of

G. Lolli (✉)
Scuola Normale Superiore, Pisa, Italy
e-mail: gabriele.lolli@sns.it

M. Panza
IHPST, CNRS, University of Paris 1 Panthéon-Sorbonne, Paris, France
e-mail: marco.panza@univ-paris1.fr

G. Venturi
FAPESP and Centre for Logic, Epistemology and the History of Science,
State University of Campinas, São Paulo, Brazil
e-mail: gio.venturi@gmail.com

formalism that was unrepresented by anyone in the active philosophical arena, these statements highly contributed to emphasize the importance of a philosophical inquiry investigating mathematicians' real work conceived as a collective enterprise, the historical development of mathematics, and the process of mathematical discovery and construction.

Lakatos's guidelines were not followed, in fact. His anti-formalism remained more a polemical slogan than a full-blown philosophical option. His reconstruction of mathematical methodology and forms of progress stayed largely short of any requirement both of a faithful historical analysis and of a fruitful philosophical perspective. His fallibilism and quasi-empiricism resulted in a quite futile negation both of the axiomatic structure of most mathematical theories and of the essential stability of mathematical achievements. His legacy contributed, however, together with other works that came after Hersh's plea, like Wilder's 1981 and Kitcher's 1984 books (Wilder 1981; Kitcher 1984), to set the stage for a philosophical reflection on mathematical practice, focusing on themes that had been underestimated both in the discussion on mathematical ontology and epistemology following Benacerraf's pivotal papers and in the technical development of logic (notwithstanding the proliferation of logical systems).

Tackling many of these themes still required resorting to logical and metamathematical tools, certainly indispensable for dealing both with abstract languages and with computational implementation. Others asked for significant historical researches or for the help of cognitive sciences and experimental psychology.

For example, historical research, aided by mathematical and logical skills, is needed for sustaining naturalistic approaches, such as Maddy's. The idea is that a philosopher of mathematics should "ask questions typically classified as philosophical" (both ontological and epistemological in nature) not "from some special vantage point outside of [...][mathematics itself], but as an active participant, entirely from within" (Maddy 2011, p. 39), and then answer these questions by giving (philosophical) voice to the autonomous methodological decisions of mathematicians. Another brand of naturalism considers, instead, that all that one has do in order to account for mathematics is to look at the cognitive competences and performances displayed in mathematical knowledge, by revealing their being rooted in or even their consisting in, cerebral mechanisms that experimental psychology or cognitive science are to shed light on (Dehaene 1997; Lakoff and Núñez 2000).

It is not necessary, however, to adhere to these forms of naturalism in order to advocate the interest of historical and cognitive inquiries for philosophy of mathematics. It is enough to accept that the latter is, or should be, a study of what mathematics actually is, that is, a human activity taking place in specific determinate circumstances and giving rise to general, abstract results only insofar as generality and abstraction are an outcome of such an activity.

If this is admitted, it becomes quite evident that philosophy of mathematics is concerned with an ever-changing subject matter. Today mathematics, understood as the activity of contemporary mathematicians, is, for example, much more strongly dependent on computational methods and algorithms (often involved in specific applications) or, at least, less pervasively going into axiomatic theories and structures than it was only some decades ago. The use of computers has brought forth, indeed, unpredictable consequences not only in the power of solving problems by tackling them from a computational or, more generally, combinatorial perspective but also in the business of proofs. This requires a fresh reflection on today's mathematical practice and on the role that new computational (and computer assisted) techniques play in it, especially on the very nature of these techniques and on the way their growing importance is changing the mathematical landscape. At the end of the 1970s, the computer-assisted proof of the Four-Color Theorem prompted a lively discussion, calling in questioning the *a priori* nature of mathematical proofs (Tymoczko 1979). The phenomenon is today so extensive as to require much more than the discussion of a single example. It is, perhaps, the very way mathematics is to be conceived that comes to be at issue, here: is it still appropriate to claim that mathematics is not, in essence, an experimental science? Or, more cautiously, how to reconcile the different parts of its multifaceted core?

A large and plural horizon of topics and approaches came thus to the fore. In 2008, Mancosu's collective volume *The Philosophy of Mathematical Practice* (Mancosu 2008) offered a comprehensive survey and a programmatic orientation for these sorts of researches, and the subsequent birth of an *Association for the Philosophy of Mathematical Practice* (http://institucional.us.es/apmp/) further contributed to foster them, though avoiding to contrast them with other approaches, still connected to metaphysical or foundational issues. It seems, indeed, to be a hopeful outcome of these researches that new approach to this issue could be thought of, according to which what they promote is less an inquiry about the ultimate grounds of mathematics, or about the existence of abstract objects, or the possibility of dispensing with them, both in mathematics itself and in its external applications, than a questioning about the way mathematics is structured and mathematical activity is carried on.

Even though the foundations of mathematics and the status of mathematical objects, if there are any, are still important subjects in the philosophical reflection on modern mathematics, it is, then, time for such a reflection to take seriously into account the actual activity of mathematicians and, consequently, the historical dimension of mathematics itself, as it results from this activity (which is properly a human activity taking place in historically determined contexts). This is not the same as rejecting logic as a privileged tool and background for conducting this reflection nor denying that mathematics is presenting crucial metaphysical problems. It is rather a way for using logical methods and techniques and metaphysical analysis to contribute to the obtaining of a more articulated and vivid picture of this activity.

As argued by Mancosu, in the introduction to the collective volume mentioned above, the philosophy of mathematics is in need of a more comprehensive perspective, appealing to different tools and methods. What is at issue, however, is not only the range of topics to be taken into account. It is also, and above all, perhaps, the nature of the mutual relations between philosophy and mathematics that is to be reassessed. That's because what makes philosophy of mathematical practice original is the kind of problems that belong to its agenda.

There are, indeed, genuinely philosophical problems that arise in the day-to-day work of mathematicians or appear when the history of mathematics is submitted to a critical look: problems that, though intimately connected with present or past mathematical activity, can be solved or appropriately stated neither through a purely mathematical treatment nor by a mere historical enquiry. Consider, for example, questions related to the purity of methods or questions about the philosophical meaning of well-established, epochal mathematical results.

On the other hand, there are mathematical questions that, though playing an internal role in the development of a certain field (or of mathematics in general), ask for a global and historically informed approach. Consider, for example, questions related to the extension of an axiomatic system—i.e., the justification of new axioms—or methodological problems that arise in considering "natural" solutions to a problem.

This is the theoretical framework of the present volume. Since it is within this framework, or better to reflect on it and to promote this way of conceiving the philosophy of mathematics, that two of us, Gabriele Lolli and Giorgio Venturi, organized the meeting *Filosofia della matematica: dalla logica alla pratica* [*Philosophy of Mathematics; from Logic to Mathematical Practice*], held at the *Scuola Normale Superiore* of Pisa, on 24–26 September 2012. The essays collected in the present volume constitute the original work of some of the scholars who took part in the conference.

One purpose was and is that of gathering together young Italian scholars in philosophy of mathematics, many of whom are working abroad. This is the reason of the common geographical origin of the authors. Two senior scholars, Paolo Mancosu and Marco Panza, also coming from Italy, but respectively working in USA and France, were also invited to the meeting as speakers. Both of them have been pleased to contribute to the volume, and the latter joined the organizers of the meeting as an editor of this volume.

Another purpose was and is that of making different analytical and conceptual tools—logical, historical, and practical in nature—interact, and to show the extent to which they are complementary in offering a philosophical account of mathematics.

The meeting has shown how active and well rooted within the international community is the youngest part of the Italian scientific community of the philosophy of mathematics. The occasion has been such an interesting opportunity to exchange different points of view and experiences as to lead to the creation of an Italian Network for the Philosophy of Mathematics (FilMat Network: http://filmat-network.com/) that is now active in promoting meetings and debates inside the Italian scientific community, strengthening its connections with the international one.

As one can see, already from the table of contents, the book deals with a wide variety of subjects and themes. Some of the papers have been coauthored by a philosopher and a mathematician or logician, and almost a third of the authors are professional mathematicians.

All papers share a fresh look at the philosophy of mathematics and take into account its recent developments, in agreement to an interdisciplinary attitude toward history, logic, and philosophy, and aim, in the very end, to promote a better understanding of the meaning and scope of a philosophy of mathematical practice. Still, though all are informed by such a spirit, they can be roughly divided in three groups, according to their main purpose and topic. Those in the first group mainly focus on the historical dimension of mathematics. Those in the second make a deeper and more detailed use of logical tools or even reflect on these tools and their capacity of contributing to a better understanding of mathematical activity. Finally, those in the third group are more concerned with questions that are typically ascribed to philosophy and its tradition.

The Historical Dimension of Mathematics

The papers composing the first part of the volume show how a philosophical reflection can arise when the history of mathematics is taken into proper account while investigating the way mathematics is structured (in the descriptions included in the present and in the two next sections, we avoid bibliographical references, since they are provided in each of the papers).

The paper by Pietro Milici, "A geometrical constructive approach to infinitesimal analysis: epistemological potential and boundaries of *tractional motion*," concerns the foundations of infinitesimal analysis. Recent approaches to this topic are essentially algebraic or computational, whereas the origins of infinitesimal analysis were essentially geometrical (though related to a reflection on the role of some algorithms in geometry), and the justification of its methods and results pertained to geometrical grounds. In some cases, as in that of the "inverse tangent problem," mechanical or better instrumental considerations were also at issue. The solutions to this problem involved certain machines, which could construct transcendental curves by appealing to "tractional motion." The main idea of Milici's paper is to come back to this idea from a modern mathematical perspective. Tractional motion is implemented as the motion of a tracing point of some ideal instruments which are formally defined and constitute the basis of a purely geometrical and finitistic axiomatic foundation for a class of differential problems. Although this research requires further development, the role of these instruments in the foundation of both computation and mathematics is deeply rooted in the history of mathematics. Think of the early-modern reinterpretation of Euclid's arguments as based on constructions by ruler and compass or of Descartes's extension of them, relying on a class of special "compasses," or, again, much more recently, of the Turing machine. Milici's paper reappraises this tradition, by suggesting, in the meantime, a new way of looking at the integral calculus.

The paper by Paolo Mancosu and Andrew Arana, "Plane and solid geometry: a note on purity of methods," deals with a methodological principle which is often at work in mathematics. Spelling out this principle would require defining purity, but this is far from simple. On a first approximation, one might say that a proof of a theorem in mathematics is pure if the conceptual tools used in the proof are already involved in the content of the theorem itself. So (broadly) conceived, this notion has played an important role in the history of mathematics—consider, for instance, the elimination of geometrical intuition from the development of analysis in the nineteenth century—and, in a way, it underlies all the investigations pertaining to conservativeness, in contemporary proof theory. A clear example of how much purity is cherished in mathematical practice is offered by the fact that Erdös and Selberg were awarded the Fields Medal for their elementary proof of the prime number theorem, which had already been demonstrated with analytical tools in the late nineteenth century. But why do mathematicians cherish purity? What is epistemologically to be gained by proofs that exclude appeal to "ideal" elements? In this note (which is based on a much longer article published in *The Review of Symbolic Logic*), Mancosu and Arana discuss an important case of essential use of three-dimensional geometry in proving results about planar geometry, namely, Desargues' plane theorem on homological triangles in projective geometry, with the aim of articulating what notion of mathematical content might best be suited for an analysis of ascriptions of purity in mathematical practice.

In "Formalization and intuition in Husserl's *Raumbuch*," Edoardo Caracciolo peruses Husserl's work on geometry that should have led to a treatment of Euclidean geometry, in the planned but unpublished second volume of the *Philosophie der Arithmetik*. Husserl started outlining a formal method that should have allowed a pure understanding of any spatial manifold; when he discovered that the analogue work of Bernhard Riemann distorted the essential features of space, inverting the precedence between flatness and curvity, he opted for dealing with spatial representation from a psychological point of view, trying to detect the intuitions grounding geometrical concepts. However, he abandoned this way when he perceived that it leads to formulating material concepts that do not match with the formal structure of pure geometry. The relation between representation, intuition, and symbolization is at the core of these early reflections of Husserl on geometry.

Looking at Mathematics Through Logic

The second part of the volume groups together contributions where a more technical use of logical tools gives the opportunity to present a detailed analysis of philosophical concepts or where these very tools are brought under scrutiny for their capacity to contribute to a philosophical account of mathematics.

The paper by Francesca Boccuni, "Frege's *Grundgesetze* and a reassessment of predicativity," focuses on the philosophical issues connected with the consistent fragments of the arithmetical system presented in Frege's *Grundgesetze der Arithmetik*, especially with respect to the predicative restrictions placed on the underlying second-order comprehension principle. Though the main aim of these consistency results may be technical, one may wonder about their possible foundational applications. On the one side, one can prove that the strongest consistent predicative fragment of Frege's system is equiconsistent with Robinson's Arithmetic **Q**, which, though not at all mathematically trivial, is still a weak system, especially if compared with Frege's original goal. On the other side, Boccuni argues that the predicative restriction imposed on the principle of comprehension leads to a radical revision also of Frege's philosophical stance toward the existence of concepts as logical entities, thus affecting both Frege's platonism and logicism all at once. She also maintains that, in order to justify Frege's platonism from a predicative perspective, a reassessment of Gödel's dichotomy between impredicativity and predicativity is required, in particular by focusing on Gödel's objections to Russell's Vicious Circle Principle (VCP). In order to achieve such a reassessment, Boccuni investigates Gödel's argument against VCP, argues against it, and suggests a different formulation of it, based on the Thesis of Arbitrary Reference, advanced by Enrico Martino, to the effect that VCP turns out to be compatible with Frege's platonism, whereas his logicism needs to be revised.

Mario Piazza and Gabriele Pulcini, in "A deflationary account of the truth of the Gödel sentence $\mathcal{G}$," address a topic discussed in recent years by, among others, Tennant, Ketland, and Shapiro. They give a negative answer to the question of whether our conviction about the truth of the Gödel sentence $\mathcal{G}$ involves a theory of truth beyond the deflationary theories. After discussing and dismissing Neil Tennant's deflationary account of incompleteness, based on the application of the so-called Local Reflection Principle, they show how a new deflationary construal of the incompletability of formal systems can be framed in the setting of Peano Arithmetic, so augmented as to include a constructive version of the ω-rule, based on the notion of prototype proof. The term "prototype," following Michael Detlefsen, has a meaning that dates back to Jacques Herbrand: "when we say that a theorem is true for all x, we mean that for each x individually it is possible to iterate its proof, which may just be considered a prototype of each individual proof."

The paper by Paolo Pistone, "Rule-following and the limits of formalization: Wittgenstein's considerations through the lens of logic," addresses the question of the justification of logical rules. In the *Tractatus* it is stated that questions about logical formatting (why just these rules?) cannot even be meaningfully formulated, since it is just the application of logical rules which enables the formulation of a question whatsoever; analogously, Wittgenstein's celebrated infinite regress argument on rule-following (it takes rules to justify rules) seems to undermine any explanation of deduction that relies on logical rules. On the other hand, important logical achievements, such as incompleteness and computational complexity, seem to expose a similar issue on formalization: does this or that proof belong to a logically correct system? Can we effectively know it? These logically motivated doubts do not concern indeed the truth of single propositions, but rather, as for Kant's celebrated *quid iuris* argument, the legitimacy of a given system of rules. By exploiting a dynamical perspective on logic (inspired by Girard's "transcendental syntax" program), in which rules are not imposed *a priori* on proofs but are rather reconstructed through the symmetries describing the interaction between proofs, such a "subjective" side of logic (fundamentally, the way we write its rules) can be unearthed: this viewpoint, made possible by the Curry-Howard bridge between proofs and programs, seems to provide technical ground (and a possible answer) to the philosophical matters on rule-following, as well as an interesting logical perspective on computational complexity.

The paper by Luca Tranchini, "Paradox and inconsistency: revising Tennant's distinction through Schroeder-Heister's assumption rules," deals with an old but still discussed problem in proof theory: Tennant's distinction, going back to Prawitz, between paradox and inconsistency. In 1982, Tennant proposed a proof-theoretic criterion of paradoxicality: a derivation of absurdity in a natural deduction system is paradoxical whenever any reduction sequence starting from this derivation eventually loops. Paradoxes are expressions governed by particular inference rules that trigger paradoxical derivations. Derivations of absurdity that do normalize are taken by Tennant as showing that the assumptions on which they rely are inconsistent. Tranchini proposes two examples that show that Tennant's formulation of the distinction is problematic. Then he precisifies the issue in terms of the extension of natural deduction proposed by Schroeder-Heister. In this setting, the notion of assumption is enriched so that rules are admitted as a special kind of assumptions alongside with sentences. According to whether the derivation is normalizable or not, the assumptions involved in the derivations will be said to be either inconsistent or paradoxical, independently of their being sentences or rules. Tranchini's analysis hints at a connection between Ekman's paradox and the relation between implications and rules and is an improvement of Tennant's analysis of paradoxes in proof-theoretic terms.

The paper by Alberto Naibo, "Costructibility and geometry," is devoted to the study of the constructive aspects of Euclid's geometry, from a formal and logical point of view. Namely, it investigates whether the intrinsic constructive nature of Euclid's geometry is captured by the standard properties of constructive logical theories, in particular the witness property conceived as the result of some deterministic computation. Starting from the analysis of concrete examples, a first negative answer is offered. This allows identifying the specific property that makes a constructive theory geometric. By focusing on deductive aspects, Naibo shows that, in such a theory, it is much more natural to work with open proofs (i.e., proofs under open assumptions), rather than with usual closed ones. But open proofs correspond to programs that do not necessarily return a value, namely, procedures. Hence, Naibo argues that one of the essential features of constructive geometry depends on the set of instructions that have to be carried out in order to perform a construction, rather than on the final result of the construction itself. In this sense, an ontological shift is pointed out: contrary to other mathematical theories, for example, arithmetic, constructive geometry can be characterized according to the actions it allows, rather than to the objects which it is about.

The paper by Michael Arndt and Laura Tesconi, "A cut-like inference in a framework of explicit composition for various calculi of natural deduction," deals with composition of proofs and the effects of cut-like inferences in various calculi of natural deduction and throws new light on the phenomenon of composition. An explicit concatenation rule is proposed, obtained by generalizations and formalization of one of the most intuitive principle of abstract reasoning, which governs the composition of abstract derivations from the left and from the right at the same time, via the mediation of control clauses that occur in the position of the major premise. Control clauses are so called because they exercise control over the manner in which the composition takes place. The sets of control clauses necessary to express various calculi of natural deduction (standard natural deduction, natural deduction with general elimination rules, bioriented natural deduction and their variants) are considered, together with the effects of its addition to these calculi. While the addition of this rule to a single-oriented control base, regardless of its directionality, does not open more possibilities of combination of abstract derivations, its addition to a bioriented control base does. In this latter case, when a formula is introduced among the assumptions by means of a certain rule, it appears as a leaf in the derivation tree, but it is not available before the application of the rule itself. Thus, it is not always possible that it be, at the same time and in that very same point of the derivation, introduced as a conclusion by some other inference. In other words, not all formulae of a derivation can be considered to be joining knots of different fragments of derivations.

In his contribution, "On the distinction between sets and classes: a categorical perspective," Samuele Maschio inquires about the relationship between sets and classes in ZF set theory by means of syntactic categories, internal categories, and algebraic set theory. He tries to clarify the way proper classes are used in mathematical practice, by means of categorical tools allowing a useful interaction between mathematics and metamathematics. Maschio claims that the peculiar nature of classes, with respect to sets, can be represented using category theory. First, he defines the category of formal classes as a full (equivalent) subcategory of the syntactic category of the theory ZF; second, using the global elements of an internal category to the category of formal classes, he manages to identify, by means of algebraic set theory, a category equivalent to the one of definable sets.

Philosophy and Mathematics

Finally the papers included in the third part of the volume deal with more classical philosophical problems that pertain to mathematics from a theoretical, practical or linguistic point of view.

The paper by Michele Ginammi, "Structure and applicability," deals with the general problem of the applicability of formal notions to physical reality and, in particular, with the issue of the effective representative power of mathematics with respect to physics. It focuses on the notion of structure and wonders whether this notions could be appealed to in order to explain mathematical effectiveness. To this purpose, the author first considers the so-called structural account, then analyzes its weaknesses, and finally suggests a way to overcome them. What is suggested is that the representative power of mathematics is strictly related to what Mark Steiner called "heuristic applicability of mathematics" in his 1998 book, that is, the capability of an appropriate mathematical representation of reality of helping in discovering new physical facts. According to Ginammi, this provides not only an improvement of the structural account, suitable for settling its difficulties, but also a clearer account of the applicability of mathematics to physics, underplaying the often mentioned "unreasonable effectiveness of mathematics."

The paper by Marina Imocrante, "Defending Maddy's mathematical naturalism from Roland's criticism: the role of mathematical depth," takes as a starting point the criticism that Jeffrey Roland moves to Penelope Maddy's naturalistic epistemology of mathematics and draws the attention to the notion of mathematical depth proposed by Penelope Maddy in her 2011 book *Defending the Axioms*. After recalling the main features of Maddy's mathematical naturalism and Roland's objections to her account, a considerable portion of the paper is devoted to an analysis of Maddy's notion of mathematical depth. Although stressing the need of certain clarifications of Maddy's notion, Imocrante suggests a possible interpretation of the facts of mathematical depth as the historical facts of mathematical practice. The peculiar role of these facts in Maddy's account is what allows Imocrante to reply to Roland's objections.

The paper by Marco Panza and Andrea Sereni, "On the indispensable premises of the Indispensability Argument," questions the often admitted dependence of the indispensability argument on confirmational holism and naturalism. The argument is generally ascribed to Quine, though its first codified version is offered by Putnam. This is, however, only a version, among others, since the term "indispensability argument" should better be intended as referring to a family of arguments sharing a common idea: that admitting the indispensability of appealing to mathematical theories in sciences, especially in physical and natural ones, must depend on ascribing some epistemic features to these theories, depending on the features ascribed to the relevant scientific theories. To make a simple example: if a scientific theory is taken to be true, and it is admitted that a certain mathematical theory is indispensable to the former, then also the latter should be taken to be true. A quite widespread opinion is that endorsing an indispensability argument requires endorsing confirmational holism (concerning the relevant scientific theories) and naturalism. Panza and Sereni's paper questions this opinion, by suggesting four basic schemas for as many subfamilies of indispensability arguments that require no premise involving any form of confirmational holism and naturalism. The authors do not intend to endorse these arguments, but to show that their weakness, if any, must depend on other, more specific assumptions, especially concerned with the notion of indispensability itself.

The paper by Luca San Mauro and Giorgio Venturi, "Naturalness in mathematics: on the statical-dynamical opposition," contains a philosophical analysis of the notion of naturalness, as it is used in the mathematical discourse. The two authors show, with the help of statistical evidence, how frequent this use is. The first part of the paper focuses on methodological issues. It investigates how a vague notion such as that of naturalness should be analyzed in mathematical contexts. San Mauro and Venturi propose a third way between an uncritical naturalism and a philosophy centered approach to mathematics. They aim to capture the autonomy of mathematical work while accounting for its truly philosophical aspects. The second part of the paper is devoted to some case studies taken from the history of set theory and computability theory, which are meant to elucidate the meaning of naturalness's ascriptions in mathematics. The main question is whether the notion of naturalness should be intended as a static or as a dynamic notion. San Mauro and Venturi favor the latter option and argue that the evidence for the former is only apparent, but still deserves a further philosophical inquiry, since it is just the apparent tension between a static and a dynamical facet of naturalness that justifies most naturalness ascriptions in mathematics.

Silvia de Toffoli and Valeria Giardino, in their paper "An inquiry into the practice of proving in low-dimensional topology," address the issue of visualization in the mathematical practice of low-dimensional topology. The authors try to clarify the epistemic value of the use of pictures in the justification of topological results. They hold that, in general, representation plays an important role in the inferential arguments of a working mathematician and that in low-dimension topology the use of pictures gains strength by their link with spatial intuition, insofar as this is connected with a faculty of "manipulative imagination," which makes mathematicians able to connect different pictures involved in a single argument. This is, however, workable only within a specific practice; hence this form of reasoning is context dependent. For this reason De Toffoli and Giardino give local criteria of validity in order to insure the soundness of arguments that in low-dimensional topology make use of forms of representation.

Concluding Remarks

Though each of these papers depends on the particular views, attitudes, interests, and competences of its authors, when taken as a whole, they illustrate the multifarious picture of the present state of philosophy of mathematics we have outlined in the opening section of the present introduction.

In particular, it seems to us that a common orientation toward a philosophy of mathematical practice is emerging, according to which this is understood not as a rigid doctrine but as a large horizon of topics and as an interdisciplinary approach.

It remains that, even on this understanding, such an orientation is still hardly identifiable as such. The danger is still there, when one refers to it, of merely gathering under a single, evocative name a mere sum, or even a mixture, of different approaches coming from general philosophy, history, logic, computer science, sociology, cognitive science, and other disciplines, including mathematics itself. The way these approaches can be held together is, indeed, an open question that the recent evolutions of a philosophy of mathematical practice are far from having settled.

In particular, it is our common opinion that a mature philosophy of mathematical practice cannot avoid coming to grips with the same traditional problems that the philosophy of mathematics has dealt with for centuries, like that of the foundation of mathematics (variously conceived) or that of the ontological and/or epistemological status of its subject matter (if there is a specific one, indeed). Being able to suggest an answer to these problems or, more soberly, a way to look at them different from those prevailing in the last decades is a crucial challenge that this orientation must face, if it is to develop into a distinctive approach. Another challenge is that of providing a unified stance for the different researches that are developing under the label "philosophy of mathematical practice," in which this new answer or way of looking at these problems can harmoniously coexist with more specific contents and inquiries.

Our volume can have the pretension neither to meet these challenges nor to dwell with them in their entire complexity. But we hope it could be, at least, a testimony of the work of a rich community of Italian scholars (many of whom are still young) working (often outside Italy, for choice or unfortunate necessity) within the philosophical framework in which those problems arise.

References

Benacerraf, Paul. 1965. What numbers could not be. *Philosophical Review* 74: 47–73.
Benacerraf, Paul. 1973. Mathematical truth. *Journal of Philosophy* 70: 661–679.
Dehaene, Stanislas. 1997. *The number sense*. New York: Oxford University Press.
Hersh, Reuben. 1979. Some proposals for reviving the philosophy of mathematics. *Advances in Mathematics* 31: 31–50.
Kitcher, Philip. 1984. *The nature of mathematical knowledge*. Oxford: Oxford University Press.
Lakatos, Imre. 1976. *Proofs and refutations*. Cambridge: Cambridge University Press.
Lakatos, Imre. 1978. *Mathematics, science and epistemology*. Cambridge: Cambridge University Press. Volume 2 of I. Lakatos, *Philosophical Papers*, Cambridge University Press, Cambridge, 1978 (2 vols).
Lakoff, George, and Núñez, Rafael. 2000. *Where mathematics comes from: How the embodied mind brings mathematics into being*. New York: Basic Books.
Maddy, Penelope. 2011. *Defending the axioms*. Oxford: Oxford University Press.
Mancosu, Paolo. (ed.). 2008. *The philosophy of mathematical practice*. Oxford/New York: Oxford University Press.
Polya, George. 1945. *How to solve it*. Princeton: Princeton University Press. 2nd ed., Garden City, NY: Doubleday Anchor Books.
Polya, George. 1954. *Mathematics and plausible reasoning*, 2 vols. Princeton: Princeton University Press.
Polya, George. 1962. *Mathematical discovery: On understanding, learning and teaching problem solving*. New York: Wiley.
Tymoczko, Thomas. 1979. The four-color problem and its philosophical significance. *The Journal of Philosophy* 76: 57–83.
Wilder, Raymond Louis. 1950. Cultural basis of mathematics, online at http://www-history.mcs.st-andrews.ac.uk/Extras/Cultural_Basis_I.html.
Wilder, Raymond Louis. 1981. *Mathematics as a cultural system*. Oxford: Pergamon Press.

Contents

Part I The Historical Dimension of Mathematics

1. **A Geometrical Constructive Approach to Infinitesimal Analysis: Epistemological Potential and Boundaries of Tractional Motion** 3
   Pietro Milici

2. **Plane and Solid Geometry: A Note on Purity of Methods** 23
   Paolo Mancosu and Andrew Arana

3. **Formalization and Intuition in Husserl's *Raumbuch*** 33
   Edoardo Caracciolo

Part II Looking at Mathematics Through Logic

4. **Frege's *Grundgesetze* and a Reassessment of Predicativity** 53
   Francesca Boccuni

5. **A Deflationary Account of the Truth of the Gödel Sentence $\mathcal{G}$** 71
   Mario Piazza and Gabriele Pulcini

6. **Rule-Following and the Limits of Formalization: Wittgenstein's Considerations Through the Lens of Logic** 91
   Paolo Pistone

7. **Paradox and Inconsistency: Revising Tennant's Distinction Through Schroeder-Heister's Assumption Rules** 111
   Luca Tranchini

8. **Constructibility and Geometry** 123
   Alberto Naibo

9	**A Cut-Like Inference in a Framework of Explicit Composition for Various Calculi of Natural Deduction** .... 163 Michael Arndt and Laura Tesconi	
10	**On the Distinction Between Sets and Classes: A Categorical Perspective** .... 185 Samuele Maschio	

Part III Philosophy and Mathematics

11	**Structure and Applicability** .... 203 Michele Ginammi	
12	**Defending Maddy's Mathematical Naturalism from Roland's Criticisms: The Role of Mathematical Depth** .... 223 Marina Imocrante	
13	**On the Indispensable Premises of the Indispensability Argument** .... 241 Marco Panza and Andrea Sereni	
14	**Naturalness in Mathematics** .... 277 Luca San Mauro and Giorgio Venturi	
15	**An Inquiry into the Practice of Proving in Low-Dimensional Topology** .... 315 Silvia De Toffoli and Valeria Giardino	

Short Presentations of the Editors

Gabriele Lolli is Professor of Philosophy of Mathematics at the Scuola Normale Superiore of Pisa after a career in Mathematical Logic. His main present interests concern the origin of set theory, mathematical proof, and the relations between mathematics and literature. Among his recent publications are *QED. Fenomenologia della dimostrazione*, Bollati Boringhieri, Turin, 2005; "Experimental methods in proofs," in R. Lupacchini, G. Corsi (eds.), *Deduction, Computation, Experiment*, Springer, 2008; *Nascita di un'idea matematica*, Edizioni della Normale, Pisa, 2013; "Mathematics according to Calvino," in M. Emmer (ed.), *Imagine Math2*, Springer, 2013.

Marco Panza is Research Director at the CNRS and member of the IHPST (CNRS and Univ. of Paris 1 Panthéon-Sorbonne). He is the author of several books and papers and the editor of several collective volumes in the domains of the history and philosophy of mathematics. For the Boston Studies in the Philosophy of Sciences, he edited, with M. Otte, *Analysis and synthesis in Mathematics. History and Philosophy* (1997). His books include *Newton et les origines de l'analyse, 1664–1666*, Blanchard, Paris, 2005, and *Plato's Problem. Introduction to Mathematical Platoinism*, Palgrave Mac Millan, Bsingstoke (UK), 2013 (coauthored with Andrea Sereni).

Giorgio Venturi is currently a postdoctoral FAPESP fellow at CLE (Campinas) and former PhD student, both in philosophy at Scuola Normale Superiore of Pisa under the supervision of Gabriele Lolli and in mathematics at the Université Paris Diderot under the supervision of Boban Velickovic. His main research field is set theory, both from a purely mathematical perspective—he coauthored Velickovic *Proper forcing remastered*, which appeared in *Appalachian set theory* (Cambridge University Press)—and from a philosophical point of view, as in *Foundation of mathematics between theory and practice*, which appeared in Philosophia scientae 18(1), 2014.

Part I
The Historical Dimension of Mathematics

Chapter 1
A Geometrical Constructive Approach to Infinitesimal Analysis: Epistemological Potential and Boundaries of Tractional Motion

Pietro Milici

1.1 Introduction

The role of infinity in modern mathematics is fundamental, but historically (since Zeno), its use has implied paradoxes. Therefore, it is quite natural to wonder when its introduction is really unavoidable and when it is possible to answer the same questions by means of other (more reachable) tools. In particular, in this chapter, we explore the foundation of (part of) Infinitesimal Analysis without infinity: from an ancient perspective, we want to consider Infinitesimal Analysis objects as generated by the motion of geometrical/mechanical ideal machines (such as the Euclidean straightedge and compass[1]), an approach adopted by Descartes to *legitimize* his algebraic curves (by tracing them with the continuous motion of appropriate geometric machines). Although some curves beyond circles and lines have been introduced since the classical period (mainly conic sections, the conchoid of Nicomedes, the cissoid of Diocles, the Archimedean spiral, and the quadratrix of Dinostratus), it was Descartes who suggested a widely accepted finer and general classification of curves, dividing geometrical from mechanical ones

---

[1]Even though Euclid's works never introduced the straightedge and compass, his axioms for planar geometry include the idea that we can draw a circle of any known radius at any known point and that we can extend any line indefinitely. These axioms, purely mathematical in nature, can also be interpreted physically by saying that the geometer has access to a compass and a straightedge, tools that were used for many purposes even before Euclid.

P. Milici (✉)
Department of Mathematics and Computer Science, University of Palermo, Palermo, Italy

UMR 8590 IHPST – Institut d'Histoire et de Philosophie des Sciences et des Techniques, Université Paris 1 Panthéon-Sorbonne, Paris, France
e-mail: p.milici@gmail.com

and considering only the first ones to satisfy the early modern metamathematical problem of geometrical *exactness* (algebra, as an analytic tool, introduced new foundational problems concerning the acceptability of solutions and constructions in the geometric paradigm) (Bos 2001). However, up to the second half of the seventeenth century, mathematicians such as Leibniz and Huygens tried to legitimize curves beyond those considered *geometric* by Descartes.

This latter attempt raised the following questions:

Q1- Is it possible to develop Cartesian machines to overtake algebra and deal with (part of) Infinitesimal Analysis? (*foundation*)

Q2- Once the allowed tools are defined, what is the class of generable functions? (*construction*)

Q3- Can this perspective offer a different (more sensible) visualization for functions (especially complex ones)? (*visualization*)

Q4- Can these tools, once actually realized, foster users' acquisition of the embodied mathematical contents? (*didactics*)

Q5- Can the behavior of ideal tools be introduced in an axiomatic theory like Euclidean geometry, so as to treat Infinitesimal Analysis from a finitistic (non-numerical) perspective? (*synthetic geometry perspective*)

The purpose of this chapter is not to answer the previous questions completely, but to ask them and perform a preliminary investigation.

In particular, while recent foundational approaches to Infinitesimal Analysis (the classical foundation or Robinson's Non-standard Analysis, the finitistic approach of Kronecker and Brouwer's intuitionism, or the Computational Analysis of Turing and Weihrauch) are essentially algebraic or computational, one of the first approaches to these problems was geometrical [Q1]. From this perspective, we may recall the seventeenth-century consideration of the "inverse tangent problem." Solutions to this problem introduced certain machines, intended as both theoretical and actual instruments, that justified the existence of certain solutions (i.e., curves) exceeding Cartesian geometry. The first documented curves constructed under tangent conditions were physically realized by the traction of a string tied to a load, which is why the study of these machines was named "tractional motion" (Bos 1988).

The main idea of this work is to deepen and further develop the analysis of tractional motion, investigating if and how these *ideal* machines (like the ancient straightedge and compass) can constitute the basis of a purely geometrical and finitistic axiomatic foundation (like Euclid's planar geometry) for a class of differential problems [Q5] (note that no axiomatization has yet been proposed). In particular, a model of these machines (i.e., the suggested components) is presented, followed by some preliminary results about the class of generable functions [Q2]. As an example of the interpretation of functions with these tools [Q3], we propose a sketch of a "tractional" planar machine embodying the complex exponential function. Finally, concerning the possible relation between these ideas and their didactical adoption through the use of concrete artifacts [Q4], we also make a suitable didactic proposal.

1.2 Brief History of Tractional Motion

In formalist mathematics, there exists every object that does not generate contradictions. This perspective is somewhat different from the Cartesian vision; in fact, the *Géométrie* in the appendix of Descartes (1637) (for a translation see Descartes 1954) *legitimized* new curves if they could be drawn by kinematic mechanical devices involving movable arms (algebraic curves).[2] In the same century, before the Infinitesimal Analysis paradigm became dominant, some geometrical ideas were developed concerning curves defined by their tangent properties.[3] Mathematicians like Huygens began to consider instruments that, like the handlebars of a bike, could guide the tangent of a curve (in analytical mechanics terms, they introduced "non-holonomic" constraints): this signaled the rise of tractional motion. Tractional motion suggested the possibility of constructing curves by imposing tangential conditions, generalizing (in a non-Cartesian way) the idea of geometrical objects, and constructing with new tools not only algebraic curves, but also some transcendental ones (seen as solutions of differential equations). In this period, the development of geometrical ideas often corresponded to the practical construction (or at least conception) of mechanical machines able to embody the theoretical properties and thus physically draw the curves.

In the early modern period, the use of algebra as an analytic tool for the solution of geometric problems posed new foundational problems, for example, what does it mean for a problem to be solved, and which constructions are legitimate? If these questions about *geometrical exactness* were so important for the early modern period, they disappeared in the eighteenth century because of the general affirmation of symbolic procedures, later considered autonomous from geometry. Hence, from our point of view, tractional motion missed the chance for its foundational potential to completely develop into a geometrical/mechanical theory. Furthermore, because of the change in paradigm, these tractional concepts were forgotten for centuries, even for practical purposes, and were independently reinvented in the late nineteenth century, when they were used to build some grapho-mechanical instruments of integration (integraphs) to analogically compute symbolically non-solvable problems. But let us start from the first curve described in a tractional way, the "tractrix."

On a horizontal plane, consider a small heavy body (subjected to the friction on the plane) tied with an ideally weightless nonelastic string, and imagine (slowly) pulling the other end of the string along a straight line drawn on the plane. Because

---

[2] In general, not every Descartes scholar would agree about this point. In fact, Descartes never solved the problem of classifying the admissible curves in an unambiguous and complete way. However, in accordance with (Bos 2001; Panza 2011), we are adopting the characterization of Descartes' geometrical curves as those that can be traced by *geometrical linkages*, that is, articulated devices basically working as joint systems, allowing a certain degree of freedom in movements between the two links they connect.

[3] This historical reconstruction is essentially based on Bos (1988, 1989) and Tournès (2007, 2009).

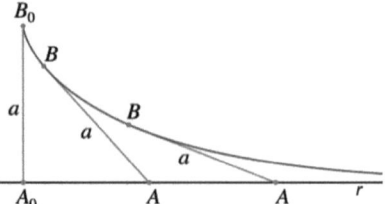

Fig. 1.1 The heavy body is B, with initial position B_0; the string is a; and the other end of the string is A, with initial position A_0. Moving A along r, B describes the tractrix (obviously, the movement is not reversible because of the non-rigidity of the string). Note how a is tangent to the curve at every point

of the friction on the plane, the body offers resistance to the pulling of the string: if the motion is slow enough to neglect inertia, the curve described by the body is called a *tractrix*. The first documented description of the *tractrix* is associated with Claude Perrault (Leibniz 1693). Examining Fig. 1.1, we can see how the curve is traced thanks to the property that the string is constantly tangent to the curve.

Christiaan Huygens (1693) deepened the theory of tractional motion and moved toward a mechanical description in order to physically build some precise instruments for tracing. In fact, the original description of the tractrix is related to at least two physical problems: the tracing plane has to be perfectly horizontal, and the heavy body, when moved, acquires inertial velocity. Huygens suggested that, abstracting the problem from its physical complexity and considering it solely in terms of tangent properties, tractional motion can be seen as a *pure geometrical movement*, independent from the motion speed (even if through the introduction of friction). This is exactly the same as the circular motion of the compass, the straight motion of the ruler, and, in general, the continuous movement considered by Descartes as the basis of analytic geometry. Furthermore, Huygens introduced a technological change in the way in which straight segments were considered: a string only works in the case of traction; on the contrary, a physical rigid bar satisfies the tangent constraint (avoiding lateral motion) not only in traction, but even in compression, making the curve realization reversible.

The foundations of tractional motion were laid, and, up to the first half of the eighteenth century, there was an improvement in related works, both in terms of practical machines (mechanical devices studied and realized to solve particular differential equations) and theoretical studies. Concerning practical machines, we may recall those made by Perks (1706, 1714) (see Fig. 1.2) for the first physical introduction of the "rolling wheel" to manage the tangent (the same solution was adopted in the nineteenth century for integraphs), whereas, concerning theoretical evolutions, we must recollect Leibniz's "universal tractional machine" (Leibniz 1693). According to him, tractional motion was the concrete realization of his vision of curves as *infinitangular* polygons, inextricably mixing the theoretical model with physical execution, each one validating the other: from a certain point of view, kinematics forms the basis that mathematics without well-defined infinitesimal

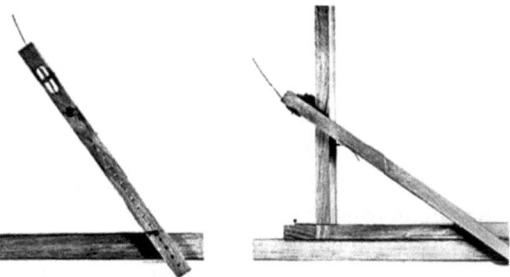

Fig. 1.2 Reconstruction of Perks' instruments for the tractrix (*left*) and for the logarithmic curve (*right*) at the Institute for History of Science, Aarhus University. Regarding the tractrix, we can see the wheel taking the place of the load: in this case, the extreme point of the fixed-length bar can freely move along a straight line. In the machine for the logarithmic curve, a horizontal fixed-length plank moves along another horizontal bar to opportunely incline the slope

entities requires. Due to its complexity, the project, so important to a single theory able to realize the quadrature of any general curve with a continuous movement, never became a real device.

A unified theory for differential equations was actually developed by Vincenzo Riccati (1752), the only complete theoretical work ever dedicated to the use of tractional motion in geometry (Tournès 2009). The Italian mathematician found geometrical proofs corresponding to those that mathematicians such as Euler derived using series, arriving at the result that "every" curve (defined in modern terminology by a differential equation $y' = f(x, y)$) can be drawn with tractional motion.[4] This result regarding transcendental curves overtook Descartes' announcement in relation to algebraic curves and developed the theory of geometric construction with simple continuous movements. One characteristic of this work is the deep interaction among algebra, geometry, mechanics, and technology to develop an abstract unified theory of differential equations based on the conception of material instruments physically drawing the integral curves. His instruments plot the integral curve of a differential equation using tractional motion:

> On a horizontal plane, one pulls one end of a tense string, or a rigid rod, along a given curve, and the other end of the string, the free end, describes during the motion a new curve which remains constantly tangent to the string. At this free end, one places a pen surmounted by a weight making pressure, or a sharp edged wheel cutting the paper, so that any lateral motion is neutralized. By suitably choosing the base curve along which the end of the string is dragged, and by suitably varying the length of the string according to a given law, one can integrate various types of differential equations. In this way of solving an inverse tangent problem, one actually materializes the tangent by a tense string and moves the string so that

---

[4]Riccati shows that, adopting modern terminology, it is possible to integrate any differential equation $y' = f(x, y)$, but he does not explicitly specify anything about the set of admissible functions f. According to the equations of the time, it is reasonable to assume that the function has to be obtained using only a finite number of algebraic operations and quadratures.

the given property of the tangents is verified at every moment. The length of the tangent is controlled at every moment by a mechanical system (a pulley or a slide channel) and by a second curve which is called the directrix of the motion. (Tournès 2004, p. 2738)

Denoting the curves traced by tractional motion as "tractorias," Riccati's work scheme was as follows: beginning from tractorias with a constant tangent (described with a constant-length string dragged along a base curve), he generalizes by allowing the integration of more and more extended classes of differential equations. The generalization supposes that the length of the string varies with the position of the tractor point (a tractoria with a variable tangent) and finally controls the length of the string by a variable directrix, whose form varies according to the position of the tractor point.

By means of such tractorias, Riccati showed that tractional motion allows the integration of any differential equation having two independent variables x and y in which the coefficients of the infinitesimal elements dx and dy are obtained using only a finite number of algebraic operations and quadratures (cf. Note 4). Under these conditions, all the auxiliary curves used by Riccati (base curves and directrix curves) are constructible by Cartesian geometry instruments extended with the possibility of quadrature. Tractional motion is therefore an additional process of construction that allows us to obtain new curves from previously known ones.

Even though this work overtakes the ancient current of geometrical problem solving by the construction of curves, and proposes a very general theoretical model to explain in a unified way the operation of a great number of tractional instruments, it was neither celebrated nor influential. The book probably arrived too late, at the end of the period of curve construction. At this time, geometry was giving way to algebra, and series were becoming the principal tool to represent solutions to differential equations, making Riccati's work almost immediately outdated (Tournès 2004).

After about 150 years of interruption, during which there was no trace of tractional motion, engineers in the late nineteenth and early twentieth centuries rediscovered, independently, the same theoretical principles and technical solutions of the eighteenth century (Pascal 1914). These solutions involved even more complicated tractional instruments, with more cutting wheels connected between them to be able to integrate differential equations beyond the first order. But after this new blooming, in the second half of the twentieth century, the technological thirst for equation-solving machines was satisfied by digital computers, which improved the performance of analog computers with very accurate error control. Therefore, because of the technological digital revolution, there was another break in the adoption of this perspective.

1.3 Tractional Motion Machines

The first goal of this study on tractional motion is to clearly define the components that can be used to obtain devices that implement certain tangent properties on a plane. We call these ideal devices "tractional motion machines" (or TMMs) (Milici 2012). As tractional motion was historically used to solve differential equations by tracing curves on the plane, we suggest TMMs should be interpreted as *function embodiers* (as we will show in Sect. 1.4, this latter interpretation allows us to make a comparison with other analog computers and to extend the results to complex values). The passage from *curve tracer* to *function embodier* is similar to the passage from curves as equations (zero loci) to parametric curves (defined as functions of an independent input). In fact, using machines, a geometric curve (interpreted as the zero locus of an equation) can be considered as the set of all planar points reachable by tracing the point of an appropriate articulated device, without the need to identify on the machine a point that generates the motion (analytically, without the need to define an independent parameter). On the contrary, to interpret a TMM as a function, we must define at once an *input* and an *output* point on the machine, so that any move in the former causes a specific move in the latter: in this way, the machine can be considered as an *operator* that directly embodies a function.

1.3.1 Components of Tractional Motion Machines

In this section, we define the mechanical components that are allowed in our interpretation of tractional motion: we adopt these because they seem to give a good compromise between the simplicity of the components (two instruments and two constraints) and that of the assembled machines (even if the proposed components are not minimal (Milici 2012, §2)). The mechanisms obtained from these components (to be considered on a plane that can be infinitely extended[5]) develop appropriate linkages, which Kempe showed to be algebraically complete (Kempe 1876) (his proof contained some flaws. For a correct proof see Kapovich and Millson 2002). Now, let us define the components of *tractional motion machines*.

We adopt infinitely extensible **rods**, and assume these have perfect straightness and negligible width. They are different from the Euclidean straight lines, because they are not statically traced objects but planar rigid bodies (mechanical entities with three degrees of freedom, two characterizing the position of a specific point and the third identifying the slope with respect to a fixed line).

It is possible to put some **carts** on a rod, each one using the rod as a rail: a cart has one degree of freedom once placed on a rod (the cart can only move up and down the rod).

The **joint** is a constraint between fixed points of two (or more) different objects (here, an object refers to the plane, a rod, or a cart). Once the joint has been applied, jointed objects can only rotate around their common point (note that, in general, the junction point does not have to be fixed on the plane).

---

[5]With regard to the plane, in general it can be substituted by any other surface in a space (as usually made in differential geometry), but the adopted surface has to be considered as given *a priori* (all we are going to construct with machines are transformations over a surface, not new surfaces). This is why we restrict ourselves to the basic case of the plane (at least for the moment).

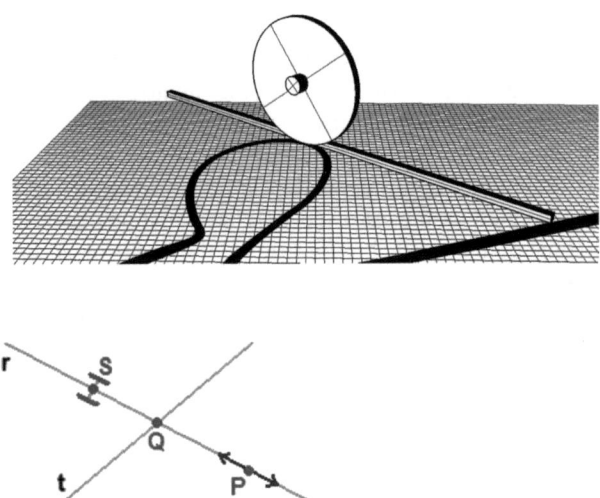

Fig. 1.3 A wheel rolling while following any regular curve has the property that its own *direction* (represented in the picture by a bar) is always tangent to the curve

Fig. 1.4 Schematic representation of the components: there are two rods (r and t) joined at Q. On r, there is also a cart P (the *arrows* stand for the possible motions the cart can have) and a wheel S (the "H" shape ideally represents the projection of a double-wheeled chariot)

Finally, we have the non-holonomic constraint, the **wheel**: once a rod r and a point S on r have been selected, we can set a wheel at S that prevents S itself moving perpendicularly to r (considering the motion of S relative to the plane). Technically, this is as if we put a fixed caster (oriented like r) at S, with its wheel rotating without slipping on the plane. As we shall evince, the avoidance of lateral motion in the rod at a point is strongly related to the tangent. If we consider the caster wheel as a disk rolling perpendicularly to the base plane, the projection of the disk surface is always tangent to the curve described by the disk contact point (see Fig. 1.3). Thus, the rod is tangent to the curve traced by the wheeled point, having the same direction as the caster wheel.

Like Kempe's linkages, our tools are assumed to be ideal (we do not care about physical inaccuracies), and we do not consider problems related to the intersection of rods or the possible collision of different carts on the same rod. Once we have specified these details, these components can be used to assemble machines whose motion on a plane is purely kinematic (just kinematic constraints, with no attention to other physical interrelations). For a diagrammatic representation of assembled components, see Fig. 1.4.

1.3.2 Some Remarks on Uniqueness

Before describing some results, it is important to deepen the "non-uniqueness" of the movement of a machine according to the motion of the input point. In fact,

Fig. 1.5 A simple TMM (the input point Z moves along a line, W rotates around)

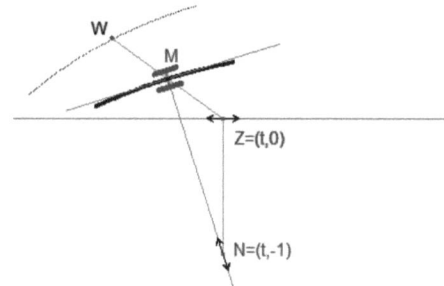

Fig. 1.6 Property of the tangent to the curve traced by a point fixed on a rolling disk

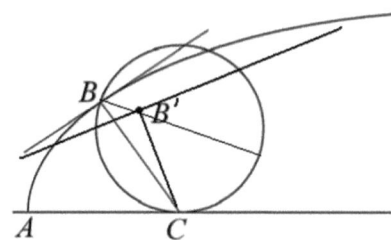

like the solution to differential equations, the motion of our machines is not always determined in a single way. In some machines, the general configuration implies that the motion of the input uniquely defines the motion of the output. However, there are some "singular" configurations in which the output motion is not uniquely determined by the input: let us introduce an explicatory machine.

Given the unitary length and an oriented rod to be used as abscissa (so that we can consider the related Cartesian plane and coordinates), a cart $Z = (t, 0)$ on the rod, and the point $N = (t, -1)$, consider the rod ZW of unitary length (W is free to turn around Z), and let M be the middle point of ZW. We can place a rod passing through M and N[6] and another one perpendicular to MN passing through M[7]: on the latter rod we can place a wheel corresponding to M, so that the tangent to the curve traced by M will always be perpendicular to MN (see Fig. 1.5).

While we move Z along the abscissa, if the absolute value of the W ordinate is strictly less than 1, W has to describe a cycloid, because of the geometrical property shown in Fig. 1.6 (in particular, we imposed the wheel on M and not on W, because, while the rod ZW rotates around Z, W can become coincident with N, leaving the rod WN undetermined). On the contrary, when W assumes coordinates $(t, \pm 1)$, the tangent to M must be horizontal, and so the motion of W can be both a cycloid and

---

[6]To constrain a rod r to pass through M and N, we first pivot r in M by a joint, then put a cart on r, and finally attach the cart with N by another joint.

[7]Concerning the construction of a rod s perpendicular in P to another rod r, we can obtain the perpendicularity by imposing the passage of s through the vertex of a right triangle with one leg on r (the right triangle can be constructed by the junction of a Pythagorean triple as segment lengths).

purely horizontal, losing the uniqueness. As we are interested (for the moment) in generating uniquely determined functions, in the following results, we take care to avoid non-uniqueness of motion when setting the machine. Furthermore, it would be interesting to find some axioms to handle redundancies and to specify when a certain machine has constraints that involve fewer degrees of freedom in particular configurations.

1.4 Results and Examples

In this section, we offer some preliminary results about the potential of TMMs. In particular, in the first subsection, we describe how to solve any real polynomial Cauchy problem. Then, we consider an example involving a complex (exponential) function, before discussing the implications for mathematics education.

1.4.1 How to Solve Real Polynomial Cauchy Problems

In this section, we propose a sketch of the proof of how to solve any real polynomial Cauchy problem (*pCp*) (Milici 2012), that is, a Cauchy problem in the form

$$\begin{cases} \underline{y}' = \underline{p}(t, \underline{y}) \\ \underline{y}(t_0) = \underline{y}_0 \end{cases}$$

where t is the free variable, $\underline{y} = (y_1, \ldots, y_n)$ is a vector function in t, and $\underline{p} = (p_1, \ldots, p_n)$ is a vector polynomial in $\underline{y}$ and t.

First of all, as an extension of Descartes' algebraic machines (Descartes 1954), TMMs can calculate real polynomials (which was an intention of Cartesian geometry), and we assume we have some TMMs (TMM$_1$, ..., TMM$_n$) that, given the values $t, y_1, \ldots, y_n$ as inputs, are able to calculate the values $p_1, \ldots, p_n$.

Once we have defined a "unitary length," we fix a rod q on the plane and assign it a direction, so that we can consider it as the abscissa of a Cartesian plane. Then, we mark on the point with Cartesian coordinates $(t_0, 0)$ on the rod. We place a cart on q at coordinates $(t, 0)$, and, by changing t, we construct the point $(t + 1, 0)$.[8]

We construct rods r and s that are perpendicular to q and pass, respectively, through $(t, 0)$ and $(t + 1, 0)$. Now, we can put n carts on r. For reasons that will become clear, we denote $y_i (i = 1, \ldots, n)$ as the ordinate of the i-th free cart. We apply the TMM$_i$ machines to t and to $y_1, \ldots, y_n$, then we report the lengths

---

[8]Note that, differently from $(t_0, 0)$, t may vary in $\mathbb{R}$. The point $(t + 1, 0)$ can be obtained by linking at $(t, 0)$ one end of a unitary length rod, whose other extremity is forced by a cart to move along q.

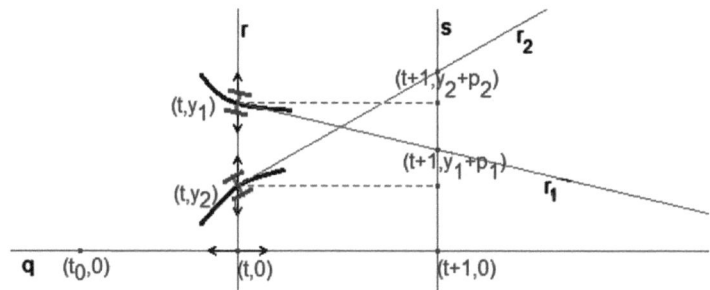

Fig. 1.7 Solving $pCps$ with a TMM (we consider y with only two components). Note that the cart at $(t, 0)$ is free to move along rod q, and along r the motion of (t, y_i) is managed by the wheels with direction r_i, where r_i connects (t, y_i) to $(t + 1, y_i + p_i)$ (y_i is an abbreviation of $y_i(t)$ and p_i denotes $p_i(t, \underline{y})$). In the figure, the TMMs used to compute p_i are not represented

Fig. 1.8 A sketch of the real exponential function TMM (cf. the right-hand picture in Fig. 1.2)

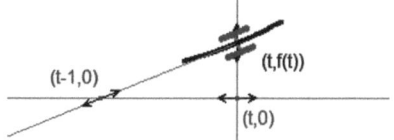

resulting from the application of the polynomial by the various TMM$_i$ on s.[9] To manage the derivative of y_i, we insert rods r_i linking (t, y_i) with the corresponding $(t + 1, y_i + p_i)$.[10] Finally, we place a wheel at (t, y_i) on every r_i: the wheels physically implement the condition $y' = p(t, \underline{y})$.

To complete the construction, we simply set the initial condition $y(t_0) = \underline{y}_0$. This can be done by moving the cart at $(t, 0)$ to $t = t_0$ and setting the y_i carts on r to the length of the i-th component of $\underline{y}_0$. Now, changing t through the motion of the cart at $(t, 0)$, we obtain the desired length $(t, y_i(t))$ (see the illustration in Fig. 1.7).

A simple example of a function that is generable by TMMs but not by Cartesian tools is the exponential function (a transcendental curve rather than an algebraic one), whose related tractional machine was first described in Perks (1706). This example, shown in Fig. 1.8, is actually simple, because we simply impose the condition that the subtangent to the exponential curve is constant (for every real t, the tangent passes through the point $(t - 1, 0)$).

Furthermore, in Milici (2012) it was proved that a TMM can generate more functions than just the solution to $pCps$. In fact, a machine has been constructed for the real function whose Cartesian graph is the cycloid (which, having cusps, is not analytic and therefore cannot be the solution to any pCp).

---

[9] Specifically, we determine the points $(t + 1, y_i + p_i)$ in function of t and the (still) free $y_1, \ldots, y_n$.
[10] To link (t, y_i) with $(t + 1, y_i + p_i)$, we place a rod r_i at (t, y_i), then we place a cart on r_i and move it to $(t + 1, y_i + p_i)$.

1.4.2 A Machine for the Complex Exponential Function

In this section, we demonstrate some new visual uses of the proposed TMM model. Although complex functions need a 4D-space to be statically represented, we can represent them through planar TMMs, merging domain and range in the same 2D-plane, so that a function is given by a machine that links the motion of the input with that of the output. A similar purpose was described in Emch (1902), although less generically, where the author showed how to perform any algebraic transformation of complex variables using only planar linkages. In this section, we want to see how TMMs can overtake algebraic operations, and so we will observe how to assemble a TMM for the complex exponential function. Recall that, in the complex case, the exponential function is the only solution to the Cauchy problem $f'(z) = f(z)$, $f(0) = 1$.

The first difference from the real case is that we have to allow the input point z to move not only along the abscissa, but across the whole plane (interpreted in Argand-Gauss coordinates). Furthermore, to represent the complex value of the function, we consider the complex value $w = z + f(z)$ as the output point, so that (for every z) $f(z)$ can be seen as the difference vector between w and z.[11]

In the real case, $f(x)$ is only 1D, whereas in the complex case we have to deal with two dimensions. Thus, instead of needing only one tangent property, we must set two. If the first can be set at the output point $w = z + f(z)$, we need another point at which to set the second. We can use Emch's result (Emch 1902): as the TMM is an extension of linkages, we can perform any complex algebraic operation, so (for every value of $f(z)$) we can dynamically multiply $f(z)$ by a complex constant k. It is at $p_k = z + k \cdot f(z)$ that we set the second tangent property to deal with the two components of the complex function.

Now, to begin handling some complex $f(z)$, let us consider $z(t)$, the trajectory of z as a function of time, adding the condition that $z'(t)$ is always not null (so z is a regular function $\mathbb{R} \to \mathbb{C}$). To investigate the tangent properties of the point $p_k = z + k \cdot f(z)$ (where k is any complex constant), we calculate its derivative (considering z as a function of t). Therefore, we have

$$\begin{aligned} p'_k(t) &= z'(t) + k \cdot f'(z) \cdot z'(t) \\ &= z'(t) \cdot (1 + k \cdot f'(z)) \end{aligned} \qquad (1.1)$$

---

[11] We have to deepen the idea of representing $z + f(z)$ instead of $f(z)$. If, at first glance, it seems so different from the representation in the real case, the main condition behind both is that the motion of the output point has to be determined by that of the input point, so it is necessary that the input *drags* the output. Mathematically, this is implemented by a vector sum (input + output), both in the real and complex cases. In the Cartesian plane, we denote the axes' unit vectors as $\hat{i}$ and $\hat{j}$, and the graph is defined as $x\hat{i} + y\hat{j}$ (so, as the domain and range axes are linearly independent, the graph of a real function "statically" represents all the information of the function). In the complex case, the domain and range have to be merged in the same planar coordinates, losing the property that any point of the plane identifies a single couple of input/output. In particular, given a real function f in x, the usual real representation on the Cartesian plane $(x, f(x))$ can be interpreted as $x + i \cdot f(x)$ on the complex plane (i is the imaginary unit).

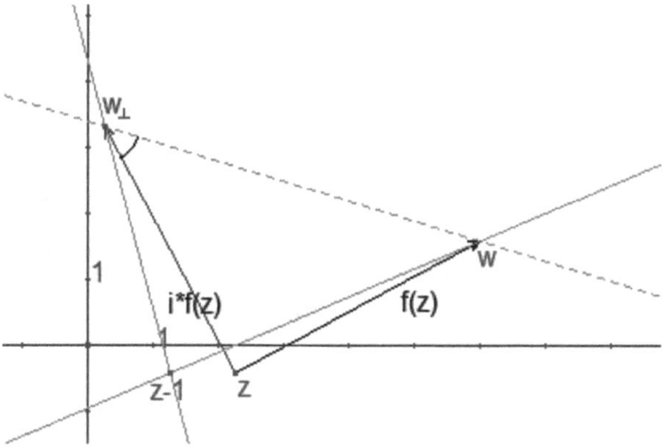

Fig. 1.9 Beginning of the construction of the complex exponential function ($w_\perp$ was constructed by rotating $f(z)$ 90° anticlockwise). At this point, no wheel has been placed and $f(z)$ does not have any constraint (it can freely move across the plane). If $w = z + e^z$, $w_\perp = z + i \cdot e^z$, and z' is real, the tangent to both w and $w_\perp$ passes through $z - 1$

Fig. 1.10 If the direction of z (while moving on its trajectory) is inclined at an angle a to the line parallel to the abscissa, the tangent at w and $w_\perp$ has an additional inclination a from the rod connecting them with $z - 1$

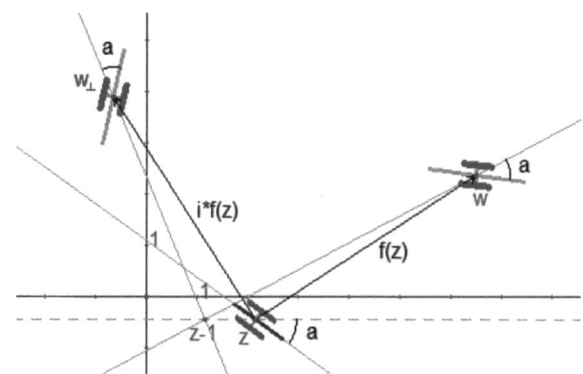

Back to the exponential function, if we set the differential condition $f'(z) = f(z)$, and (for the moment) consider $z' = 1$, the tangent to the curve $p_k(t)$ always passes through the point $z - 1$ (in Fig. 1.9, we have considered the points $w = p_1$ and $w_\perp = p_i$).[12]

Until now, we have considered the case $z' = 1$; this can be extended to any not-null real value of z'.[13] In fact, in general, (1.1) implies that the tangent to p_k has direction $(1 + k \cdot f'(z))$ rotated by an angle $\theta_{z'}$ (so the slope does not depend on $\rho_{z'}$): this can be set by adding the angle $\theta_{z'}$ to the inclinations of the rods connecting w and $w_\perp$ to $z - 1$, as shown in Fig. 1.10. This addition of angles is possible, for example, through the linkage constructions named *angle adders* (Emch 1902). Once the tangents have been identified, we can set the wheels at w and $w_\perp$.

---

[12]Even though we do not encounter them in this chapter, there are problems when $p_k = k \cdot f(z) = -1$ (the tangent is not defined, because p_k is not moving). To overcome this, we would need to construct not only p_{k_1} and p_{k_2} but also p_{k_3} (with k_i different from each other), so that there would be at least two well-defined tangent conditions everywhere.

[13]As we shall evince, the tangent to p_k depends on the argument of z'. Concerning the polar form (ρ, θ) of any complex value z, the argument θ is determined if and only if $z \neq 0$.

In summary, given z and z', we set the tangent properties at w and $w_\perp$. This defines the exponential function, because we have two constraints guiding the two components of the complex output value. However, even if the position of z' is visible on the plane, z' is not accessible: to identify it, or more precisely its inclination $\theta_{z'}$, we simply add a rod r at z, and put a wheel on r at point z. The rod r is free to rotate around z, but the wheel prevents any lateral motion of z with respect to the direction of r, which means that, when z moves, r is the tangent to the trajectory of z.

In this way, if we move the input point z from the initial condition $f(0) = 1$ and change the inclination of the rod r to make z move along a certain trajectory, then the complex vector $w - z$ is exactly e^z.

1.4.3 Some Educational Implications

Hitherto, we have explored the TMM theoretical model, but can its transposition into concrete artifacts be useful? We think it can, especially for didactical purposes. Indeed, the actual manipulation of an artifact can help students to experience and internalize the underlying mathematical contents, if suitably introduced into educational pathways (as suggested in "Theory of the Semiotic Mediation" (Bartolini Bussi and Mariotti 1999), which focuses on the use of artifacts to transmit mathematical knowledge).

As an elementary introduction to tractional machines in mathematics education, we must mention the pathway described in Di Paola and Milici (2012), which introduces a very simple concrete artifact to deepen the tangent concept (a TMM composed of just a rod and a wheel: making the wheel move along a curve, the rod outputs the tangent to the curve). What we are going to observe in this section is the possible use of a more complex artifact, one that was ideated and designed by the author and realized in collaboration with Benedetto Di Paola (this artifact was presented in a workshop at the *64th Conference of the International Commission for Study and Improvement of Mathematics Education – CIEAEM 64* (Milici and Di Paola 2012) and mathematically deepened in Salvi and Milici (2013)). According to the way we assemble its components, this artifact generates two different curves, one algebraic and one transcendental,[14] as shown in Fig. 1.11.

---

[14]In particular, the fact that two functions, one transcendent and the other algebraic, can be constructed through similar devices of equal complexity is an epistemological point, in contrast with the Cartesian dualism between the different legitimization of geometrical (algebraic) and mechanical (transcendental) curves. Concerning this, we may mention the letter that Poleni wrote to Hermann in September 1728 (published in Poleni 1729), in which the author wondered about the nature of tractional curves. With a simple modification to the exponential tractional machine (just changing an angle, which is essentially the same thing we did, as shown in Fig. 1.11), the author had realized that tractional machines draw curves defined by differential equations in a uniform way, regardless of their algebraic or transcendental nature.

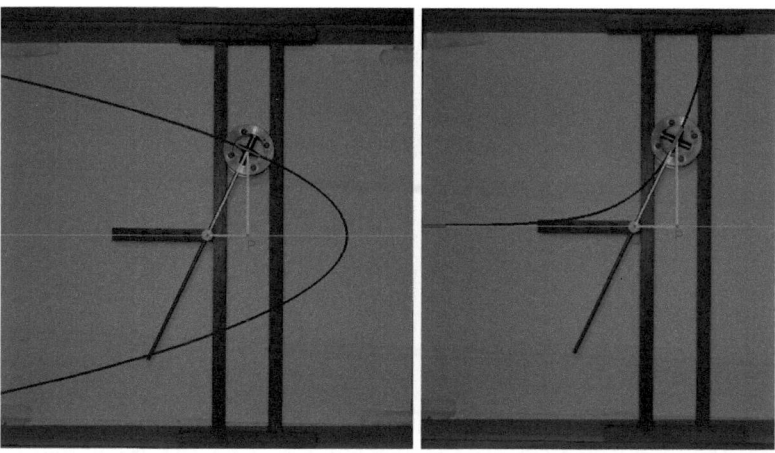

Fig. 1.11 The artifact generating a parabola (*left*) and an exponential curve (*right*), according to the inclination of the wheel. It was ideated, designed, and realized by the author and Benedetto Di Paola for the G.R.I.M. (Research group of teaching/learning mathematics, University of Palermo)

The proposed artifact was designed to permit a double use: a first explorative approach (from machine to mathematics) and a second constructive one (from formula to machine). In particular, being interested in developing a pathway involving the field of Infinitesimal Analysis, instead of focusing on curves, we want to focus on the generated functions. Thus, we propose the exploration of a machine embodying the square root and the construction of a machine for the exponential curve (both machines can be obtained by assembling the same components in a different manner).

We choose these functions because, though very different from the usual didactic perspective, they are markedly similar in their TMM interpretation. The real exponential function has been discussed in both Sects. 1.2 and 1.4.1 (Figs. 1.2 and 1.8), and the choice of the square root function has been made because of its nature, which, though simple, reveals many significant aspects that can be highlighted in the geometrical/mechanical interpretation. In particular, the related didactical pathway focuses on a specific "new" reading of the mathematical concepts of tangents (geometrical and analytical approach with the derivative), continuity, real function asymptotic behaviors, and differential equations.

Although the function $f(x) = \sqrt{x}$ is algebraic, we do not interpret it as the inverse of x^2. Specifically, the machine (Fig. 1.12) solves the differential equation $f'(x) = \frac{1}{2f(x)}$ with the initial condition $f(1) = 1$, whose single solution for positive values is the square root.[15]

---

[15] This definition is solved by the square root only for the real values; it does not apply in the complex extension.

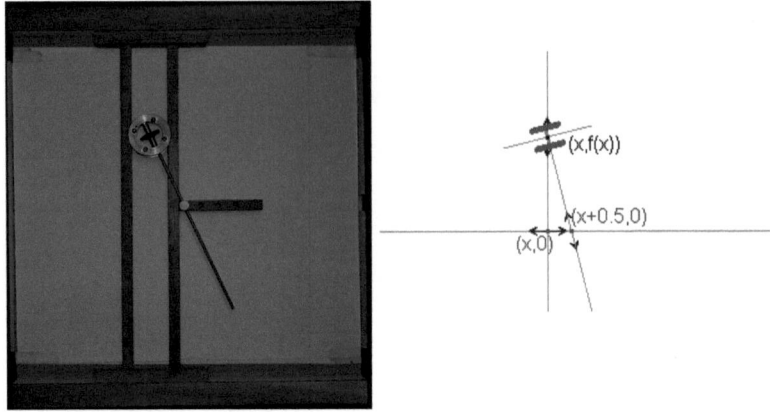

Fig. 1.12 The actual artifact (compared to the left picture of Fig. 1.11, the machine has been rotated 180°) and the related TMM for $f(x) = \sqrt{x}$. To work, it must be shifted horizontally along the "basis cathetus" (the segment with extremes $(x, 0)$ and $(x + \frac{1}{2}, 0)$). The wheel at $(x, f(x))$ implements the condition that the tangent at $(x, f(x))$ must be perpendicular to the hypotenuse

Table 1.1 Square root function: analytical and geometrical/mechanical properties

Analytical register	Geometrical/mechanical register
Domain: $\mathbb{R}^+$	Here there is a great difference in comparison to the analytical register. The artifact in Fig. 1.12, if not physically dragged on the plane, *does not allow us to evaluate the domain* (because the abscissa values are used in a dynamic way). On the other hand, it is possible to realize how the artifact becomes stuck when $f(x) = 0$ (the wheel becomes perpendicular to the horizontal motion)
$f(x) \geq 0$	Knowing that the artifact becomes stuck when $f(x) = 0$ and that $f(1) = 1$, the function is always nonnegative as a result of its continuity
$f'(x) > 0$	The tangent has to be perpendicular to the *hypotenuse*, so the derivative is positive when f is not negative (in the whole domain)
$\lim_{x \to +\infty} f(x) = +\infty$	Since it is increasing, f cannot oscillate. By *reductio ad absurdum*, suppose that f converges, so f' tends to 0. Mechanically, this implies that the *hypotenuse* tends to be parallel to the ordinates (even if this can never physically happen), and this occurs only if f tends to infinity. Hence, the absurdum (f had to converge)
$\lim_{x \to +\infty} f'(x) = 0$	Once the divergence has been observed, according to the previous *reductio ad absurdum*, the tangent tends to be parallel to the abscissa

Considering Table 1.1, we get a sense of the possible translation from the analytical to the geometrical/mechanical semiotic register[16] (Duval 1993) and the manner in which we dynamically analyze the artifact and its mechanical components (constraints, rods, and so on).

---

[16]TMMs (and real artifacts) are not only able to visualize some properties (like in dynamic geometry) but also prove them in a specific register. Unfortunately, this register is currently not

1.5 Conclusions and Future Perspectives

Although this research into TMMs requires further development, the role of formalized ideal machines in the foundation of both computation and mathematics is deep rooted: we may think about the Euclidean straightedge and compass, or about the extension through the Cartesian machines, up to the Turing machine (even if it is set on a digital/arithmetic paradigm instead of a geometrical/mechanical one). In this way, after the study of relative boundaries, it would be interesting to construct an axiomatic system for TMMs to reinterpret (part of) the classical real and complex analysis from a constructive and geometrical perspective (probably starting with the axiomatization of the behavior of specific tractional tools, such as the equiangular compass proposed in Milici and Dawson 2012).

As previously announced, we have not yet been able to define the class of functions that are generable by the proposed model. At first glance, it seems that, with these means, we will never be able to generate classical functions such as the Weierstrass function (continuous but nowhere differentiable) or to draw the Koch snowflake and fractals. The situation of the Euler Γ-function (at the basis of fractional calculus[17]) appears even more complicated. In fact, being differentiable but not a solution to any pCp, we do not yet know whether or not Γ can be generated by a TMM. In both cases, it would be interesting to know, so that we may understand the limits and potentials of the proposed geometrical interpretation of the derivative as tangent (furthermore, if generable, it would also be interesting to investigate the tractional machine of some functions, such as Γ and Riemann-ζ, for both real and complex values).

Another perspective concerns the comparison with some models of computation. In particular, in Milici (2012), a comparison with Shannon's *General Purpose Analog Computer* (Shannon 1941) was proposed. This pathway still has to be deepened, and another long-term aim is to investigate whether and how a relation between digital and analog computation can be traced (especially in relation to the *Computable Analysis* paradigm).

---

autonomous, and we are still trying to define a suitable theory to highlight the primitive concepts for the construction and functioning of the machines. Meanwhile, as visible in Table 1.1, we had to use some analytic properties in the geometrical/mechanical register (properties about continuous or monotonic functions) to obtain some informal proofs. In addition, the geometrical and mechanical registers, even if different, have been summarized in the same column for simplicity.

[17]Fractional calculus is the study of an extension of derivatives and integrals to non-integer orders (for further reading, see, e.g., Ross 1975).

References

Bartolini Bussi, M.G., and Mariotti, M.A. 1999. Semiotic mediation: From history to the mathematics classroom. *For the Learning of Mathematics* 19(2): 27–35.

Bos, H.J.M. 1988. Tractional motion and the legitimation of transcendental curves. *Centaurus* 31(1): 9–62.

Bos, H.J.M. 1989. Recognition and wonder: Huygens, tractional motion and some thoughts on the history of mathematics. *Tractrix* 1: 3–20. Reprinted in, Bos, H.J.M. 1993. *Lectures in the history of mathematics*. Providence: American Mathematical Society, 1–21.

Bos, H.J.M. 2001. *Redefining geometrical exactness: Descartes' transformation of the early modern concept of construction*. New York: Springer.

Descartes, R. 1637. *Discours de la méthode pour bien conduire sa raison & chercher la varité dans les sciences plus la diotrique, les meteores, et la geometrie, qui sont des essais de cete methode*. Leyde: Maire.

Descartes, R. 1954. *The geometry of René Descartes*. New York: Dover. Includes a facsimile of the appendix of the original French edition Descartes (1637).

Di Paola, B., and Milici, P. 2012. Geometrical-mechanical artefacts mediating tangent meaning: The tangentograph. *Acta Didactica Universitatis Comenianae – Mathematics* 12: 1–13.

Duval, R. 1993. Registres de représentation sémiotique et fonctionnement cognitif de la pensée. *Annales de didactique et de sciences cognitives* 5: 37–65.

Emch, A. 1902. Algebraic transformations of a complex variable realized by linkages. *Transactions of the American Mathematical Society* 3(4): 493–498.

Huygens, C. 1693. Letter to H. Basnage de Beauval, February 1693. *Œuvres de Huygens* 10: 407–422. Printed in *Histoire des ouvrages des sçavants* (or *Journal de Rotterdam*), 244–257 (1693).

Kapovich, M., and J. J. Millson. 2002. Universality theorems for configuration spaces of planar linkages. *Topology* 41(6): 1051–1107.

Kempe, A.B. 1876. On a general method of describing plane curves of the nth degree by linkwork. *Proceedings of the London Mathematical Society* 7: 213–216.

Leibniz, G.W. 1693. Supplementum geometriæ dimensoriæ seu generalissima omnium tetragonismorum effectio per motum: similiterque multiplex constructio lineæ ex data tangentium conditione. *Acta Eruditorum* 385–392 [Translated from Math. Schriften 5: 294–301].

Milici, P. 2012. Tractional motion machines extend GPAC-generable functions. *International Journal of Unconventional Computing* 8(3): 221–233.

Milici, P., and Dawson, R. 2012. The equiangular compass. *The Mathematical Intelligencer* 34: 63–67.

Milici, P., and Di Paola, B. 2012. Geometrical-mechanical artefacts for managing tangent concept. In *Proceedings of the 64th conference of the international commission for study and improvement of mathematics education*, Rhodes, Greece, 23–27 July 2012.

Panza, M. 2011. Rethinking geometrical exactness. *Historia Mathematica* 38(1): 42–95.

Pascal, E. 1914. *I miei integrafi per equazioni differenziali*. Napoli: B. Pellerano.

Perks, J. 1706. The construction and properties of a new quadratrix to the hyperbola. *Philosophical Transactions* 25: 2253–2262.

Perks, J. 1714. An easy mechanical way to divide the nautical meridian line in Mercator's projection, with an account of the relation of the same meridian line to the curva catenaria. *Philosophical Transactions* 29(338–350): 331–339.

Poleni, J. 1729. *Epistolarum mathematicarum fasciculus*. Typographia Seminarii: Patavii.

Riccati, V. 1752. *De usu motus tractorii in constructione æquationum differentialium commentarius*. Bononiæ: Ex typographia Lælii a Vulpe.

Ross, B. 1975. A brief history and exposition of the fundamental theory of fractional calculus. In *Fractional calculus and its applications*, 1–36. Berlin, Springer.

Salvi, M., and Milici, P. 2013. Laboratorio di matematica in classe: due nuove macchine per problemi nel continuo e nel discreto. *Quaderni di Ricerca in Didattica (Mathematics)* 23: 15–24.

Shannon, C.E. 1941. Mathematical theory of the differential analyzer. *Journal of Mathematics and Physics MIT* 20: 337–354.

Tournès, D. 2004. Vincenzo Riccati's treatise on integration of differential equations by tractional motion (1752). *Oberwolfach Reports* 1: 2738–2740.

Tournès, D. 2007. La construction tractionnelle des équations différentielles dans la première moitié du XVIIIe siècle. In *Histoires de géométries: texte du séminaire de l'année 2007*, ed. D. Flament. Paris: Fondation Maisons des Sciences de l'homme.

Tournès, D. 2009. *La construction tractionnelle des équations différentielles*. Paris: Blanchard.

Chapter 2
Plane and Solid Geometry: A Note on Purity of Methods

Paolo Mancosu and Andrew Arana

2.1 Introduction

Traditional geometry concerns itself with planimetric and stereometric considerations, which are at the root of the division between plane and solid geometry. When one raises the problem of the relationship between these two areas, one encounters epistemological, ontological, semantical, and methodological problems. In addition, other issues related to psychology and pedagogy of mathematics emerge naturally. In this note (based on Arana and Mancosu 2012), we will focus on a methodological aspect: purity of methods (see Detlefsen 2008 and Detlefsen and Arana 2011). After a few historical remarks concerning the role played by solid geometry in the development of plane geometry, we will move on to the analysis of a specific case, Desargues' theorem on the plane (which we will call "Desargues' plane theorem"). This theorem was proved by Desargues by making use of metric notions (congruence principles) that were key to a theorem that played a central role in the demonstration, namely, Menelaus' theorem. However, the development of geometry in the nineteenth century led to the analysis of the foundations of projective geometry and to the attempt to eliminate as much as possible from this discipline non-projective notions such as congruence or measure. Desargues' theorem played a crucial role in this type of investigation. A purely projective proof of this theorem had already been given in 1822 by Poncelet. Poncelet had shown how a version of Desargues' theorem in space (we will call it "Desargues' solid theorem") provided, as a simple corollary,

---

P. Mancosu (✉)
University of California, Berkeley, CA, USA
e-mail: mancosu@socrates.berkeley.edu

A. Arana
University of Illinois at Urbana-Champaign, Champaign, IL, USA

a projective demonstration of Desargues' plane theorem. The appeal to congruence in Desargues' original proof for the plane theorem was thus eliminated through the introduction of spatial notions. One can however ask whether this appeal to space is legitimate and necessary. The legitimacy question originates from considerations related to purity of methods. The issue about necessity is tied to logical considerations. One had to wait until the works of Peano and Hilbert to obtain an (affirmative) answer to the latter question. Moreover, these results are at the basis of a more articulate discussion of the legitimacy problem (namely, the purity problem). These considerations will be developed in the final part of the note.

2.2 Historical Notes on the Relationship Between Plane and Solid Geometry

In ancient geometry, we encounter few interesting applications of solid geometry to plane geometry (of course, solid geometry requires plane geometry). Euclid's *Elements* present us with a sharp separation between plane geometry and solid geometry (with the latter relegated to the last books of the *Elements*), a division that will have a lasting impact on the presentation of elementary geometry until the end of the nineteenth century. There are however, already in Greek times, some advanced directions of research in which techniques of solid geometry are applied to the study of problems in plane geometry. We can mention, for instance, the quadrature of the circle provided by (Pappus 1876–1878) which is obtained generating the quadratrix curve on the plane through a projection of the cylindrical helix. It is also important to note that the distinction between plane, solid, and linear problems given by Pappus is orthogonal to that between plane and solid geometry. Pappus' taxonomy concerns the types of curves required for the solution of problems (line and circle for plane problems, conic sections for solid problems, and "more complex" curves for the linear problems). Euclid's solid geometry ends up classified as "plane" in Pappus' taxonomy, and conversely, problems stated in plane geometry, such as the trisection of an arbitrary angle, are classified as "solid." While Pappus criticizes the use of curves that do not correspond to the nature of the problem (such as the use of conic sections for solving "plane" problems), we are not aware that any Greek mathematician (or philosopher) ever raised objections to the use of solid geometry in investigations of problems of plane geometry.

In the seventeenth century, one notices a lively interest for the application of solid geometry in the solution of problems in plane geometry. Consider for instance the example of Evangelista Torricelli. In his treatise, *De quadratura parabolae* (1644; see Torricelli 1919–1944), Torricelli presented 20 different proofs of the quadrature of the parabola (a theorem of plane geometry proved for the first time by Archimedes) and classified them according as to whether they were proved by "classical" demonstrations (using techniques by reductio ad absurdum) or with demonstrations obtained through the geometry of indivisibles of Cavalierian inspiration. What is striking in this treatise is Torricelli's attention vis-à-vis the use of solid geometry in the proofs of theorems of plane geometry. All the most important results of Euclidean and Archimedean stereometry are appealed to and Torricelli showed how one can obtain the quadrature of the parabola from each one of these stereometric theorems, either using exhaustion techniques (reductio ad absurdum) or indivisibilist arguments. Of course, not a single one of these stereometric results can be considered necessary for the proof of this plane theorem, for Archimedes had already given a proof that only appeals to concepts and results of plane geometry. Torricelli does not raise any methodological problems concerning the use of solid geometry in investigating problems of plane geometry.

With the development of projective geometry in the nineteenth century, the use of spatial techniques in the study of problems of plane projective geometry begins to show its fruitfulness. Monge's school, in particular, made extensive use of the interaction between planar and spatial notions. In his famous 1837 *Aperçu*, Chasles described the school of Monge by means of its propensity towards the use of three dimensions in the proof of plane theorems.

We conclude these brief historical remarks by recalling that within elementary geometry, the separation between plane and solid geometry was challenged seriously for the first time in the work of the Italian geometer Riccardo de Paolis in *Elementi di geometria* (1884). In this book, de Paolis emphasized the importance of analogies between plane and solid geometry (angles-diedra, polygons-polyhedra, etc.) as well as the importance of using space for the understanding and the simplification of theorems of plane geometry. This "fusionist" position, namely, the request that plane and solid geometry be developed together, was at a source of the debate known as "fusionism" which saw the involvement of Italian, French, and German geometers. The debate between those who advocated "fusionism" and their opponents led to discussions concerning the legitimacy as well as the necessity of using space in proofs of theorems in plane geometry. But in order to seriously tackle such issues, one had to wait for the foundational works of Peano and Hilbert, which we will discuss below.

2.3 The Foundations of Projective Geometry

In the early nineteenth century, geometers set out to develop the foundations of projective geometry, independently of Euclidean geometry. Some, for instance, Möbius and Plücker, sought to develop an analytic projective geometry, analogous to Cartesian analytic geometry for Euclidean geometry. Others, for instance, Steiner, sought a coordinate-free development of projective geometry that had the same power as the new analytic projective geometry. In these research programs, these geometers freely used metric considerations. They appealed either to the Euclidean distance metric or to principles of proportionality or congruence. However, these are not projectively invariant.

Beginning with his *Geometrie der Lage* (1847), von Staudt sought to eliminate these metric considerations from projective geometry, on purity grounds:

> I have tried in this work to make the geometry of position into an independent science that does not require measurement.

Though there were gaps concerning continuity that were later filled by others, von Staudt's work yielded a means of defining projective coordinates by purely projective means. The key to his accomplishment was a particular construction ("quadrilateral construction"), which provides a way, given any three collinear points, to find uniquely a fourth point on that line with a certain relation to the other three points; it is then said that the four points together form a "harmonic range" (a notion that we do not need to define here). The uniqueness of the fourth harmonic point can be shown by metric considerations. Following his aim of purifying projective geometry of metric considerations, von Staudt proved the uniqueness of the fourth harmonic point by purely projective means, in particular by appealing to Desargues' theorem whose statement does not involve any metric notion.

Desargues' Plane Theorem If two triangles lying in the same plane are such that the lines connecting their corresponding vertices intersect at a point, then the intersections of their corresponding sides are collinear.

As we have already mentioned, Desargues' original proof appeals to metric notions, since it appealed to congruence by way of Menelaus' theorem. Von Staudt's aim was to purify projective geometry of metric considerations and, in particular, to define projective coordinates by purely projective means. The key to doing so was Desargues' theorem.

His aim would only have been satisfied if he had a nonmetrical proof of Desargues' theorem. However, Desargues had also stated a solid version of the result.

Desargues' Solid Theorem If two triangles lying in different planes are such that the lines connecting their corresponding vertices intersect at a point, then the intersections of their corresponding sides are collinear.

The planar Desargues' theorem can be proved by "projecting" the solid version into the plane, as Poncelet showed in his *Traité des propriétés projectives des figures* (1822). This proof is purely projective, avoiding metrical considerations (all one needs to observe is that two planes intersect at a line and that the lines that connect the vertices of the triangles lying on different planes can only meet in the line of intersection of the two planes). Hence, von Staudt was able to achieve his aim by using this proof. However, this proof draws on considerations from solid geometry, despite the fact that Desargues' theorem concerns just triangles in the plane.

We have thus reached the point where the fusionist debate began. That fusionism had to be a necessity in the foundations of projective geometry was also the conclusion reached by Felix Klein in his article *Über die sogenannte Nicht-Euklidische Geometrie* (1873).

In an influential 1891 lecture (Wiener 1892) remarked, without proof, that Desargues' theorem cannot be proved by purely planar projective considerations, observing that "this area of geometry is not self-contained." Peano and Hilbert took up this metamathematical question shortly thereafter.

2.4 Peano and Hilbert

The key to Klein's observation on the necessity of appealing to space in the foundations of projective geometry is Beltrami's theorem (1865), which says that a "smooth" (i.e., Riemannian) surface has constant curvature if and only if it can be mapped to a plane so that the geodesics of that surface are mapped to straight lines in that plane. The result applies to the Euclidean plane and even to the projective plane. Klein understood Beltrami's theorem as asserting that a Riemannian surface of nonconstant curvature cannot be represented on a plane so that the geodesics of that surface "behave" like straight lines in the plane.

In 1894 Peano developed this suggestion, sketching a proof that his axioms of planar geometry have models in which Desargues' theorem fails by appealing to Riemannian surfaces of nonconstant curvature. Hence, his planar axioms are provably insufficient for proving Desargues' theorem. Once solid axioms are added, Peano's axioms prove Desargues' theorem as expected.

In lectures delivered in 1898–1899, Hilbert developed his own axiomatization for geometry, dividing his axioms into classes I (incidence), II (order), III (parallel), IV (congruence), and V (continuity). He observed that Desargues' theorem is provable in this system using spatial axioms, or alternately using axioms of congruence. He then showed that Desargues' theorem cannot be proved in plane geometry (in fact, from axioms I 1–2, II, III, IV 1–5, and V), by presenting explicitly a model in which these axioms are satisfied but Desargues' theorem is not. Hence, it follows that his planar axioms (I 1–2) are provably insufficient for proving Desargues' theorem.

In his lectures of 1898–1899 (Lectures on Euclidean Geometry), Hilbert commented upon the result by emphasizing the importance for the issue of purity of methods:

> This theorem gives us an opportunity now to discuss an important issue. The content [Inhalt] of Desargues' theorem belongs completely to planar geometry; for its proof we needed to use space. Therefore we are for the first time in a position to put into practice a critique of means of proof. In modern mathematics such criticism is raised very often, where the aim is to preserve the purity of method [die Reinheit der Methode], i.e. to prove theorems if possible using means that are suggested by [nahe gelegt] the content of the theorem. (Hallett and Majer 2004, pp. 315–316)

What is critical for a proof's being pure or not, then, is whether the means it draws upon are "suggested by the content of the theorem" being proved. Since the "content of Desargues' theorem belongs completely to planar geometry," solid considerations would not appear to be "suggested by the content of the theorem," and therefore, it would seem that Hilbert judged solid proofs of Desargues' theorem impure. Hilbert also showed that if a planar geometry satisfies axioms I 1–2 (the planar incidence axioms), II (the order axioms), and III (the parallel axiom), then Desargues' theorem is necessary and sufficient for that planar geometry to be an element of a spatial geometry satisfying all the incidence axioms I in addition to the axioms of II and III. That is, a plane satisfying axioms I 1–2, II, and III, and also satisfying Desargues' theorem, will also satisfy the spatial incidence axioms I 3–7. Hilbert proved this by showing, firstly, how, in a planar geometry satisfying axioms I 1–2, II, III, and Desargues' theorem, to construct an algebra of segments that is an ordered division ring and, secondly, how this ordered division ring can be used to construct a model of axioms of I, II, and III, that is, a model of spatial geometry. (Order is inessential here.)

Here is how Hilbert summarized the situation in his 1898–1899 lectures:

> Then the Desargues Theorem would be the very condition which guarantees that the plane itself is distinguished in space, and we could say that everything which is provable in space is already provable in the plane from Desargues. (Hallett and Majer 2004, p. 240)

In other words, Desargues' theorem can be used as a replacement for Hilbert's solid axioms: it has the same provable consequences as those axioms in Hilbert's axiomatic system (see Hilbert 1899, 1971).

2.5 The Problem of Content

In a recent article (Hallett 2008) and in his introductions to the Hilbert's lectures on geometry published in the first volume of the Hilbert Editions (Hallett and Majer 2004), Michael Hallett has drawn some interesting consequences, which in our opinion are questionable, on the notion of the content of Desargues' theorem and on the issue of purity of methods. Hallett writes:

> What this shows is that the Planar Desargues's Theorem is a sufficient condition for the orderly incidence of lines and planes, in the sense that it can be used to generate a space. We thus have an explanation for why the Planar Desargues's Theorem cannot be proved from planar axioms alone: the Planar Desargues's Theorem appears to have spatial content. (Hallett 2008, p. 229)

Moreover, in his introduction to Hilbert's 1898–1899 lectures, Hallett writes that Hilbert's work "reveals that Desargues' planar Theorem has hidden spatial content, perhaps showing that the spatial proof of the Planar Theorem does not violate 'Reinheit' after all" (pp. 227–228). Thus, Hallett believes that Hilbert's work should cause us to revise our judgments of what counts as a pure proof of Desargues' theorem. While solid considerations would seem "at first sight" to be impure for proving Desargues' theorem, Hallett infers from Hilbert's work reveals that they are not, for Desargues' theorem is in fact a theorem with (hidden) solid content.

This position defended by Hallett appeals to the notion of "hidden higher-order content" developed by Dan Isaacson in the context of some articles aimed at providing an interpretation of Gödel's incompleteness results for Peano arithmetic (Isaacson 1987). In our paper (Arana and Mancosu 2012), we develop a detailed analysis both of Isaacson's notion of "hidden higher-order content" as well as the consequences drawn from it by Hallett with respect to the issue of purity of methods.

The central aspect of the issue is that the notion of content proposed by Hallett, on the basis of the Hilbertian analysis of Desargues' theorem, is based on the deductive role played by this theorem with an axiomatic context. This notion is very close to that of content as deductive equivalence (within an axiomatic system) that had been proposed by Carnap. Hallett sees in Desargues' plane theorem a statement with (hidden) solid content exactly because, within a certain axiomatic theory, Desargues' theorem plays the same inferential role as the space incidence axioms.

Our criticism to Hallett's position is based on the following five objections, which we simply state here without giving any arguments (for which we refer to Arana and Mancosu 2012):

(a) If the content of Desargues' theorem were spatial, it would seem to follow that an investigator with no beliefs or commitments concerning space (such as a character of Flatland) could not understand Desargues' theorem, which seems implausible.

(b) To claim that Desargues' plane theorem has a solid content on account of the inferential role it plays in Hilbert's axiomatic system requires a deep metatheoretical analysis such as the one carried out by Hilbert. But what to say then of statements that have not yet been subjected to such a deep metatheoretical analysis or, worse, for which we don't know whether they are true or false (such as the twin prime conjecture)? Intuitively, we understand the content of the twin prime conjecture even though we have no metatheoretical analysis of it.

(c) Hallett's view implies a radical contextualism regarding the content of statements like Desargues' theorem. The inferential role of Desargues theorem within metrical geometries is quite different than its inferential role within projective geometry; in the former, spatial considerations are unnecessary, while in the latter they are necessary. If Hallett were correct, the content of Desargues' theorem would change dramatically depending on which axiomatic context we use it in, without its formulation changing at all.

(d) Hallett infers that the spatial content "revealed" by Hilbert's work belongs specifically to Desargues' theorem, when Hilbert's work shows only that Desargues' theorem added to the planar axioms of classes I, II, and III has the same spatial consequences as the spatial axioms of those classes. Even if it is reasonable to maintain that the planar axioms plus Desargues' theorem have tacit spatial content on account of their shared inferential role (which we have contested), it is illicit to single out that content as belonging to Desargues' theorem. For those spatial consequences belong only to the axiomatic system as a whole, not to Desargues' theorem alone. While it is true that without Desargues' theorem these spatial consequences are not ensured, it is also true that Desargues' theorem alone does not ensure them. Hence it would be more accurate to say that these spatial consequences are partly the result of the planar axioms and partly the result of Desargues' theorem. Indeed, Hallett's argument would just as well establish that one of the planar axioms, say I.1, has tacit spatial content.

(e) From the analysis of the notion of content defended by Hallett, it follows that every theorem has a pure proof. It seems to us implausible that this can be true *a priori*, simply as a consequence of the analysis of the notion of content. Purity would end up being trivialized.

We conclude that the notion of content offered by Hallett can be of interest for other theoretical goals but not for the clarification of the ascription of purity that are often found in mathematical practice. The notion of content that in our opinion is useful for clarifying judgments about purity of proofs must be tied to the understanding of the meaning of the statement of a theorem and not to its inferential role within an axiomatic system. Moreover, our position on Desargues' theorem seems to us to be identical to the one defended by Hilbert: Desargues' plane theorem does not have a pure proof in a projective context.

Acknowledgements This note is a translation of a note we first published in Italian (Mancosu and Arana 2012). We would like to thank Abel Lassalle Casanave for having given us the stimulus to write the note in Italian and Marco Panza for useful suggestions and comments. In addition, we would like to thank Gabriele Lolli and Giorgio Venturi for having invited the first author to a meeting at the Scuola Normale Superiore in Pisa (on September 23, 2012) and for having suggested to include a translation of our Italian note into English for publication in this volume.

References

Arana, Andrew, and Paolo Mancosu. 2012. On the relationship between plane and solid geometry. *The Review of Symbolic Logic* 5(2): 294–353.
Chasles, Michel. 1837. *Aperçu Historique sur l'Origine et le Développement des Méthodes en Géométrie*. Bruxelles: M. Hayez.
De Paolis, Riccardo. 1884. *Elementi di Geometria*. Torino: Loescher.
Detlefsen, Michael. 2008. Purity as an ideal of proof. In *The philosophy of mathematical practice*, ed. Paolo Mancosu, 179–197. Oxford: Oxford University Press.
Detlefsen, Michael and Arana, Andrew. 2011. Purity of methods. *Philosophers' Imprint* 11(2)
Hallett, Michael, and Ulrich Majer (eds.). 2004. *David Hilbert's Lectures on the Foundations of Geometry, 1891–1902*. Berlin: Springer.
Hallett, Michael. 2008. Reflections on the purity of method in Hilbert's *Grundlagen der Geometrie*. In *The philosophy of mathematical practice*, ed. Paolo Mancosu, 198–255. Oxford: Oxford University Press.
Hilbert, David. 1899. *Grundlagen der Geometrie*. Leipzig: B.G. Teubner.
Hilbert, David. 1971. *Foundations of Geometry*. La Salle: Open Court. English translation of Hilbert 1899.
Isaacson, Daniel. 1987. *Arithmetical truth and hidden higher-order concepts*. In Logic Colloquium '85. Amsterdam: North Holland: 147–169.
Klein, Felix. 1873. Über die sogenannte Nicht-Euklidische Geometrie. *Mathematische Annalen* 6: 112–145.
Mancosu, Paolo (ed.). 2008. *The Philosophy of Mathematical Practice*. Oxford: Oxford University Press.

Mancosu, Paolo and Arana, Andrew. 2012. Geometria Piana e Solida: una nota sulla purezza del metodo [Plane and solid geometry: A note on purity of methods]. *Notae Philosophicae Scientiae Formalis* 1: 89–102. http://gcfcf.com.br/pt/revistas/vol1-num1-maio-2012/

Pappus. 1876–1878. *Collectionis quae supersunt*, 3 vols. Berlin: Weidman.

Peano, Giuseppe. 1894. Sui Fondamenti della Geometria. *Rivista di Matematica* 4: 51–90.

Poncelet, Jean-Victor. 1822. *Traité des Propriétés Projectives des Figures*. Paris: Bachelier.

Torricelli, Evangelista. 1919–1944. *Opere di Evangelista Torricelli*. Faenza: Stabilimento Tipografico Montanari.

von Staudt, Karl Georg Christian. 1847. *Geometrie der Lage*. Nürnberg: F. Korn.

Wiener, Hermann. 1892. Über Grundlagen und Aufbau der Geometrie. *Jahresbericht der Deutschen Mathematiker-Vereinigung* 1: 45–48.

Chapter 3
Formalization and Intuition in Husserl's *Raumbuch*

Edoardo Caracciolo

3.1 The *Philosophie der Arithmetik* and the Origins of the *Raumbuch*

The *Philosophie der Arithmetik* was published in 1891, but it marks the convergence point among contrasting theoretical issues dating from 1886, when Husserl moved to Halle and started collaborating with Carl Stumpf in order to obtain his *Habilitation*. During this time, Husserl's ideas start to diverge from those inherited by his former master Karl Weierstrass.[1] Husserl, indeed, declares his intention to carry on Weierstrass' work using the theoretical tools that he inherited from his second master, Franz Brentano.[2] So Husserl dedicated the first part of the *Philosophie der*

This paper deals with the methodological issues that Husserl encountered when he was developing his first space theory. In particular, the present paper tries to throw some light on the impact of intuition and formalization on early Husserl's geometrical studies. In *Il problema dello spazio nel primo Husserl* ("Rivista di Filosofia," vol. CIV, n. 2. agosto 2013), instead, I offer an historical overview on the *Raumbuch*, mostly focusing on the psychological issues, like Husserl's critique of Helmholtz's space theory.

[1] Cf. Miller, J.P. 1982. *Number in presence and absence. A study of Husserl's philosophy and mathematichs*, 11. The Hague/Boston/London: Nijhoff.

[2] Cf. HUA XII pp. 294–295. The *Philosophie der Arithmetik* reflects on the foundation of arithmetic that, in the first part of the book, is defined as "science of number," a subject based on the concept of positive integer. Indeed, this definition is originated from Weierstrass's studies. Claudio Majolino explains in which way Brentanian psychology answers to an exigency of intuitive researches that Weierstrass left unfulfilled. Cf. Majolino, C. 2004. Declinazioni dello spazio, sul rapporto tra spazialità percettiva e spazialità geometrica nel primo Husserl. *Paradigmi* XXII(64/65): 223–238.

E. Caracciolo (✉)
Dipartimento di Filosofia e Scienze dell'Educazione, Università degli studi di Torino, 10124, Italy
e-mail: edoardo.caracciolo@unito.it

Arithmetik to develop a set of psychological analysis aiming to discover the intuitive roots of the concept of number. Actually, after a few chapters, it appears obvious that this inquiry layout rests on a theory of representation that cannot handle those numbers lacking of corresponding intuition (e.g., imaginary and irrational numbers).

After the failure of this first research trend, the second part of the *Philosophie der Arithmetik* contains the ideas that Husserl developed during his lectureship as *Privatdozent* in Halle.[3] Here, the arithmetical method is presented as a computational technique (*Rechenkunst*) that "can break completely free of the conceptual substrata" focusing on the mode of relation (*in der Weise der Beziehung*). In this sense, it is a "formal processing method, i.e., algorithmic," that is

> a system of formal rules by means of which mathematical problems can be solved in purely mechanical operations, i.e. we can find unknown numbers and numerical relations starting from known ones.[4]

The *Rechenkunst* is a valuable method because, filtering all kinds of numbers within the same algorithmic system, it can deal with all conceivable types of numbers: so, it avoids the *impasse* affecting the first section of the book.[5]

In the preface to *Philosophie der Arithmetik*, Husserl alludes to a second volume that should contain logical researches on the arithmetical algorithm and a philosophical theory of Euclidean geometry, both sharing the same principles (*Grundgedanken*).[6] It is possible that he devises to study the applicability of algorithms to the geometrical field, being interested in a frame of research connecting theory of geometry, formal arithmetic, and manifold theory.[7]

In the same time, Husserl plans "[...] to communicate more detailed investigations concerning symbolic representations and the methods of cognition grounded thereon"[8] in an appendix to the second volume. Both algorithms and general psychological representation fall within the domain of symbolic representations

---

For an overall view on Husserl's juvenile years cf. Rollinger, R.D. 1999. *Husserl's position in the school of Brentano*, Phaenomenologica, 15–21. Dordrecht: Kluwer.

[3] Douglas Willard hypothesize that this second perspective was influenced by Schröder's algebra of logic: indeed, Husserl was writing a (negative) review on his *Vorlesungen über die Algebra der Logik* during the composition of the *Philosophie der Arithmetik*, last chapter. (cf. D. Willard, D. 1984. *Logic and the objectivity of knowledge. A study in Husserl' early philosophy*, 109. Athens: Ohio University Press.).

[4] HUA XII p. 132; cf. also pp. 258, 346. For example, numbering is a mechanical operation that "[...] substitutes the names for the concepts, and then by means of the systematic of names and a purely external process, derives names from names, in the course of which there finally issue names whose conceptual interpretation necessarily yelds the result sought" (HUA XII p. 239). On this matter, cf. also Sinigaglia, C. 2000. *La seduzione dello spazio*, 64–66. Milano: Unicopli.

[5] Cf. HUA XII p. 283.

[6] Cf. HUA XII pp. 7–8.

[7] Cf. HUA XXI pp. 244–249, 396. Cf. also Sinigaglia, C. *La seduzione dello spazio*, 61, op. cit.

[8] HUA XII p. 193.

because they are both inauthentic representations.⁹ The link between arithmetic and psychology becomes even more clear considering that *arithmetica universalis* is part of *formal logic*; the latter is still conceived as a Brentanian *Kunstlehre* – the art that detects the proper judgment on the base of psychological categories.¹⁰

The convergence between arithmetical and psychological studies appears quite clearly in geometry because space can be described by special algorithms or analyzed as a psychological representation. At first, Husserl will deal with the formal side of the space problem elaborating a critique of analytical geometry. The psychological perspective, instead, will be developed in the later *Raumbuch*. For these reasons, the *Raumbuch* can be regarded as the last outcome of the philosophical theory of Euclidean geometry that should have been presented in the second volume of the *Philosophie der Arithmetik*.¹¹

3.2 The *Raumbuch* Affair

Before the nineteenth century, geometers believed that Euclidean geometry was based on intuitive space: they only argued on the origin of space representation. Husserl gives a brief account of the old *Raumproblem* (space problem) in some notes dated back to 1893: he distinguishes the apriorist faction – according to which space concepts are already present before any experience (i.e., Kant, König, Baumann, Sigwart), from the empiricist one, according to which space concepts are abstractions or idealizations of empirical spatial figures (i.e., Comte, Mill, Taine, Beneke).¹²

Then, the spread of non-Euclidean geometries complicated the connection between geometrical and intuitive space, showing that geometrical concepts may not be grounded on intuition.¹³ This situation urged to perform a new philosophical

---

⁹In *Zur Logik der Zeichen*, Husserl explores the wide range of symbolic representations. Among them he numbers the artificial signs (*Künstliche Zeichen*) of general arithmetic and those conceptual second class signs (*symbolichen Vorstellungen der Zweiten Klasse*) standing for things that cannot be properly represented; cf. HUA XXI pp. 349–350, 354–356.

¹⁰Cf. HUA XXI p. 248. On the relation between Husserl and the brentanian logic, cf. De Boer, T. 1978. *The development of Husserl's thought*, 91–93. Den Haag: Nijhoff.

¹¹Cf. Argentieri, N. 2008. Matematica e fenomenologia dello spazio. In *Forma e materia dello spazio, dialogo con Edmund Husserl*, ed. P. Natorp, 246, edited by N. Argentieri. Napoli: Bibliopolis, Corrado Sinigaglia proposes a close analysis of the relations between the *Philosophie der Arithmetik* and the *Raumbuch*; cf. Sinigaglia, C. 2001. La libera variazione delle forme. Husserl lettore di Riemann. In *Logica e politica. Per Marco Mondadori, Fondazione Arnoldo e Alberto Mondadori*, edited by M. D'Agostino, G. Giorello, and S. Veca, 377–403. Milano: il Saggiatore.

¹²Cf. HUA XXI pp. 285–286.

¹³Non-Euclidean geometries deny the parallel postulate. This postulate states that if a straight line falling on two straight lines makes the interior angles on the same side less than two right angles, then the two straight lines, if produced indefinitely, meet on that side on which the angles are less

foundation of geometry, since it offered new perspectives on the issue, therefore updating the classical dispute on the *Raumproblem*. Indeed, non-Euclidean ideas were born within a mathematical frame of discussion but reverberated through the scientific world stimulating new interpretation of space representation from biological, psychological, and philosophical points of view. Thus, some philosophers dealt with non-Euclidean geometries, opening the way to a common research field for geometry and philosophy. For example, both Hermann Lotze's and Hermann von Helmholtz's investigations conclude that our space representation does not reflect the external space. According to Lotze, space is the form through which mind perceives the external space acting upon the mind itself: since that form coincides with Euclidean space, a non-Euclidean intuition is just impossible.[14] According to Helmholtz too, space representation is consistent with Euclidean law since it originates from the properties of an Euclidean world affecting our nerves. Nevertheless, we can suppose that a stimulus generated by a non-Euclidean world could induce a non-Euclidean intuition.[15]

Having this debate in the background, Husserl has been interested in geometry at least since 1886, when he writes a note on homogeneous and heterogeneous *continua*. During the 1889–1890 winter semester, he also delivers lectures on the *Grundproblem der Geometrie*, so at the time, he is already dealing with space from a mathematical point of view.[16] Between 1892 and 1894, he writes down a collection of notes in which he deals with the psychological/philosophical side of the *Raumproblem*, outlining his first theory of space. This work is approximately planned in a brief draft called "Spacebook diary" (*das Tagebuch zum Raumbuch*); following this clue, Ingeborg Strohmeyer gathered the recommended passages and shaped them in a quite organic treatise that has been published in Husserliana XXI volume.[17]

---

than the two right angles. Cf. M. Kline, M. 1972. *Mathematical thought from ancient to modern times*, vol. I, p. 59, vol. III, p. 865. New York/Oxford: Oxford University Press. On non-Euclidean geometries paternity, cf. Kline, M. *Mathematical thought from ancient to modern times*, vol. III, 869–870, op. cit.

[14]Cf. Torretti, R. 1984. *Philosophy of geometry from Riemann to Poincaré*, 285–291. Dordrecht: Reidel. On the fracture between things and representation, cf. Lotze, H. 1899. *Microcosmus: An essay concerning man and his relation to the world*, 344–353, 573–578. Edinburgh: T. & T. Clark.

[15]Cf. Helmholtz, H. 1867. *Handbuch der Physiologischen Optik*, 194. Leipzig: Voss; Helmholtz, H. 1876. The origin and the meaning of geometrical axioms. *Mind* 1(3): 316–318. On the relation between Lotze and Helmholtz, cf. Gehlhaar, S. 1991. *Die Frühepositivistische (Helmholtz) und phänomenologische (Husserl) Revision der Kantischen Erkenntnislehre*, 30. Cuxhaven: Junghans-Verlag.

[16]The note can be found in Ms. K I 50/47a. The *Grundproblem der Geometrie* is published in HUA XXI pp. 312–347.

[17]Cf. HUA XXI pp. 262–311. The *Raumbuch* structure is presented in a note published in HUA XXI pp. 402–404. For an historical panorama on the *Raumbuch* birth and on the *Tagebuch zum Raumbuch*, cf. the Textkritische Anmerkungen published in HUA XXI pp. 469, 485–486; HUA D. I pp. 36–37; Mohanty, J. N. 1999. The development of husserl's thought. In *The Cambridge*

3.4 Mereology, Material *a Priori*, and Idealization: The Other Way

Several passages of the *Raumbuch* witness the essential role that intuitive qualities play in the descriptive analysis of space – an aspect shared with Stumpf's *Über den psychologischen Ursprung der Raumvorstellung*.[39] And just like Stumpf, Husserl conceives space as a multisided whole.

Husserl, in fact, notices that the "space of the world" (*Weltraum*) is composed by a variety of places connected through a network of symbolic cross-references[40]: for instance, the observation of the wall naturally leads to perceive the room; the exploration of the room reminds that this last one is part of the flat that, in turn, is a portion of the house located in the neighborhood and so on until the process reaches the space that contains every places, i.e., the space of the world. This connection between places is not only possible but also necessary because each place neither can *be*, nor can *be perceived*, nor can *be thought* without surrounding places. Therefore, each intuitive representation of space contains a symbolical reference to the surrounding space.[41]

The smallest part of this spatial mosaic is "spatiality" (*das Räumliche*), a basic extent that is the "abstract substratum" of every intuitive quality.[42] This argumentation was previously deploid by Stumpf against the Kantian thesis according to which space would be the form of sensibility organizing phenomenical multiplicity.[43] According to Stumpf and Husserl, the concept of space simply highlights the structure of real space, the organization of raw empirical data instead of shaping it. For instance, both extent and visual/tactile qualities display their own configuration: they are "abstract elements" or "grounded contents" because we cannot conceive

---

[39]C. Stumpf, C. 1873. *Über den psychologischen Ursprung der Raumvorstellung*. Leipzig: Verlag von S. Hirzel.

[40]Cf. HUA XXI p. 281.

[41]This argument may remind a Kantian thesis, but, actually, the *Raumbuch* displays an anti-Kantian perspective on space. Kant wanted to prove the priority of spatial form over spatial object showing that object cannot be displayed without a surrounding space, whereas space itself can be conceived as object-free. Instead, Husserl speaks in terms of extension: the single fraction of space is an extension as well as the world space. Obviously, the first extension is part of the second one, but – here it is the difference from kantianism – the single place cannot be conceived without conceiving its surrounding places as well as the world space cannot be conceived without its composing parts. On this subject, cf. Kant, I. *Kritik der reinen Vernunft*, A24, B39. This thesis anticipates, in a different theoretical context, an idea that Husserl will elaborate in the *Logischen Untersuchungen*. There he notices that every representation has both intuitive and symbolical sides, each one contributing in a different degree to the whole representation. Cf. HUA XIX pp. 610–614.

[42]HUA XXI p. 276.

[43]Cf. Kant, I. *Kritik der reinen Vernunft*, A 99, 107, 120n; B 201-2n, 218-9, 129–130, 134–135. Victor Popescu highlights the subtle differences between Stumpf's and Husserl's mereologies: cf. Popescu, V. 2003. Espace et mouvement chez Stumpf et Husserl, une approche méréologique. *Studia Phaenomenologica* III(1–2): 115–133.

extent without color in such a way "that the suppression of the former implies the suppression of the latter."⁴⁴ Besides, extent and qualities are connected in a subtle way. For example, colors fade or shine depending on surface illumination and according to *a priori* material laws; on the other hand, when color is obscured, surface disappears as well. These relations do not concern formal consistence between parts (*contra* Kant) and they are neither grounded on habits (*contra* Empiricism). Instead, they express an objective configuration that does not change depending on the intentional subject.⁴⁵

At this basic stage of perception, we can only sense sides of things. In order to perceive an object, we have to synthesize the separated extent into a stable composition of visual sides. Each perceived side contains symbolical references pointing to the other side; the synthetic act binds these symbolic references to the first intuition and crystallizes them into an object (e.g., the room is composed following the references to the adjacent walls that are contained in the perception of the first wall). This gradual composition is a "teleological process" because it aims to give a complete representation of the object, i.e., to "accomplish" this object perception linking all its sides into a complete whole.⁴⁶

Sensible objects exhibit intuitive qualities in a greater or lesser degree of perfection: e.g., a straight line could be more or less straight and a point could have more or less extent. We can appreciate qualitative differences because each quality value is disposed on a teleological scale leading to an ideal perfection, to an unperceivable "limit" (straightness, redness). As Husserl will point out in the *Ideen* – within a different theoretical context – geometrical concepts as "ideas in a Kantian way" express "something invisible."⁴⁷ In order to conceive these concepts, we should execute an "idealization" (*Idealisierung*) – a process that reiterates endlessly an "almost induction" (*Quasi-Induktion*) and accentuates a material content until perfection.⁴⁸ Indeed, idealization "starts from what is intuitively given and implied

---

[44]Cf. HUA XXI pp. 281, 307. Cf. also Stumpf, C. *Über den psychologischen Ursprung der Raumvorstellung*, 107–109, op. cit. This distinction will be further developed in the *Psychologische Studien zur elementaren Logik* (cf. HUA XXII pp. 97–98) and in the *Logische Untersuchungen* (cf. HUA XIX pp. 231–240, 272–274). Cf. Kaiser-El-Safti, M. *Fenomenologia trascendentale versus iletica. Psicologia e fenomenologia in Husserl e Stumpf*, 236, op. cit.; Majolino, C. *Declinazioni dello spazio, sul rapporto tra spazialità percettiva e spazialità geometrica nel primo Husserl*, 230–231, op. cit.

[45]According to Stumpf, those judgments describing objective relations are necessary by nature and universally valid. Starting from those judgments, we can develop a set of *a priori* material laws. Cf. Stumpf, C. 1982. Psychologie und Erkenntinistheorie. In *Abhandlung der Königlich Bayerischen Akademie der Wissenschaften*, I Classe, 19, 2, München, 494–495. On this subject cf. De Palma, V. 2001. *L'a priori del contenuto. Il rovesciamento della rivoluzione copernicana in Stumpf e Husserl*. In: *Carl Stumpf e la fenomenologia dell'esperienza immediata*, edited by S. Besoli and R. Martinelli, Discipline Filosofiche, XI, 2, 316–318. Macerata: Quodlibet.

[46]HUA XXI p. 284. Pursuing this strand of research, in the *Dingvorlesung*, Husserl will deal with the problem of the tridimensional circularity of the real object.

[47]Cf. HUA III/1 p. 138.

[48]Cf. HUA XXI p. 286.

into the nature of a thing"[49] and simply enhances the material content. This happens, for example, when we detect a median point between two points that are gradually getting closer: when these two points became indistinguishable, the process can be further protracted beyond "the limits of the operating potentiality of our measuring instruments."[50] In this way, we gain the "best conditions of sight,"[51] and since our intentional activity has been freed from any subjective defect, we can finally conceive the geometrical concept in its ideal and universal objectivity. Thus, for instance, we conceive the concept of point subtracting extent to the point until it becomes a dimensionless geometrical entity. The real point and the concept of point are both dimensionless to different degree: the real object is linked to the concept through a shared content (e.g., being dimensionless). This shared content legitimizes the bond of continuity between object and concept, and therefore, it highlights the intuitive roots of the concept. This kind of link is further confirmed by the continuity of the idealization process: indeed, idealization connects concept to intuition by an uninterrupted and iterative induction. For example, the concept of point is a product of an uninterrupted process that subtracts extent to the point.[52] Thanks to this double line of continuity, intuition and concept are so "similar" that "intuitions symbolize concepts, the former are not the object of concepts but symbols, more precisely hieroglyphs of the concepts."[53]

The symbolic relation between concept and intuition makes conceptual work simpler since it allows to translate a conceptual problem into intuitive terms. Nevertheless, intuitive evidence is not exact as formal evidence and an intuitive demonstration is not as rigorous as a formal demonstration: by interpreting topics of pure geometry in terms of intuitive figures we may oversimplify the issue.[54] Thus, in notes dated 1894, Husserl reconsiders the differences between the analytic and the "other way," and this time, he underlines the merit of the analytic side. According to the "other way," intuition and concept should be reconnected by idealization – a procedure enhancing similarities: actually, many passages clearly deny there is such a similarity. For instance, as noted down in 1893, sensible space and ideal space have totally different features since, while we can perceive the former, we can only think the latter. To be more precise, pure geometry is a formal domain of contentless objects that "has to expels errors from the same foundations by a purely

---

[49]HUA XXI p. 308.

[50]HUA XXI p. 296. This passage reminds Lobačevsky's *New principles of geometry*: "[...] it will be possible to form any body by means of composition, reaching an identity degree beyond which our senses stop perceiving imperfections. [...] although we get our first concepts from it [the nature] we owe the rigor of the former to our senses imperfection" (Lobačevsky, N. *Neue Anfangsgründe der Geometrie mit einer vollständigen Theorie der Parallellinien*, p. 81, op. cit.).

[51]HUA XXI p. 287.

[52]*Ibidem.*

[53]Cf. HUA XXI pp. 289–290, 294. Not only single objects but the entire intuitive space may be used as a symbolic surrogate of pure geometrical space.

[54]Cf. HUA XXI p. 295.

formal procedure and rigorous axioms and [has to] show the intuitive procedure in its own limits [...]."[55] It deals with a pure concept of space that shares almost nothing with the empirical concept of space studied by physical geometry: we cannot subsume the latter under the former because there is no continuity between a formal concept, devoid of any contents, and a sensible concept, still characterized by material contents.[56] Husserl further develops this idea, and in a letter he writes to Natorp in 1897, he numbers three concepts of space differing in formal purity. The spatial manifold is the most formal concept; from it we deduce the tridimensional Euclidean manifold by formal determination. The third and less formal concept is the concept of intuitive space that cannot be derived by formal determination because it is enriched by material contents. Thus, there are two kinds of space concept – the formal one and the material one. Moreover, these two kinds of concept cannot be linked through a single act of mind – neither formal determination nor idealization.[57] This last process, in fact, reveals its uselessness when it pretends to conceive formal concept – devoid of any content – by a continuous enhancement of intuitive contents.

3.5 Representation, Intuition, and Symbolization

Such a methodological issue implies that when Husserl was planning the *Raumbuch*, he did not clearly distinguish between formal and material concepts.[58] First, he needed to clarify which kind of intellectual act could conceive concepts, that is he needed to discover that we can visualize some kinds of concepts through an intuition (*Anschauung*).[59] In the *Raumbuch*, instead, he still relies on a slightly modified version of the theory of representation introduced in the *Philosophie der Arithmetik*: he makes a few distinctions, but he still contrasts concept and intuition.

> [...] we should ask ourselves if the real representation that each time we have, has the character of intuition or symbolization (*Repräsentation*) and, in this last case, if it has the character of an intuitive or non-intuitive symbolization (proper or improper) of what we call space. In the case of non-intuitive representations we have to investigate if they have [...] the character of conceptual representation, which relation they have with corresponding intuitions, if they can be grounded on these last ones or if [...] they necessarily lack of corresponding intuition.[60]

---

[55]HUA XXI pp. 271, 295–296.

[56]Cf. HUA XXI p. 296. This passage anticipates the distinction between physical and pure geometry in the *Prolegomena*; cf. HUA XVIII p. 251.

[57]Cf. HUA D. III/5 pp. 53–54.

[58]Cf. Brisart, R. *Le Général et l'abstrait: sur la maturation des Recherces Logiques de Husserl*, 39–40, op. cit.

[59]This idea shows up in the *Psychologische Studien zur elementaren Logik*, cf. HUA XXII p. 104.

[60]HUA XXI p. 262.

Husserl still defines intuition and symbolization according to the guidelines of the *Philosophie der Arithmetik*:

> If a content is not directly given that which it is, but it is only indirectly given through signs that univocally characterize it, then, instead of having an authentic representation, we have a symbolic representation of it [...].[61]

So, because of symbolization mechanics, we can represent concepts through intuitions standing for them: for example, a real point may stand for the concept of point because watching the former we catch a symbolic link to the latter. This connection implies that intuitions and concepts are both different and similar in a way that Husserl does not further clarify.

Furthermore, symbolization (whose content *is not* directly given to us) is defined as a mere negation of intuition (whose content *is* directly given to us), and therefore, its representational domain is reduced to what *is not* intuitive. As a consequence, symbolization has not an autonomous representational status.[62] Such a feeble demarcation of the conceptual domain can be interpreted as a symptom of veiled and impending psychologism. If "what is intuitive" is determined by subjective configurations and if symbolization is simply "what is not intuitive," then these psychological faculties will define "what is not intuitive" too. Thus, psychological faculties circumscribe the symbolical domain (and the conceptual one within it) by defining what they are not. For example, empirical concepts are presented as what is beyond "the limits of the operating potentiality of our measuring instruments."[63] Leaving aside the classical debate about Husserl's supposed psychologism, it seems that the domain of symbolization shrinks depending on the extent of the intuition domain.[64] Nevertheless, we find various *Raumbuch* passages implying that concepts have an objective and defined status. For instance, an impossible concept cannot be represented, no matter the subject; in another note, he says that a concept of space based on material *a priori* determines the conditions of possibility of experience; elsewhere, idealization is presented as a procedure purifying the psychic process from subjective defects.[65] In the end, it seems that the *Raumbuch* representational theory is quite fuzzy.

---

[61] XII p. 193. This definition has many similarities with the one that Husserl gives in HUA XXI p. 272. It is worth noticing a minor semantic sliding: the "proper representations" in the *Philosophie der Arithmetik* are named "intuitions" in the *Raumbuch*.

[62] Cf. HUA XXI pp. 295–296. Husserl will define the representational status of concepts when he will deal with the categorial intuition in the *Logische Untersuchungen*. There he also dismantles the intuition/symbolization dichotomy that structures the *Raumbuch* representational theory.

[63] HUA XXI pp. 295–296.

[64] That reminds an idea from the *Philosophie der Arithmetik*, where arithmetic is presented as a tool dealing with sets that cannot be intuited because of subjective inability; consequently, since powerful subjectivities, like angels or God, need not to develop arithmetic to handle large sets, then their arithmetical domain (objects and procedures) is almost empty. At the end, "results" are the same: both man and angel represent the same large sets, but the former uses a tool (arithmetic), whereas the latter need it not. Cf. HUA XII pp. 191–192.

[65] Cf. HUA XXI pp. 262, 296, 287.

Moreover, the internal distinctions between the various kinds of symbolization are not even always respected, at least not during the psychological analysis of the "everyday life space." According to Husserl, the "everyday life space" is an ideal formation: a everyday life object, in fact, is an "ideal object" formed by an intellectual synthesis binding together all the sides of the object.[66] No object, indeed, can show us all its sides at the same time, but, nevertheless, we intend the complete object with all its sides when we conceive the object. For example we know that a dice has six faces but we cannot *see* six faces at the same time. We know, however, that the three manifest faces hide other three faces and that all those faces together form the dice. When Husserl says that every real object is an ideal object, Husserl confuses the symbolic link between the intuitive symbol (the manifest three sides) and the symbolized object (the hidden three sides) with a symbolic link between the intuitive symbol (the three sides) and the symbolized concept (the concept of dice).

In general, this theory lacks of balanced composition of its founding concepts. For instance, non-intuitive symbolization (concept) is defined as alternative to intuition, despite non-intuitive symbolization (concept) has a direct contrary, i.e., intuitive symbolization (intuitive signs). Such a definition has two consequences. First, we do not know what symbolization is *per se*, but we only know that it is not intuition. Second, the various species of symbolization are not defined as reciprocal alternatives but, all together, as alternatives to intuition.

The conceptual representation *depends* on intuition because Husserl tries to ground concepts into sensible experience relying on immature psychological methods. This effort becomes evident when he analyzes real bodies following a method inspired by solid geometry. He adopts the technical terminology of geometry when he explains how "division" (*Teilung*) decomposes the "physical body" (*Körper*) into geometrical entities as "surfaces" (*Flächen*), "lines" (*Linien*), and "points" (*Punkte*).[67] The transition from the esthetical to the geometrical dimension is witnessed by a synonymical overlap of the terms "physical body" and "figure" (*Gebilde*) – an overlap justified by the fact that we can extract "forms" (*Formen*) and "corporeal figures" (*körperlichen Figuren*) from every physical body.[68] At the base of this consideration, there is a major confusion between external experience (physical/esthetical body) and internal one (geometrical form). Thus, the *Raumbuch* betrays the first rule of immanentism according to which, since external experience data are untrustworthy, analysis should be focused only on the immanence of conscience – where the features of intentional objects can be ascertained once

---

[66]Cf. HUA XXI pp. 281–283.

[67]Cf. HUA XXI pp. 278–279.

[68]Cf. HUA XXI pp. 278–279, 286. Stumpf explains that many space theories of his time incorrectly mix two strands of research that should be kept separated: epistemology, focusing on immediately evident truths, merges with descriptive psychology, focusing on the genesis of concepts. Thus, the researches on the origin of spatial representation overlap the studies on the nature of geometrical axioms. For this reason, the spatial analysis of the early Husserl displays a geometrical nuance. Cf. Stumpf, C. *Psychologie und Erkenntinistheorie*, 484, op.cit.

for all.[69] In order to achieve an optimal description of the spatial representation, analysis should keep spatial sensations and spatial things separated; moreover, this separation should be actively maintained. Husserl will satisfy these two requisites by developing the *epoché* in the *Ideen*. This procedure neutralizes the presuppositions threatening the purity and the independence of phenomenological analysis: for instance, we should "put between brackets" the belief in the existence of the natural world in order to focus on sensations rather than on things, unlike what happen within the *Raumbuch* frame.[70] Moreover, a radical philosophical investigation should avoid concepts, methods, and practices that have been derived from other sciences. A pure psychological analysis should also start from the bottom, highlighting those primordial structures that ground the edifice of science: in this sense, we should not aim to justify a scientific idea since that would adjust analysis and distort results. Actually, the *Raumbuch* analysis is influenced both by implicit geometrical categories and by a previously established aim, i.e., the justification of Euclidean geometry.[71]

Thus, because of a methodological immaturity, the first Husserlian theory of space collapses within a few years. The *Raumbuch* is one of those experiences pushing Husserl to conceive the *epoché* – a phenomenological method that will redefine the relations between space, geometry, and experience:

> If the province of phenomenology were presented with such an immediate obviousness as the province pertaining to the natural attitude in experiencing, or if it became given in consequence of a simple transition from the latter to the eidetic attitude as, for example, the province of geometry becomes given when one starts from what is empirically spatial, then there would be no need of circumstantial reductions with the difficult deliberations which they involve.[72]

3.6 Conclusion

Although just after the publication of the *Philosophie der Arithmetik* Husserl plans to develop a formal approach to geometry, in the notes from the early 1890s, he criticizes the formal method of analytical geometry for not being able to grasp the essence of space. So he chooses to investigate the intuitive side of space through

---

[69]Husserl inherits immanentism from Brentano, and when he is working on the *Raumbuch*, he still adopts this intentional theory. For example, he stresses the distinction between immanent object (*immanente Objekt*) and real object; he defines the metaphysical space – i.e., the real space – as transcendent space (*transzendent Raum*). Cf. HUA XXI pp. 262, 265–266, 270, 305. Paradoxically, he makes the same mistake that he highlights in Helmholtz's empiricist space theory: according to him, Helmholtz confuses the inner psychological experience with the real external one. Cf. HUA XXI p. 309.

[70]Cf. HUA III/1 pp. 59–60, 108, 115–116.

[71]Cf. HUA III/1 pp. 112–115. It is no accident that several sciences presuppose the axiomatic method called *mathesis universalis* – whose first model was Euclidean geometry.

[72]HUA III/1 pp. 115–116.

the *Raumbuch* psychological analysis; anyway, in the *Raumbuch* last notes, he approves formalization again. Such a theoretical mutability may be explained if we consider that in the early 1890s, Husserl is still defining the representational status of conceptual representations; as a consequence, he has not conclusively established if geometry is a formal or a material science or both. For these reasons, he cannot formulate an ultimate description of the geometrical method.

The first step towards a solution can be found in the *Prolegomena* where Husserl clarifies the relation between intuitive geometry as "phenomenical space science" and formal geometry as "categorial form of geometrical theory."[73] Although this distinction has already been sketched in the *Raumbuch* note dealing with the differences between physical and pure geometries, in the *Prolegomena*, it is associated with a pondered reflection about methods that redefine the various kinds of concepts. This process is completed in the *Ideen*, where Husserl coins two procedures that elaborates formal and material concepts, each one within its own research field. The first procedure is "formalization" and replaces contents with contentless variables in order to reduce a material field to a manifold whose formal objects are defined solely by the form of their connections with other objects.[74] The second procedure is called "generalization" and explores all the possible manifestations of a material content by means of imagination; it investigates which features are essential and which ones are not, until the eidetic essence of the content emerges as the invariable core of every possible manifestation.[75]

This methodological reorganization redefined the relations between geometry and space theory. The phenomenological space theory, as developed in the *Dingvorlesung*, abandons geometrical categories and adopts esthetical ones: it primarily focus on the transcendental constitution of the real space rather than on our geometrical representations. In the same time, geometry deals no more with intuitive space, contents and qualities: as specified in the *Ideen*, geometry should be an axiomatic-deductive system dealing with exact concepts. Once its proper object has been detected, geometry finally finds a place among the other sciences as an "exact science" based on formalization.[76]

References

Argentieri, N. 2008. Matematica e fenomenologia dello spazio. In *Forma e materia dello spazio, dialogo con Edmund Husserl*, ed. P. Natorp, 187–294, edited by N. Argentieri. Napoli: Bibliopolis.
Brisart, R. 2003. Le Général et l'abstrait: sur la maturation des Recherces Logiques de Husserl. In *Aux origines de la phénoménologie*, edited by D. Fisette e S. Lapointe. Paris: Vrin.

---

[73]Cf. HUA XVIII p. 252.
[74]Cf. HUA III/1 pp. 133–136; HUA XVII pp. 79–80; HUA XVIII pp. 247–248.
[75]Cf. HUA III/1 pp. 131–132.
[76]Cf. HUA III/1 pp. 133–139.

De Boer, T. 1978. *The development of Husserl's thought.* Den Haag: Nijhoff.
De Palma, V. 2001. *L*'a priori *del contenuto. Il rovesciamento della rivoluzione copernicana in Stumpf e Husserl.* In: *Carl Stumpf e la fenomenologia dell'esperienza immediata,* edited by S. Besoli and R. Martinelli, Discipline Filosofiche, XI, 2, 309–352. Macerata: Quodlibet.
Gehlhaar, S. 1991. *Die Frühepositivistsche (Helmholtz) und phänomenologische (Husserl) Revision der Kantischen Erkenntnislehre.* Cuxhaven: Junghans-Verlag.
Grassman, H. 1878. *Die lineale Ausdehnungslehre.* Leipzig: Otto Wigand.
Hartimo, M.H. 2007. Towards completeness: Husserl on theories of manifolds 1890–1901. *Synthese* 156: 281–310.
Hartimo, M.H. 2008. From geometry to phenomenology. *Synthese* 162: 225–233.
Helmholtz, H. 1867. *Handbuch der Physiologischen Optik.* Leipzig: Voss.
Helmholtz, H. 1876. The origin and the meaning off geometrical axioms. *Mind* 1(3): 301–321.
Helmholtz, H. 1921. Über die Thatsachen, die der Geometrie zu Grunde liegen (1868). In *Schriften zur Erkenntnistheorie,* edited by M. Schlick and P. Hertz. New York: Springer.
Husserl, E. 1970. *Philosophie der Arithmetik mit erganzenten Texten, (Husserliana* XII), edited by L. Eley. Den Haag: Nijhoff.
Husserl, E. 1973. *Ding und Raum Vorlesungen 1907, (Husserliana* XVI), edited by U. Claesges. Den Haag: Nijhoff.
Husserl, E. 1975. *Logische Untersuchungen. Erster Teil, (Husserliana* XVIII), edited by E. Holenstein. Den Haag: Nijhoff.
Husserl, E. 1976. *Ideen zu einer reinen Phänomenologie und phänomenologischen Philosophie, (Husserliana* III/1 e III/2), edited by W. Biemel. Den Haag: Nijhoff.
Husserl, E. 1979. Psychologische Studien zur elementaren Logik. In *Aufsätze und Rezensionen (1890–1910), (Husserliana* XXII), edited by B. Rang, 92–123. Den Haag: Nijhoff.
Husserl, E. 1983. *Studien zur Arithmetik und Geometrie, (Husserliana* XXI), edited by I. Strohmeier. Den Haag: Nijhoff.
Husserl, E. 1984. *Logische Untersuchungen. Zweiter Teil, (Husserliana* XIX), edited by U. Panzer. Den Haag: Nijhoff.
Husserl, E. 1994a. *Briefwechsel – Die Brentanoschule, (Husserliana Dokumente* III/1), edited by K. Schuhmann. Deen Haag: Kluwer.
Husserl, E. 1994b. *Briefwechsel – Die Neukantianer, (Husserliana Dokumente* III/5), edited by K. Shuhmann. Den Haag: Kluwer.
Kaiser-El-Safti, M. Fenomenologia trascendentale versus iletica. Psicologia e fenomenologia in Husserl e Stumpf. In *Carl Stumpf e la fenomenologia dell'esperienza immediata,* edited by S. Besoli and R. Martinelli, Discipline Filosofiche, Anno XI, numero 2, 231–260. Macerata: Quodlibet.
Kant, I. 1966. *Kritik der reinen Vernunft,* edited by Ingeborg Heidemann. Stuttgart: Philipp Reclam Jun.
Kline, M. 1972. *Mathematical thought from ancient to modern times,* vol. I–III. New York/Oxford: Oxford University Press.
Lobačevsky, N. 1898. Neue Anfangsgründe der Geometrie mit einer vollständigen Theorie der Parallellinien. In *Zwei geometrische abhandlungen aus dem russischen uebersetzt, mit anmerkungen und mit einer biographie des verfassers,* edited by F. Engel and P. Stäckel, 67–236. Leipzig: Teubner.
Lotze, H. 1899. *Microcosmus: An essay concerning man and his relation to the world.* Edinburgh: T. & T. Clark.
Majolino, C. 2004. Declinazioni dello spazio, sul rapporto tra spazialità percettiva e spazialità geometrica nel primo Husserl. *Paradigmi* XXII(64/65): 223–238.
Miller, J.P. 1982. *Number in presence and absence. A study of Husserl's philosophy and mathematichs.* The Hague/Boston/London: Nijhoff.
Mohanty, J. N. 1999. The development of Husserl's thought. In *The Cambridge companion to Husserl,* edited by B. Smith and D. W. Smith, 45–77. Cambridge: Cambridge University Press.
Mulligan, K. 1999. Perception. In *The Cambridge companion to Husserl,* edited by B. Smith and D. W. Smith, 168–238. Cambridge: Cambridge University Press.

Parrocchia, D. 1994. La forme générale de la philosophie husserlienne et la théorie des multiplicités. *Kairos* 5: 133–164.
Riemann, B. 1868. Über die Hypothesen, welche der Geometrie zugrunde liegen. In *Abhandlungen der Königlichen Gesellschaft der Wissenschaften in Göttingen*, vol. XIII, 133–152. Göttingen.
Rollinger, R.D. 1999. *Husserl's position in the school of Brentano*, Phaenomenologica, vol. 150. Dordrecht: Kluwer.
Schuhmann, K. 1977. *Husserl Chronick, (Husserliana Dokumente I)*. Den Haag: Nijhoff.
Sinigaglia, C. 2000. *La seduzione dello spazio*. Milano: Unicopli.
Sinigaglia, C. 2001. La libera variazione delle forme. Husserl lettore di Riemann. In *Logica e politica. Per Marco Mondadori, Fondazione Arnoldo e Alberto Mondadori*, edited by M. D'Agostino, G. Giorello, and S. Veca, 377–403. Milano: il Saggiatore.
Stumpf, C. 1873. *Über den psychologischen Ursprung der Raumvorstellung*. Leipzig: Verlag von S. Hirzel.
Stumpf, C. 1892. Psychologie und Erkenntinistheorie. In *Abhandlung der Königlich Bayerischen Akademie der Wissenschaften*, I Classe, 19, 2, München.
Torretti, R. 1984. *Philosophy of geometry from Riemann to Poincaré*. Dordrecht: Reidel.
Willard, D. 1984. *Logic and the objectivity of knowledge. A study in Husserl' early philosophy*. Athens: Ohio University Press.

Part II
Looking at Mathematics Through Logic

Chapter 4
Frege's *Grundgesetze* and a Reassessment of Predicativity

Francesca Boccuni

It is well known that Frege's *Grundgesetze der Arithmetik* is inconsistent. The inconsistency is due to the coexistence of two assumptions within Frege's formal system, namely, the impredicative second-order comprehension axiom and unrestricted Basic Law V.[1] Still, it is also known that there are consistent fragments of *Grundgesetze*. In the 1980s, Peter Schroeder-Heister and Terence Parsons provided, respectively, a syntactic and a semantic proof of consistency for the first-order fragment of Frege's *Grundgesetze*. At the time, Parsons conjectured that any extension of the first-order fragment to some second-order system of *Grundgesetze* would result in an inconsistent set of axioms. Nevertheless, in 1996, Richard Heck proved that the *predicative* second-order fragment of *Grundgesetze* has a model. A few years later, Kai Wehmeier proved the consistency of the Δ_1^1-fragment of *Grundgesetze*.[2] This article will concern mainly Heck (1996). Heck's result shows that the fragment of *Grundgesetze* resulting from predicatively restricting the comprehension axiom,[3] while maintaining Basic Law V unrestricted, is consistent. Though Heck (1996) focuses on achieving a technical goal, one may wonder what the possible foundational applications of his result are. In particular, one may be

---

[1] In *Grundgesetze*, one finds the so-called substitution rule, which is nevertheless equivalent to the usual second-order unrestricted comprehension axiom from second-order logic. For a matter of perspicuity, I will discuss the unrestricted comprehension axiom. Basic Law V is the renowned Frege's axiom according to which extensions α and β are identical if, and only if, their corresponding concepts F and G are coextensive.

[2] See also Burgess (2005) and Ferreira and Wehmeier (2002).

[3] By this restriction, no bound second-order variables are allowed on the right-hand side of the axiom's biconditional.

F. Boccuni (✉)
Faculty of Philosophy, University Vita-Salute San Raffaele, 20132, Milan, Italy
e-mail: boccuni.francesca@unisr.it

interested in whether such a consistent fragment could provide a formal core for revising Frege's foundational programme.

Nevertheless, in order to be feasible at all, any such revision of Frege's logicism has to deal with two important issues: first, the issue of what the mathematical strength of such a revision is as compared to Frege's original programme of a logicist foundation of arithmetic; secondly, the issue of whether such a revision implies some radical modifications of Frege's philosophical assumptions.

As far as mathematical strength is concerned, the predicative fragment of *Grundgesetze* is known to be quite weak, since it is equi-interpretable with Robinson arithmetic Q.[4] Provided that Frege's logicism really is the claim that arithmetic is derivable from purely logical basis, such a system should be strengthened so that it recovers, in the best case scenario, full second-order Peano arithmetic. Nevertheless, my main interest in what follows will concern the issue of the possible revisions of Frege's *philosophical* assumptions in a predicative setting. In particular, since the predicative restriction on the comprehension axiom affects, first and foremost, Frege's view on *concepts*, I will investigate whether there is some possible interpretation of predicativity that is compatible with Frege's philosophical view on them.[5] This will be achieved through a general reassessment of the notion of *predicativity*, as it is first motivated by Gödel (1944). Gödel's objection to the use of predicativity in mathematics relies on ontological considerations. The reassessment of predicativity I propose detaches the acceptability of a predicative approach from ontological preferences and connects it to logico-mathematical reasoning. On these grounds, I will finally investigate where things stand as for Frege's philosophical view on concepts in a predicative setting, and I will conclude that such a view may be at least partially retained.

4.1 Predicativity and Predicativism: Russell's VCP

Predicativity is just a syntactic means by which the definition of a mathematical entity by quantification over a totality it belongs to is disallowed and some large portions of mathematics can or cannot be recovered. Predicativity in mathematics has to be distinguished from predicativism, which is the cluster of philosophical

---

[4] See Burgess (2005), Ferreira and Wehmeier (2002), Ganea (2007), and Heck (1996).

[5] Ferreira and Wehmeier (2002) shows that the Δ_1^1-comprehension axiom augmented by unrestricted Basic Law V is consistent. Δ_1^1-comprehension allows only for second-order existential formulæ that are provably equivalent in the system to second-order universal formulæ to appear on the right-hand side of the biconditional. Still, in what follows I will focus on Heck (1996). Δ_1^1-comprehension with Basic Law V, in fact, though very interesting mathematically because of its consistency proof, is still mathematically quite weak, since it is taken to interpret just Robinson arithmetic Q, like Heck (1996). So, if we take the recovery of portions of mathematics larger than Q as one of the two important issues any revisions of Frege's logicism should tackle, then Δ_1^1-comprehension with Basic Law V falls short of being an alternative to Heck (1996) as for broader foundational purposes.

claims motivating predicativity.[6] These claims are usually taken to be captured by Russell's *vicious circle principle* (VCP). Gödel (1944) is possibly the most classical article where a detailed discussion about VCP takes place. It has also been possibly the most influential view ever since on the divide between predicativity and impredicativity, and their philosophical implications. It is just fair, then, that VCP and Gödel's criticism are among the main topics of this article. In particular, I will defend a form of VCP against Gödel's criticisms as they are presented in Gödel (1944).

It is well known that Russell offers several formulations of VCP.[7] Gödel (1944) points out that these formulations boil down to three VCPs, respectively, formulated in terms of *definability*, *presupposition*, and *involvement*:

Definability VCP If, provided a certain collection had a total, it would have members only definable in terms of that total, then the said collection has no total.[8]

Presupposition VCP Given any set of objects such that, if we suppose the set to have a total, it will contain members which presuppose this total, then such a set cannot have a total.[9]

Involvement VCP Whatever involves all of a collection must not be one of the collection.[10]

Famously, Gödel (1944) focuses mainly on a critical appraisal of Definability VCP. Nonetheless, Jung (1999) shows that Presupposition VCP is the most basic formulation of the principle,[11] and it is quite plausible that Russell had this formulation in mind all along. Also, Gödel (1944) points out that Presupposition VCP (as well as Involvement VCP) is a more plausible principle than Definability VCP. In the reminder of this section, then, I will focus on Presupposition VCP, and in closing I will provide further motivation for viewing Definability VCP just as a formulation of it.

In order to spell out Presupposition VCP, the notion of presupposition has to be tackled. Nevertheless, such a notion is admittedly rather vague. In the literature, different notions of presupposition may be found.[12] Each of them gives rise to a

---

[6]See, for instance, Hellman (2004).

[7]See, for instance, Jung (1999) for a detailed survey on them.

[8]Russell (1908, 63).

[9]Russell B. and Whitehead A., *Principia Mathematica*, vol. 1, p. 37.

[10]Russell B. and Whitehead A., *Principia Mathematica*, vol. 1, p. 63.

[11]In fact, Jung (1999, 69–74) shows that both Definability and Involvement VCPs follow from Presupposition VCP.

[12]See, for instance, Fine (1995) and Correia (2008). See also Linnebo (forthcoming). Hellman (2004) claims that there is also an epistemic justification for predicativism, namely, that rational beliefs in mathematics extend only to predicatively definable objects. Epistemic predicativism is indeed a possible interpretation of Russell's VCP. Nevertheless, I will not investigate it in this article, though it is worth mentioning that epistemic VCP may be interesting to anyone working on

formulation of Presupposition VCP. These notions of presupposition, and the related formulations of VCP, may be formulated as follows:

Presupposing for existence: An entity **A** existentially presupposes an entity **B** just in case **A** cannot exist unless **B** does. Consider, for instance, sets. A non-empty set x exists only if its members exist.[13]

> **Ontological VCP:** No entity can presuppose for its existence a totality it belongs to.

Presupposing for essence: An entity **A** essentially presupposes an entity **B** just in case **A** cannot be what it is unless **B** is. For instance, what set x is presupposes what members it contains.[14]

> **Metaphysical VCP:** No entity can essentially presuppose a totality it belongs to.

Presupposing for specification: An entity **A** presupposes an entity **B** for its specification just in case **A** cannot be specified unless **B** is. For instance, whether we can specify a set x presupposes that we are able to specify its members.[15]

> **Specifiability VCP:** No entity can presuppose for its specification a totality it belongs to.

Both essential presupposition and presupposition for specification are tightly connected with identity (or at least equivalence). It may be argued that, after all, they come down to the same notion. Nevertheless, even though they both indeed involve some requirements for identity, I believe they should be carefully separated. We may

---

some Platonist response to Benacerraf's dilemma. Benacerraf's dilemma claims that the Platonist has to face severe epistemic problems as for the accessibility of the entities she takes to exist mind-independently. To this extent, epistemic predicativism seems to support Benacerraf's view.

[13]On existential presupposition, see Fine (1995) and Correia (2008).

[14]On essential presupposition, see Fine (1995) and Correia (2008).

[15]See, for instance, Linnebo (forthcoming), which is nevertheless focused on investigating *first-order* impredicativity in abstraction principles. An abstraction principle has the form $\S F = \S G \leftrightarrow R_E(F, G)$, where $\S$ is an abstraction operator mapping a given collection of entities into a collection of entities of different sort and R_E is an equivalence relation. Well-known examples are the so-called Hume's Principle and Basic Law V. The impredicativity Linnebo investigates concerns the fact that the entities introduced on the left-hand side of the biconditional can be among the values of the first-order variables appearing on the right-hand side (consider, for instance, Basic Law V: $\{x : Fx\} = \{x : Gx\} \leftrightarrow \forall x(Fx \leftrightarrow Gx)$). Thus, if abstraction principles serve the purpose of *individuating* or *specifying* the entities introduced on the left-hand side by the equivalence relation on the right-hand side, their impredicativity would imply that the entities the terms on the left-hand side refer to are individuated or specified on the basis of a totality they belong to. Nevertheless, I am here analysing the impredicativity underlying second-order logic, which originates from the comprehension axiom and concerns the specification of the second-order entities the left-hand side of the biconditional refers to.

in fact argue that there are examples of essential presupposition that do not apply to specifiability. Consider, for instance, the power set of ω. We may argue that such a set is the set it is, exactly because of the members it contains. Nevertheless, not all subsets of ω are specifiable. To this extent, should the power set of ω be specifiable at all, it would not be so on the basis of the presupposition of the specifiability of its members, even though it would still be the set it is because of the members it contains.

There is a further notion of presupposition I would like to mention, namely, referential presupposition. It is worth mentioning that referential presupposition, and its related VCP, will be the main focus of the paper. It may be surprising that in this section I set it aside right after I state it, but I will profusely come back to it in Sects. 4.4 and 4.5. All that is in between is propaedeutic to a motivation for referential presupposition and its related VCP, and their relation to the issue of second-order predicativity in a Fregean setting.

Presupposing for reference: The possibility of referring to an entity **A** presupposes the possibility of referring to an entity **B** just in case the possibility of referring to **A** makes ineliminable use of the possibility of referring to **B**. Let me mention one example to illustrate what I have in mind. Consider tropes. If I say 'Your smile is like no other', I am apparently referring to your smile. Now, could I refer to it, in case I were not able to refer to *you* in the first place? Hardly so. To this extent, the possibility of referring to your smile presupposes the possibility of referring to you.

Referential VCP: No entity can presuppose for its reference to a totality it belongs to.

In the opening of this section, I mentioned the three formulations of VCP available in Russell's writings, and in passing I mentioned Gödel's article on Russell's mathematical logic. In that article, Gödel claims that it is Definability VCP that is of interest in mathematics, since it disallows impredicative definitions and thus undermines the derivation of most of classical mathematics[16]:

> (...) the vicious circle principle (...) applies only if the entities involved are constructed by us. In this case, there clearly must exist a definition (namely the description of a construction) which does not refer to a totality to which the object defined belongs, because the construction of a thing can certainly not be based on a totality of things to which the thing to be constructed itself belongs. If, however, it is a question of objects that exist independently of our constructions, there is nothing in the least absurd in the existence of totalities containing members, which can be described (i.e., uniquely characterized) only by reference to this totality.[17]

A clear example of this is the definition of the set of the natural numbers, which is provided in terms of all inductive sets. If impredicative definitions are claimed to

---

[16]See Gödel (1944, 455–459).

[17]Gödel (1944, 456).

be impermissible, such a definition cannot be recovered. Nevertheless, according to Quine (1969, 242–3, emphasis added):

> No question of legitimacy can arise in connection with definition, so long as a mechanical procedure is provided for expanding the new notation in all cases uniquely into old notation. Now what Poincaré criticized is not the definition of some special symbol as short for '$\{x : x \notin x\}$', but rather the very assumption of the existence of a class y fulfilling '$(x)(x \in y \leftrightarrow x \notin x)$'. We shall do better to speak not of impredicative definitions but of *impredicative specification* of classes, and, what is the crux of the matter, impredicative assumptions of class existence.

Here, definability plays no role at all, since, independently of the possible underlying ontological assumptions concerning mathematical entities, definitions are just a matter of introducing a new notation that is always eliminable in terms of the old one. The best way to see this is to consider comprehension axioms, which are actually what Quine seems to have in mind in the quotation above. Comprehension axioms are not definitions, but they indeed provide a means to specify entities through predicative or impredicative formulæ. But if definitions are unproblematic, independently of the underlying ontology, how are we to make sense of Gödel's objection to Definability VCP? One possible way out is the one envisaged by Quine in the previous quotation, namely, to leave definability out of the picture and rather consider *specifiability* and thus rephrase Definability VCP as Specifiability VCP.[18] To this extent, specifiability may be considered as the rather general notion of 'singling out' an entity from other entities. A further formulation of VCP can be provided in terms of *individuation*. Interestingly, Quine (1985, 166–7) disallows impredicative individuation. In connection with individuation of events, he says:

> For my own, I welcome impredicative definitions. I have remarked that there is nothing wrong with identifying the most typical Yale man by averaging measurements and tests of all Yale men including him. But now we observe that impredicative definitions are no good for individuation. Here a difference between the impredicative and the predicative emerges which is significant quite apart from any constructivist proclivities. We can define impredicatively but we cannot individuate impredicatively.

This quotation calls for a distinction between impredicative individuation as impermissible, as opposed to impredicative specification as permissible. What would be Quine's reasons for allowing impredicative uses of the latter, while banning impredicative uses of the former? I take it that in Quine's view individuation goes through some identity criterion. Impredicative individuation has to be disallowed since if individuation presupposes identity, then the individuation of an object must not be performed by any identity statements concerning that very object. On the other hand, impredicative specification might be allowed since it is a weaker notion than individuation. In particular, it does not require an identity criterion: for instance, in the case of impredicative second-order comprehension for concepts, the singling out of a concept by a condition does not require, at least *prima facie*, an identity criterion for concepts. Though I acknowledge the different import of

---

[18]See also Jung (1999, 59) on this point.

impredicative individuation, on the one hand, and impredicative specification, on the other, I believe that none of them is permissible, at least as far as concepts are concerned. In Sect. 4.3, I provide reasons for equating specification of concepts to reference to them, on the grounds that, because of their intensional nature, their specification can only be performed by language. To this extent, specifiability of a concept and reference to it are equivalent notions. By bearing on this equivalence, in Sects. 4.4 and 4.5 I will provide motivations for banning, *pace* Quine, impredicative specification of concepts on the basis of Referential VCP.

4.2 Gödel's Criticism of VCP

VCP was subjected to some radical criticisms, especially by Gödel (1944) but also by Quine and Ramsey. So, it is quite important to go through these criticisms, and try to provide some counter-objection, to the aim of making the use of predicativity compatible with Frege's view on concepts. In this section, I will sum up Gödel's (or some broadly Gödelian) criticism of VCP.

Gödel's main objection to Russell's VCP is based on some philosophical considerations he puts forward: if a Platonist view of mathematical entities holds, then one is entitled to impredicativity, since VCP in a Platonist perspective is either easily rejected or trivial; on the other hand, if VCP holds and by this one has solid motivation for predicativity, one is also committed to some form of mathematical Constructivism. Thus, in Gödel's view, the divide between predicativity and impredicativity, and the acceptance or the rejection of VCP, ultimately hinges on ontology.

Let us see how Gödel's argument may be reconstructed. Consider a Platonist attitude towards the existence of the entities of a certain domain. To this extent, these entities exist mind-independently, and none of them presupposes the totality it belongs to for its existence, because they are just there from the start. Consequently, in a Platonist framework, Ontological VCP is trivial:

> Such a state of affairs[19] would not even contradict (...) the third form[20] if 'presuppose' means 'presuppose for the existence'.[21]

Let us consider now Specifiability VCP, which, under the revision of Gödel's objections to Definability VCP, is the real target of a Gödelian objection. From a Platonist perspective, Specifiability VCP is false, since 'there is nothing in the least absurd in the existence of totalities containing members, which can be described

---

[19]That is, Platonism.

[20]That is, Presupposition VCP.

[21]Gödel (1944, 456). The same argument goes as for Metaphysical VCP. Consider, for instance, sets. A set x is the set it is because of the members it contains, not because of the totality it belongs to. Thus, we may argue against Metaphysical VCP that it is trivial as much as Ontological VCP, under a Platonist stance of the universe of sets. In such a view, in fact, Metaphysical VCP would hold by default. From now on, then, I will not consider Metaphysical VCP anymore.

(i.e., uniquely characterized) only by reference to this totality'.[22] And in fact Quine's quotation from above continues:

> And what now of the vicious circle? (...) [I]mpredicative specification of classes (...) is hardly a procedure to look askance at, except as one is pressed by the paradoxes to look askance at something or other. For we are not to view classes literally as created through being specified - hence as dated one by one, and as increasing in number with the passage of time. (...) The doctrine of classes is rather that they are there from the start. This being so, there is no evident fallacy in impredicative specification. It is reasonable to single out a desidered class by citing any trait of it, even though we chance thereby to quantify over it along with everything else in the universe. Impredicative specification is not visibly more vicious than singling out an individual as the most typical Yale man on the basis of averages of Yale scores including himself.[23]

Finally, if the Platonist view on the entities of a given domain holds, also Referential VCP is false. Gödel, in fact, claims that

> [i]t is demonstrable that the formalism of classical mathematics does not satisfy the vicious circle principle in its first form,[24] since the axioms imply the existence of real numbers definable in this formalism only by reference to all real numbers. Since classical mathematics can be built up on the basis of *Principia* (including the axiom of reducibility), it follows that even *Principia* (...) does not satisfy the vicious circle principle in its first form (...).
>
> I would consider this rather as a proof that the vicious circle principle is false than that classical mathematics is false, and this is indeed plausible on its own account. For, first of all one may, on good grounds, deny that reference to a totality necessarily implies reference to all single elements of it or, in other words, that 'all' means the same as an infinite logical conjunction.[25]

If Platonism is the best ontological account of classical mathematics on the market and we hold it, then we may indeed quantify over all the entities of a totality without presupposing reference to each and every one of them. Consider, for instance, real numbers. If we help ourselves to a Platonist view of the reals, we think of their collection as a totality of some sort. But quantifying over the totality of the real numbers does not imply that we have as many individual constants as the reals (and in fact we have not) and that we can attach referents to them.

Even more so, a further criticism of Referential VCP may target the principle on the basis of the relation between reference, on the one hand, and existence and specifiability, on the other. In fact, without existence, we may not specify, let alone refer. But even if we grant existence with no specification, can we actually refer to an entity without being able to single it out by some means? According to a common intuition, in order to refer by a term, the referent of that term must exist and it has to be, at least, specifiable in some way. In particular, Referential VCP may be criticised from a further angle, which is connected with its relation to Ontological

---

[22]Gödel (1944, 456).
[23]Quine (1969, 242–3).
[24]That is, Definability VCP.
[25]Gödel (1944, 454–455).

and Specifiability VCPs. It may be argued, in fact, that Referential VCP is hardly news as compared to Ontological and Specifiability VCPs. These objections will be addressed in a few sections.

4.3 Predicativity and Presupposition Versus Frege's Platonist Logicism About Concepts

If Gödel's argument holds, *prima facie* none of the aforementioned formulations of VCP is compatible with Frege's Platonist view on concepts. Accordingly, if we try to motivate a predicative restriction on second-order comprehension in a Fregean setting and we take predicativity to be motivated by some notion of presupposition, the consistent predicative fragments of *Grundgesetze* do *not* seem to provide a philosophical basis to revive Frege's foundational programme.

Previously, I claimed that specifiability and reference may be seen as distinct though connected notions. As for concepts, however, the situation is different. Whether the underlying ontological assumption for either of those principles is Platonism or Constructivism, Specifiability VCP and Referential VCP boil down to the very same principle. Specification means to single out a concept from the other concepts. But then again, how can we specify an intensional entity such as a concept? Given their intensionality, a rather natural way to specify a concept is via some linguistic expression expressing it, like a *predicate*. To this extent, the specification of a concept comes down to providing a way to refer to it via an appropriate linguistic means. To this, it may be objected that a possible alternative way of specifying a concept is via thought. For instance, in order to specify the concept 'being red', we can picture red things or maybe just think 'red'. In this case, specifiability presupposition would not come down to referential presupposition. Nevertheless, consider that I am interested in second-order languages in a Fregean setting and in Frege's view on concepts. In Frege's perspective, mental representations are to be avoided in the philosophical investigation on mathematics and logic. The Fregean, thus, would not be troubled with the above objection to the equivalence between specifiability presupposition and referential presupposition for concepts. In a Fregean perspective, it is quite natural to view concepts as specified by language.[26] But then again, to this extent, specifiability presupposition and referential presupposition turn out to be equivalent notions. If then a Platonist view of Fregean concepts holds, Specifiability and Referential VCPs will be both rejected, on the grounds that on that view Referential VCP is false.

There seems to be a further difficulty with predicativity in a Fregean perspective. In Frege's view, concepts are logical entities whose existence is guaranteed by the laws of thought alone, i.e. the laws of the True. According to Frege, these laws are

---

[26]Consider, for instance, the Fregean view that predicates, i.e. the linguistic items standing for concepts, are obtained by extrapolation of singular terms from sentences.

rigorously formalised in classical logic as it is presented in the *Begriffsschrift* as a higher-order logical system. Frege is not just a Platonist about concepts; he is a *logicist* Platonist: if we grant Frege that the very laws of thought are exhaustively captured by higher-order logic, then in his view higher-order logic is *the* logic of the laws of thought. In a Fregean perspective, conceptual Platonism is not just an ontological stance on concepts independent of his logicism; it is instead strongly motivated by, or at the very least deeply connected with, his logicism. In this view, unrestricted comprehension for concepts is both motivated by Frege's Platonist view on concepts, but even more so it entails, given his logicism, that unrestricted comprehension is a fundamental tool to capture the laws of the True.

In what follows, I will argue that there is a motivated way to make the consistent predicative fragment of *Grundgesetze* compatible with Frege's view. This motivation hinges on a revision of Gödel's dichotomy between predicativity and impredicativity and their philosophical implications. To this aim, in the following sections, I will propose to detach the acceptance of predicativity from ontological considerations and to connect it to considerations concerning logical and mathematical reasoning. On these grounds, I will finally claim that, in the light of these considerations, it is possible to make Frege's Platonist logicism about concepts compatible with a predicative revision of his foundational programme. A word of caution is here needed, though. I will argue that Frege's conceptual Platonism is indeed compatible with second-order predicativity. As for Frege's logicism, it will turn out that Frege's claim that arithmetic can be reduced to logic alone has to be revised. Nevertheless, this revision will not necessarily make Frege's original view a complete nonstarter. It will turn out that a predicative restriction on second-order comprehension will indeed provide a starting point for a revision of Frege's logicism. Before that, though, I will have to take a slight detour through the notion of *arbitrary reference*.

4.4 TAR

Let us recall the formulation of referential presupposition from above: the possibility of referring to an entity **A** presupposes the possibility of referring to an entity **B** just in case the possibility of referring to **A** makes ineliminable use of the possibility of referring to **B**. First of all, in order to make sense of it and provide some counter-objections to Gödel's arguments against Referential VCP, the notion of *possibility of reference* needs to be clarified.

According to Martino (2001, 2004), the possibility of directly referring, at least ideally, to any entity of a universe of discourse is presupposed both by logical and mathematical reasoning, even when non-denumerable domains are concerned. Martino (2001) labels this claim the *Thesis of Arbitrary Reference* (TAR).[27] Such

---

[27]More suggestively, Martino (2004) calls this claim the *Thesis of Ideal Reference*. In what follows, it will become clear why.

a possibility of direct reference is very well expressed by the crucial role *arbitrary reference* plays both in formal and informal reasoning. Its cruciality lies in that arbitrary reference exhibits two different logical features that make it essential for performing proofs, i.e. *arbitrariness* and *determinacy*. Through arbitrary reference, we may consider *any* object a of a universe of discourse. Consequently, the arguments about a retain their general validity. At the same time, though, within the arguments about it, 'a' is required to denote a determinate object, distinct from all the other objects in the domain it belongs to. Typically, when a derivation is completed, we may detach 'a' from the individual we attached it to and reuse it in a different derivation. But within a derivation on a, 'a' has to refer to a determinate individual. The same argument holds, *mutatis mutandis*, also for reference to concepts via second-order free variables. In fact, every argument in favour of the genuine referentiality of arbitrary reference put forward in this section applies both to first- and second-order arbitrary reference.

In order to motivate TAR, an account of the *genuine* referentiality of arbitrary reference and its *directness* has to be provided. It may be argued, in fact, that arbitrary reference is not genuinely referential, since parameters and free variables do not refer at all.[28] Evidence in favour of the genuine referentiality of arbitrary reference may be found in Breckenridge and Magidor (2012) and Martino (2001, 2004). In this section, I will provide a more general argument to this aim. My claim is that the soundness of arguments in mathematical and logical reasoning requires the underlying assumption of the genuine referentiality of arbitrary reference. The relation between soundness and referentiality will be accounted for in terms of sameness and determinacy of reference. Though my argument will be spelled out in terms of reference to individuals, nothing will prevent its application to reference to concepts as well.

Usually, a parameter 'a' is used to refer to the *same* individual a within a derivation on a. A crucial reason for this is to be found in the requirement of soundness we want to impose on some valid argument schemas. If sameness of reference were not a basic ingredient of derivations, soundness would be in jeopardy.[29] Consider the rule of existential elimination in natural deduction. When we pass from a premise of the form $\exists x \phi x$ to the auxiliary assumption $\phi(a)$, 'a' has to be an unused parameter, or at least it has not to appear in any of the assumptions which $\exists x \phi x$ depends upon. Consider now the following (invalid) deduction:

(1) $\exists x H x$ $\mathscr{A}$
(2) $\exists x \neg H x$ $\mathscr{A}$
(3) Ha $\mathscr{A}$
(4) $\neg Ha$ $\mathscr{A}$
(5) $Ha \wedge \neg Ha$ 3,4 intr. $\wedge$

---

[28] See, for instance, Pettigrew (2008).

[29] A further argument to this aim, from the uniformity of substitution of predicate and individual letters in argument schemas, may be found in Boccuni (2010).

Invalidity stems out from that, in eliminating the existential quantifiers, respectively, from (1) and (2), we use the very same parameter in (3) and (4).[30] Say that H is the property of being even and x varies over the natural numbers: (1) and (2) say, respectively, that there is at least a number which is even and there is at least a number which is not. Both these sentences are true in the standard model of Peano arithmetic. Nevertheless, if we use the same parameter to perform existential elimination in the derivation above, in (3) and (4) we, respectively, say that a number is even and that *the very same number* is not, from which the contradiction in (5) arises. For this reason, using an already used parameter in (4) cannot be allowed.

In order to explain the invalidity of the derivation (1)–(5), 'a' must be referring to the same, though arbitrary, individual both in lines (3) and (4). Thus, in order to achieve soundness in the previous example, in line (4) we have to use a different parameter than 'a', because we need to express that a *different* individual than a is $\neg H$ within the same derivation, in accordance with the restrictions imposed on universal introduction and existential elimination. But then again, in order to distinguish between a and any other arbitrary individual that is $\neg H$, we have to assume that a is a *determinate*, though arbitrary, individual of the domain. A similar argument can be provided as for second-order existential elimination. Say, for instance, that you have two second-order existential assumptions such as $\exists F \phi F$ and $\exists F \neg \phi F$, which are both true in second-order arithmetic, for ϕ being the formula 'is finite'. Indeed, there are finite concepts in second-order arithmetic as there are infinite ones, so the two existential assumptions are true. But now if we use the same parameter to eliminate both of them, we incur in the contradictory conclusion that the same concept satisfies a formula and its negation.

The motivation for this requirement is very nicely explained by Suppes:

> (...) ambiguous names,[31] like all names, cannot be used indiscriminately. The person who calls a loved one by the name of a *former* loved one is quickly made aware of this. (...) Such a happy-go-lucky naming process is bound to lead to error, just as we could infer a false conclusion from true facts about two individuals named 'Fred Smith' if we did not somehow devise a notational device for distinguishing which Fred Smith was being referred to in any given statement. The restriction which we impose to stop such invalid arguments is to require that when we introduce by existential specification an ambiguous name in a derivation, that name has not previously been used in the derivation.[32]

---

[30] See Suppes (1999, 82) for this example.

[31] That is, parameters like 'a'.

[32] Suppes (1999, 82). Of course, it is not always the case that using the same parameter leads to invalidity nor that different parameters always have to refer to different entities. For instance, consider using 'a' for eliminating the quantifiers both from $\forall x F x$ and $\forall x G x$ in the same derivation, where x varies over the natural numbers and both formulæ have a model in Peano arithmetic. Or consider using 'a' and 'b' for eliminating, respectively, the first quantifier and the second, where a and b may be the same individual. In none of these examples, sameness of reference seems to lead to invalidity, but such an innocuousness does not by itself speak against the genuine referentiality of 'a' or the importance of sameness of reference to derivations. It rather testifies that there are contexts in which the co-referentiality of all the occurrences of 'a' (or of 'a' and 'b', for that matter) is not problematic.

The reasons for restricting the rules of introduction and elimination of quantifiers in natural deduction are semantic: in derivations, we perform a semantic reasoning that we want to be captured by deductive rules and restrictions on them. Such a reasoning is crucially based on sameness and determinacy of reference of parameters. But then again, in order to make sense of sameness and determinacy and consequently of the requirements we impose on deductive rules for the sake of soundness, we have to assume the genuine referentiality of parameters. Genuine referentiality is a necessary condition for soundness. This relation can be highlighted by investigating the role that sameness and determinacy of arbitrary reference have in derivations. In fact, if 'a' were not referential at all, how could we account for a being the same individual throughout an argument? Those who support the non-referentiality of arbitrary reference should provide some argument for explaining how formal and informal reasoning functions in the way it does (for instance, by certain constraints on introduction and elimination of quantifiers).[33]

A further issue concerns the connection of arbitrary reference with existence. In fact, arbitrary reference may be used within derivations in which one of the assumptions leads to contradiction. Consider the example 'Let r be the set of all the sets that are not members of themselves'. It is well known that a contradiction is derived from the assumption of the existence of the Russell set. We readily conclude that r does not exist. But then again 'r', under the assumption of the genuine referentiality of arbitrary reference, has to genuinely refer to an (arbitrary) individual throughout the derivation on it. Is then 'r' genuinely referential after all, i.e. in the derivation of Russell's paradox does it refer to an individual that does not exist? Reference in arguments by *reductio ad absurdum* may be explained as follows. Throughout the derivation leading to the contradiction, we temporarily assign an individual as the value of 'r'. But then we find out that the individual we picked does not, and in fact cannot, satisfy the condition 'set of all the non-self-membered sets'. Since the individual chosen is an arbitrary individual, that the Russell condition is not satisfied holds of all the individuals of the domain. But in the process, 'r' has indeed been referential: it referred to an arbitrary individual that does not satisfy the Russell condition. The derivation of the contradiction does not just say that there is nothing 'r' refers to, but that for every assignment of a referent to 'r', r does not satisfy the defining formula.

The directness of arbitrary reference may be appreciated by considering its relation with quantification. Consider once again the rule of existential elimination. As Martino (2004) points out, the possibility of passing from a purely existential assumption such as $\exists x \phi x$ to the consideration of an arbitrary object a such that ϕa

---

[33] A further issue concerns the semantics that better captures the genuine referentiality of arbitrary reference. To the best of my knowledge, there are two competing options on the market: Kit Fine's view according to which arbitrariness is a property of some special kind of objects, namely, those referred to by parameters, and an epistemic view, championed by Breckenridge and Magidor (2012) and Martino (2001, 2004), according to which arbitrariness is an epistemic feature of our reasoning – a is a determinate individual, and 'a' determinately refers to it, but we do not know *which* individual a is.

is guaranteed by the rule of elimination of the existential quantifier which allows to substitute the given existential assumption with the auxiliary assumption ϕa. If the rules of inference that govern the use of the logical constants in natural deduction are justified by the meaning of the constants themselves, the meaning of the existential quantifier presupposes the possibility of singularly referring, at least ideally, to any individual of a domain, and consequently existential quantification logically presupposes such a possibility of reference.[34] Thus, before we simultaneously consider several entities through quantification, we are required to be able to refer to each of them, at least ideally: quantification logically presupposes the ideal possibility of referring to each and every element of a domain, before we consider those elements through generalisation.[35] In this perspective, no individual can be referred to only on the basis of reference to a totality it belongs to, as an ideal way to directly refer to it is required in order not to violate TAR: Referential VCP is a corollary of TAR.[36]

4.5 Gödelian Criticism to Referential VCP

Let us recall that Referential VCP may be criticised with respect to two issues: (1) the fact that quantification does not imply an infinite conjunction or disjunction and (2) Referential VCP being hardly news over Ontological and Specifiability VCPs. In this section, I will tackle these two issues.

According to TAR, quantification presupposes arbitrary reference: if the meaning of the quantifiers is governed by their rules of introduction and elimination in natural deduction, then quantification presupposes the possibility of reference, at least in principle. In this respect, under certain restrictions, we may pass from the consideration of an arbitrary individual a such that ϕ to the generalisation that every x is such that ϕ.[37] In this regard, one is reminded of Gödel's objection to quantification as presupposing reference to each and every member of a domain. Nevertheless, Gödel's objection apparently does not take into consideration the cruciality of arbitrary reference, but is limited to considering that quantification

---

[34] Analogously as far as the rule of introduction for universal quantification is concerned. See Martino (2004, 110).

[35] For further justification and applications of arbitrary reference, see also Breckenridge and Magidor (2012).

[36] See Martino (2004, 119). Notice that Referential VCP follows from TAR also when non-denumerable domains are concerned. Even though a language cannot display non-denumerably many names, TAR still holds, as the ideal possibility of directly referring to each and every individual in a non-denumerable domain may be performed via arbitrary reference, as in the case of, e.g. 'let a be an arbitrary real number'. Also, Martino (2001, 2004) provides a special semantics, the acts of choice semantics, in order to make sense of how the directness of arbitrary reference should work.

[37] And analogously as for existential quantification.

does not imply infinite conjunctions or disjunctions of formulæ where the bound variables is substituted by a constant. Referential VCP as implied by TAR, though, does not necessarily concern reference through constants, either first- or second-order, but at the very least concerns arbitrary reference. The restriction to arbitrary reference is really all that is needed to make a case for predicativity concerning Fregean concepts, because the predicativity that is employed to that aim is motivated by a reflection on the relation between quantification and the ideal possibility of reference. To this extent, TAR *does* imply that quantification involves arbitrary reference, but it does *not* imply that quantification involves reference to each member of a totality by an infinite conjunction or disjunction.[38] According to TAR, quantification implies reference to *any* individual of that totality, by arbitrary reference: if we refer to a totality, then we must be able, at least ideally, to refer to each and every member of it through arbitrary reference prior to the consideration of the totality itself. In this respect, the first objection to Referential VCP is a nonstarter.

What now of the criticism that, given the relation between reference, on the one hand, and existence and specifiability on the other, Referential VCP is hardly news over Ontological and Specifiability VCP? I take this objection to come down to whether there are cases in which specifiability and ontological presupposition do not hold, whereas referential presupposition does. As for Specifiability VCP, arbitrary reference is such that specification and reference are detached. In order to fix the reference of 'a', we do not need to be able to specify a. In a sense, our ignorance of *which* individual a is justifies the possibility of introducing universal quantification in reasoning (given the appropriate restrictions on universal introduction). For instance, we are not required to specify an arbitrary real number r, in order to genuinely refer to it via an arbitrary name 'r'. Furthermore, there are also cases in which ontological presupposition does not hold, whereas referential presupposition does. Consider, for instance, the empty set $\{x : x \neq x\}$. Its existence does not depend upon the existence of non-self-identical individuals, as in fact there are none. Nevertheless, the possibility of referring to $\{x : x \neq x\}$ seems to depend upon the possibility of referring to the individuals of the domain through x in the formula $x \neq x$, if only to acknowledge that no individual is non-self-identical. This holds on the grounds of TAR and in particular on the grounds that parameters and free variables are genuinely referential. Consider, furthermore, a view on concepts

---

[38] What if we wanted to extend Referential VCP to reference via individual constants? In this case, the relation between reference as involved in Referential VCP and arbitrary reference in TAR should be further motivated. Brandom (1996) suggests a way to deal with this issue. While arbitrary reference – which he calls 'parametrical' – and reference via individual constants are both genuinely referential, we may explain their relation by saying that (i) either they convey different notions of reference (ii) or arbitrary reference embodies the only notion of reference there is, and either reference via individual constants is built up from it or it is reducible to it. I discuss option (i) in the main text. In the case of (ii), arbitrary reference would be primitive, so Referential VCP would concern it by default and would easily follow from TAR. I sincerely thank an anonymous reviewer for pressing me on this issue.

quite like Frege's. In such a view, the existence of concepts does not depend upon the existence of predicates expressing them. Nevertheless, since concepts can be specified and referred to only via language, the possibility of referring to a concept depends upon the possibility of referring to a predicate expressing it. At least a few cases, in which referential presupposition holds, are not covered by ontological presupposition. To this extent, referential presupposition and Referential VCP are indeed news.

4.5.1 Frege's Platonist Logicism About Concepts

I shall now turn to the issue of whether there is a way to make the use of predicativity compatible with Frege's view on concepts. I mentioned earlier that concepts are intensional entities and, as such, the only way to specify and refer to them is via language, i.e. through formulæ which express them. I also claimed that specification of and reference to concepts boil down to the very same notion. Consider the impredicative comprehension principle in second-order logic: $\exists F \forall x (Fx \leftrightarrow \phi)$. Considering the concept F means considering the formula ϕ, under the intended interpretation. But then again, the concept F cannot be referred to by quantification over the domain it belongs to, since quantification would require that we are able to refer to F prior to specifying it through a predicate. In ϕ, therefore, there cannot be any bound second-order variables, on pain of violating TAR and Referential VCP.

Now, what about ontology and in particular Frege's Platonism on concepts? The relation of logical presupposition of quantification from arbitrary reference provides a motivation for formulating Referential VCP *independently* of the underlying ontology. Such a relation in fact holds no matter what ontological preferences one has. It is just a matter of a relation between two crucial logical tools, i.e. (arbitrary) reference and quantification, and as such it is impervious to ontological considerations. One thing is the ontological assumptions underlying the consideration of a domain of entities; another matter is the consideration of how reference works. The underlying ontological assumptions, though inevitably connected with reference, are not the only considerations playing a role in the referential picture. TAR provides independent reasons for the dependence of quantification on reference, to which the underlying ontological preferences are irrelevant. TAR is a claim concerning how language works *logically*: ontology entering the picture is merely accidental. To this extent, holding a Platonist stance towards the existence of the entities of a certain domain does not affect Referential VCP. Given that Referential VCP is based on TAR and TAR just embodies a claim of logical dependence of quantification from arbitrary reference, a rejection of Referential VCP does not follow from a Platonist attitude towards existence. To this extent, the Fregean may still be a Platonist about concepts, even though she holds Referential VCP. Moreover, all concepts may exist: the only restriction TAR imposes is on the concepts which are expressible within a language. These latter may well exist, depending on the ontological preferences one has; as for the ones which are not expressible, they too may well exist, even though the language cannot talk about them.

What about logicism, finally? As for Frege's idea that impredicative higher-order logic captures the very laws of thought, it has to be drastically revised. Since TAR is a claim concerning logic and in particular the relation between arbitrary reference and quantification, I here hold the strong claim that logic as the body of the laws of thought should be inherently predicative. Frege's logicism has to be restricted to the extent that it finally takes into account the relation of logical presupposition of quantification from arbitrary reference. Is this bad news to the Fregean? Yes and no. It is bad news to the extent that a predicative second-order fragment of Frege's *Grundgesetze* like Heck (1996) cannot interpret more than Robinson arithmetic Q, which is a rather poor, though not trivial, fragment of mathematics, especially as compared to Frege's original attempt to found (full second-order) arithmetic on pure logic. In this respect, the objection from mathematical strength to a predicative revision of Frege's logicist programme hits the target.

The limitation that TAR imposes on the underlying logic, though, may still be good news to the Fregean. First of all, TAR provides some independent motivation, other than the mere search for consistency, for restricting the underlying second-order logic. Secondly, given its compatibility with a Platonist stance towards the existence of concepts, TAR provides some room to manœuvre for claiming that the predicative fragments of *Grundgesetze* may indeed provide a consistent formal core for reviving Frege's foundational programme. Most of all, the limitation may just imply that higher-order logic *as the Fregean might interpret it*, namely, as a theory of concepts, has to be revised, but there may be further extensions of Heck (1996) that may both count as logic and recover second-order Peano arithmetic.[39]

4.6 Concluding Remarks

In this paper, I tried to show that the consistent predicative fragments of *Grundgesetze*, though *prima facie* at odds with Frege's philosophical stance on concepts, may indeed provide some philosophical grounds for revising Frege's programme. I proceeded through an independent reassessment of the very notion of predicativity, not on ontological grounds as Gödel does, but on some logical basis. It turns out that this reassessment is compatible with Frege's Platonist view on the existence of concepts but requires a drastic revision of his logicist view. In spite of this pessimistic conclusion, I suggest to look on the bright side of the matter: once both consistency and Frege's conceptual Platonism are secured by TAR on independent philosophical grounds, the predicative fragments of *Grundgesetze* may provide a stepping-stone for further investigations towards a (possibly partial) revival of Frege's philosophical views on mathematics.

---

[39] See, for instance, Boccuni (2010).

Acknowledgements I am most thankful to the Institute of Philosophy (School of Advanced Study, London) for sponsoring a visiting fellowship during which I worked on this paper. I also wish to thank Øystein Linnebo, Salvatore Florio, Sean Walsh, Barry Smith, Jönne Speck, Neil Barton, Toby Meadows, Alexander Bird, Anthony Everett, Leon Horsten, Samir Okasha, Christopher Clarke, Mark Pinder, the audience of the Pisa conference 'Filosofia della matematica: dalla logica alla pratica: Giovani studiosi a confronto', and two anonymous reviewers for useful comments on earlier versions of this paper.

References

Boccuni, F. 2010. Plural *Grundgesetze*. *Studia Logica* 96(2): 315–330.
Brandom, R. 1996. The significance of complex numbers for Frege's philosophy of mathematics. *Proceedings of the Aristotelian Society* 96: 293–315.
Breckenridge, W., and Magidor, O. 2012. Arbitrary reference. *Philosophical Studies* 158(3): 377–400.
Burgess, J.P. 2005. *Fixing Frege*. Princeton: Princeton University Press.
Correia, F. 2008. Ontological dependence. *Philosophy Compass* 3(5): 1013–1032.
Ferreira, F., and Wehmeier, K.F. 2002. On the consistency of the Δ^1_1-CA fragment of Frege's *Grundgesetze*. *Journal of Philosophical Logic* 31: 301–311.
Fine, K. 1995. Ontological dependence. *Proceedings of the Aristotelian Society, New Series* 95: 269–290.
Ganea, M. 2007. Burgess' PV is Robinson's Q. *Journal of Symbolic Logic* 72(2): 619–624.
Gödel, K. 1944. Russell's mathematical logic. In *The philosophy of Bertrand Russell*, ed. P.A. Schilpp, 123–153. Evanston/Chicago: Northwestern University; in Benacerraf and Putnam (eds.) 1983 *Philosophy of mathematics: Selected readings*. Cambridge: Cambridge University Press.
Heck, R. 1996. The consistency of predicative fragments of Frege's *Grundgesetze der Arithmetik*. *History and Philosophy of Logic* 17(1): 209–220.
Hellman, G. 2004. Predicativism as a philosophical position. *Revue Internationale de Philosophie* 229: 295–312.
Jung, D. 1999. Russell, presupposition, and the vicious-circle principle. *Notre Dame Journal of Formal Logic* 40(1): 55–80.
Linnebo, Ø. forthcoming. Impredicativity in the Neo-Fregean programme. In *Abstractionism in mathematics: Status belli*, ed. P. Ebert and M. Rossberg. Oxford: Oxford University Press.
Martino, E. 2001. Arbitrary reference in mathematical reasoning. *Topoi* 20: 65–77.
Martino, E. 2004. Lupi, pecore e logica. In *Filosofia e logica*, ed. M. Carrara and P. Giaretta, 103–33. Catanzaro: Rubettino.
Pettigrew, R. 2008. Platonism and aristotelianism in mathematics. *Philosophia Mathematica* 16(3): 310–332.
Quine, W.O. 1969. *Set-theory and its logic*, revised ed. Harvard: Harvard University Press.
Quine, W.O. 1985. Events and reification. In *Actions and events: Perspectives on the philosophy of Davidson*, ed. E. Lepore and B. McLaughlin. Oxford/New York: Blackwell.
Russell, B. 1908. Mathematical logic as based on the theory of types. *American Journal of Mathematics* 30(3): 222–262.
Suppes, P. 1999. *Introduction to logic*. New York: Dover.

Chapter 5
A Deflationary Account of the Truth of the Gödel Sentence $\mathcal{G}$

Mario Piazza and Gabriele Pulcini

5.1 Introduction

According to deflationism, truth is a metaphysically thin property, redundant and dispensable, but useful as generalisation device as in 'whatever the oracle told you is true'. Philosophers of course disagree whether deflationism about truth is true. Some objections to it are more penetrating than others, but none of them seems to be decisive. Yet some authors such as Stewart Shapiro and Jeffrey Ketland have argued that the refutation of deflationism can be, to some extent, a *mathematical* task: it is a matter of showing that our conviction about the truth of the independent Gödel sentence $\mathcal{G}$ involves 'a theory of truth which *significantly transcends the deflationary theories*' (Ketland 1999, p. 88; Shapiro 1998). More specifically, Shapiro and Ketland maintain that a truth-theoretic extension of a given arithmetical formal system such as Peano Arithmetic **PA** is deflationarily licit only when it satisfies the *conservativeness* requirement, i.e. when '[it does] not allow us to prove anything in the original language that we could not prove before we added the truth predicate' (Shapiro 1998, p. 497). Thus, as $\mathcal{G}$ is independent of **PA** albeit expressed within its language—so the anti-deflationary argument goes—any truth-theoretic extension allowing us to prove the truth of $\mathcal{G}$ *must* be nonconservative and so deflationarily illicit.

M. Piazza
Department of Philosophy, University of Chieti-Pescara – Via dei Vestini 31, 66013 Chieti, Italy
e-mail: mpiazza@unich.it

G. Pulcini (✉)
Centre for Logic, Epistemology and History of Science, State University of Campinas – Cidade Universitária "Zeferino Vaz", CEP 13083-859 Campinas, SP, Brazil
e-mail: gab.pulcini@gmail.com

This view has provoked Neil Tennant's reply on behalf of deflationism, according to which the Gödel sentence $\mathcal{G}$ can actually be recognised to be 'true', in the sense of 'assertable', without deploying or invoking a 'thick' concept of truth, i.e. avoiding the semantical notion of model or a Tarski's style truth predicate. In particular, Tennant proposes a way of deflationarily achieving the proposition $\mathcal{G}$ by means of reflective extensions of formal arithmetic which augment the deductive apparatus of PA with a suitable version of the reflection principle (Tennant 2002, 2010). In this way, much current debate about deflationism and Gödel phenomena has been subsumed under the discussion about the justificatory status of reflective statements without appeal to nonconservative truth-theoretic extensions of PA (Field 1999; Halbach 2001, 2011; Ketland 2005; Tennant 2005, 2010; Cieśliński 2010).

Generally speaking, a natural way to be deflationist in mathematics is to equate truth with proof. However, the received view of incompleteness is spontaneously *inflationary* holding that the constitutive element of the First Incompleteness Theorem is the *independence* of truth from proof in PA, so that this theorem would prove the existence of arithmetical sentences which are *true* but *unprovable*. The current orthodoxy therefore favours an anti-deflationary stance: truth can be easily conceived of as a substantial, not a deflationary, property of some sentences, if there is a discontinuity between their truth and their proof. As the anti-deflationist Ketland puts it, our understanding of the significance of the First Incompleteness Theorem is primarily a matter of sensitivity to the *proof-transcending* truth of $\mathcal{G}$ (Ketland 1999, p. 91).

Admittedly, the inflationary reading of incompleteness is as old as Kurt Gödel's own informal argument in the very first paragraph of his 1931 article. This argument, indeed, incorporates a commitment to a 'thick', primitive, concept of truth. Let PA denote the formal system of first-order arithmetic and let

$$\mathcal{G} \Leftrightarrow \mathcal{G} \text{ is not provable in } \mathsf{PA}$$

Let us now ask whether $\mathcal{G}$ is provable or not in PA:

- Suppose that $\mathcal{G}$ is provable in PA. Then, *for what it literally says of itself*, it is a false statement. This means that PA is unsound inasmuch as it allows a false statement to be proved. Hence, if PA is sound, then $\mathcal{G}$ is unprovable in it.
- Suppose instead that $\mathcal{G}$ is unprovable in PA. Then, it is a true statement and its negation $\neg\mathcal{G}$ is false. Again, if PA is sound, then $\neg\mathcal{G}$ is unprovable.

Therefore, PA is *syntactically incomplete*: there exists a statement $\mathcal{G}$ such that neither $\mathcal{G}$ nor its negation $\neg\mathcal{G}$ is provable in PA. Moreover, since $\mathcal{G}$ is a true statement, it follows that PA is also *semantically incomplete*, i.e. there exists a true statement that PA cannot prove (Gödel 1965).

Yet this semantical argument that Gödel launches as a sort of guide for the perplexed performs the function of a heuristic insight. (It should also be noted in passing that Gödel qualifies the independent statement as 'richtig' not as 'wahr'.) The insight in question, engaging as it is, departs from the effective logical meaning of $\vdash_{\mathsf{PA}} \mathcal{G} \leftrightarrow \neg Theor_{\mathsf{PA}}(\ulcorner \mathcal{G} \urcorner)$, which states that $\mathcal{G}$ and $\neg Theor_{\mathsf{PA}}(\ulcorner \mathcal{G} \urcorner)$ are mutually

5 A Deflationary Account of the Truth of the Gödel Sentence $\mathcal{G}$

interchangeable with regard to provability within PA. So the gap between $\mathcal{G}$ and its supposed translation in terms of natural language cannot by any means be filled by a precise logical argument. However, in the sequel of his article Gödel sets the scene for incompleteness in purely syntactical (and moreover intuitionistic) terms so that the core of his construction properly involves a syntactical sensitivity rather than a semantical one.

On the other hand, the fact that proof and arithmetical truth in PA do not co-travel should be viewed only as a *consequence* of the First Incompleteness Theorem *under a classical view of truth*: insofar as $\nvdash_{PA} \mathcal{G}$ and $\nvdash_{PA} \neg\mathcal{G}$, one grants that either $\mathcal{G}$ or $\neg\mathcal{G}$ must be true. But the *irrelevance* of bivalence from a *mathematical* point of view suggests a rationale for a deflationary approach to incompleteness. This point can be illustrated by means of an example. Consider Σ_1-completeness in its contraposed form: since $\neg\mathcal{G}$ is a Σ_1-statement such that $\nvdash_{PA} \neg\mathcal{G}$, then $\neg\mathcal{G}$ is false in the standard model $\mathcal{N}$ and, therefore, $\mathcal{G}$ is true in $\mathcal{N}$ (see Corollary A.19). In other words, the Σ_1-completeness allows us to achieve the truth of $\mathcal{G}$ in the standard model by the very independence of $\mathcal{G}$ from PA. Now take Goldbach's conjecture: *for all* $n \in \mathbb{N}$, *if n is greater than 2, then it can be expressed as the sum of two primes*. Like $\mathcal{G}$, Goldbach's conjecture is a Π_1-statement. Let $\pi(x)$ be the predicate of being a prime number; the conjecture can actually be formalised as follows:

(GC) $\quad \forall x(x > 2 \rightarrow \exists y < x \; \exists z < x(x = y + z \wedge \pi(y) \wedge \pi(z)))$.

Imagine then that someone were to prove that GC is independent from PA and so, by an argument analogous to that for $\mathcal{G}$, that GC is true in the standard model $\mathcal{N}$; again, the truth of GC is yielded by its independence proof. We think it is safe to notice that mainstream number theorists would feel inclined to seek a *counterexample* of Goldbach's statement in elementary number theory (or a proof of it in higher systems).

Another source of epistemological worries about inflationary (model-theoretical) demonstrations of $\mathcal{N} \vDash \mathcal{G}$ concerns the inegalitarianism about models, that is, the choice of the standard model $\mathcal{N}$ as the official platform for establishing the truth of $\mathcal{G}$. A general line of reasoning runs as follows:

1. The central aim of a formal system of arithmetic is to give a formal account for elementary number theory as faithful as possible.
2. But any model of the expanded theory $PA \cup \{\neg\mathcal{G}\}$ (i.e. the theory where $\neg\mathcal{G}$ is assumed to be true) must be a *non-standard* one.
3. Therefore, the exclusion of the standard model $\mathcal{N}$ from the range of possible mathematical structures verifying the axioms of the theory would flout our intuition, because $\mathcal{N}$ expresses the *intended* structure of natural numbers, i.e. it does not include heterogeneous entities like non-standard numbers.

As Michael Dummett points out in his famous paper on Gödel's Theorem, this argument is epistemologically plagued by the puzzling loop caused by the notion of standard model (Dummett 1963). In Crispin Wright's words, 'as soon as it is granted that any intuitively sound system of arithmetic *merely* partially describes the subject matter to which it answers, an explanation is owing of *how* the subject matter in question can possess a determinacy transcending complete description' (Wright 1994). In practice, the intended structure of elementary number theory $\mathcal{N}$ is the theory we want to show as being free from contradictions, and to this end we try to characterise faithfully the class of its truths by means of the property of being a theorem of PA. But our judgement that $\mathcal{G}$ is true though unprovable in PA depends for its acceptability on the *assumption* that the structure $\mathcal{N}$ is clear enough to regulate the deductive behaviour of PA.

In this chapter, we articulate a new deflationary construal of incompleteness where the concept of truth has no substantial role to play in our conviction that the independent sentence $\mathcal{G}$ should be asserted. Moreover, we focus on the actual conceptual core of the Gödelian construction, namely, the *deductive inexhaustibility* of PA. Indeed, whereas inflationary theories of truth tend to 'complete' PA while proving $\mathcal{G}$, current deflationary accounts dismiss the problem of the incompletability of formal systems. Our syntactical path leading to the achievement of $\mathcal{G}$ avoids the heavy commitment to reflection principles, so transparently ad hoc. As a consequence, we deny the thesis about the mathematical refutability of deflationism about truth via Gödel's Theorem while bypassing as inessential the anti-deflationary demand for the derivability of the reflection principles without truth-theoretic principles (Feferman 1991).

The plan of the chapter is as follows. In the next section, we discuss a whole set of problems emerging from the deflationary approach to incompleteness proposed by Tennant in Tennant (2002). In Sect. 5.3, we indicate a syntactical, deflationary route to $\mathcal{G}$. In particular, we pursue a reading of incompleteness in terms of the (constructive) ω-rule and the notion of *prototype proof* in Jacques Herbrand's sense of the term. The constructive ω-rule is shown to have a deflationary character, so as not only to avoid (necessarily nonconservative) semantical justifications but also to overcome the very inexhaustibility phenomenon. Finally, we briefly discuss our deflationary proposal in relation to the procedure sketched by Dummett for achieving the truth of $\mathcal{G}$. In order to make the chapter as self-contained and readable as possible, the appendix provides notations and basic notions as well as the proofs of the theorems involved.

5.2 Against Tennant's Deflationary Reading of Incompleteness

In order to provide a proof for $\mathcal{G}$ in an augmented formal theory including PA, Tennant suggests extending PA with a reflection principle in Feferman's spirit (Feferman 1962)—i.e. an axiom schema that disquotes the truth-predication coming from theoremhood. Under this principle, called the *principle of uniform primitive recursive reflection*, he intends to show 'that there is a 'deflationary way' of faithfully carrying out the semantical argument for the truth of the independent Gödel sentence' (Tennant 2002, p. 557).

5 A Deflationary Account of the Truth of the Gödel Sentence $\mathcal{G}$

For ease of exposition, we sketch his argument by taking into account the formal system PA^{Rfn}, that is, PA expanded to include the *local reflection principle*:

$$Rfn : \forall \alpha, Theor_{\mathsf{PA}}(\overline{\ulcorner \alpha \urcorner}) \to \alpha.$$

PA^{Rfn} is called the *soundness extension* of PA for the reason that the *Rfn* principle is taken into account to represent the formal counterpart to the metalogical property of soundness. Now, from the soundness of PA with regard to the standard model $\mathcal{N}$, follows in particular a proof of $\mathcal{N} \vDash \mathcal{G}$, so that it is possible to formalise a PA^{Rfn} proof of $\mathcal{G}$. We clearly have both

$$\vdash_{\mathsf{PA}^{Rfn}} Theor_{\mathsf{PA}}(\overline{\ulcorner \mathcal{G} \urcorner}) \to \mathcal{G}$$

and

$$\vdash_{\mathsf{PA}^{Rfn}} \neg Theor_{\mathsf{PA}}(\overline{\ulcorner \mathcal{G} \urcorner}) \to \mathcal{G}.$$

Therefore, by stressing the classical tautology

$$((\alpha \to \beta) \wedge (\neg \alpha \to \beta)) \to \beta$$

we can easily conclude

$$\vdash_{\mathsf{PA}^{Rfn}} \mathcal{G}.$$

The following three questions introduce as many objections that can be brought against Tennant's deflationism:

1. *Does the reflection principle actually express the soundness property?*
2. *Does the consistency extension lack good philosophical motivations?*
3. *Does the reflection principle enable us to fill the gap between provability and truth?*

Let us approach these questions in turn.

5.2.1 Does the Reflection Principle Actually Express the Soundness Property?

Tennant's argument depends on the assumption that the *Rfn* axiom schema expresses a *syntactical*, and so genuinely deflationary, rendition of the metatheoretical property of soundness within PA. Such a translation is meant to preserve at least the intensional meaning of the soundness property, that is, the fact that if α is a theorem of PA, then α can be *accepted* as an arithmetical truth; this allows us to derive the epistemological justification for it from the very belief in the soundness of the theory. Now, the soundness of PA with respect to the standard model yields a model-theoretical proof of the *inflationary* truth of $\mathcal{G}$ (Dummett 1963). Analogously, Tennant claims that a deflationary rendition of the soundness property will allow the achievement of a proof of $\mathcal{G}$ in a *deflationary* way.

But a fundamental difficulty comes into view by acknowledging that PA is sufficiently strong to prove the so-called *provable Σ_1-completeness* (see Rautenberg 2000), i.e. the fact that

$$\forall \alpha \in \Sigma_1, \vdash_{PA} \alpha \to \text{Theor}_{PA}(\ulcorner \alpha \urcorner).^1$$

Σ_1-completeness brings with it the fact that any independent Π_1-formula is recognised to be true by $\mathcal{N}$. Specifically, it is possible to provide an easy proof for $\mathcal{N} \vDash \mathcal{G}$. Indeed, $\mathcal{G} \in \Pi_1$ and consequently $\neg\mathcal{G} \in \Sigma_1$; hence, by the contrapositive of the Σ_1-completeness, we obtain $\mathcal{N} \nvDash \neg\mathcal{G}$, i.e. $\mathcal{N} \vDash \mathcal{G}$. Accordingly, such a proof undermines the above defence of deflationism. Indeed, if the provable Σ_1-completeness actually represented its metatheoretical counterpart, then the sentence $\mathcal{G}$ would be provable within PA itself! So, the fact that the provable Σ_1-completeness can be understood on a disquotational parallel with the *Rfn* principle erases any trust in the intensional character of the correspondence between *Rfn* and the soundness property.

5.2.2 Does the Consistency Extension Lack Good Philosophical Motivations?

Tennant quickly dismisses the *consistency extension* of PA, obtained by adding to PA axioms the sentence asserting the syntactical consistency of PA, $Cons_{PA}$. The motivation for this dismissal lurks in the cryptic metaphor that consistency extension is 'an uninformative hammer with which to crack the independent walnut' (Tennant 2002, p. 573). But it is hard to make sense of this from the very perspective he embraces. First of all, as Dummett points out in his contribution on the Gödel's Theorem (Dummett 1963), the truth of $\mathcal{G}$ is a truth *under the assumption of the consistency of* PA. Now, it is well known that through the Second Incompleteness Theorem, this dependency can be strengthened and formalised within PA, $\mathcal{G}$ and $Cons_{PA}$ being two provably equivalent propositions: $\vdash_{PA} Cons_{PA} \leftrightarrow \mathcal{G}$. In this way, a proof of $\mathcal{G}$ can be very simply obtained as the result of a *modus ponens* with the new axiom $Cons_{PA}$. Secondly, the consistency extension takes

---

[1] Note that this formulation can be depurated from any residual metatheoretical feature, simply by individuating a recursive predicate $\mathfrak{S}(x)$ such that $\vdash_{PA} \mathfrak{S}(\overline{n})$ if and only if n is the Gödelian coding of a Σ_1-formula. In such a way, the provable Σ_1-completeness turns to be condensed into the following axiom schema:

$$\forall \alpha, \vdash_{PA} \mathfrak{S}(\ulcorner \alpha \urcorner) \to (\alpha \to \text{Theor}_{PA}(\ulcorner \alpha \urcorner)).$$

advantage of being *weaker* than the soundness extension. In general, the soundness of an arithmetical theory T implies its consistency (otherwise T would prove false statements), but the converse does not hold given the existence of unsound consistent theories (Isaacson 2011). Indeed, since $Cons_{PA}$ is provably equivalent to $\mathcal{G}$, $PA \cup \{Cons_{PA}\}$ is a *minimal* deductive extension allowing us to prove $\mathcal{G}$ (similarly to Tennant's extension based on the *uniform primitive recursive reflection*). From an epistemological standpoint, belief in the soundness of our arithmetical theory implies belief in its consistency to the extent that the cognitive act of recognising a certain cluster of axioms as intuitively true runs under the implicit assumption of their reciprocal consistency. Could anyone recognise as evidently true a pair of axioms contradicting each other? In this way, belief in consistency turns out to be the very first step towards belief in the soundness of the theory. This is the real reason why the proof of the soundness of PA is regarded as uninformative with respect to the consistency of the system: it cannot prove a property silently assumed by the proof itself (Piazza & Pulcini 2013). To conclude, the point is not the lack of *good* reasons for setting aside the consistency extension *in favour* of the local reflection principle. Rather we say that an allegation against the consistency extension cannot be justified without compromising the acceptance of the soundness extension itself.

5.2.3 Does the Reflection Principle Enable Us to Fill the Gap Between Provability and Truth?

The First Incompleteness Theorem establishes much more than the syntactical incompleteness of PA: it shows that *any* first-order formal system capable of faithfully representing a certain amount of elementary number theory is deductively *inexhaustible*. In the specific case of PA, if we attempt to fill the deductive hole $\mathcal{G}$ by extending its deductive power, the deductive hole replicates itself through a new independent proposition $\mathcal{G}'$. Such a phenomenon is due to the fact that the undecidable proposition $\mathcal{G}$ constructed by Gödel involves the predicate $Theor_{PA}(x)$ which is, by definition, strictly dependent on the set of axioms of PA, so that it gives rise to an independence phenomenon which is intrinsically insurmountable.[2]

In the case of the specific extension PA^{Rfn} studied by Solomon Feferman and advocated by Tennant, we can iterate Gödel's construction so as to produce another independent proposition $\mathcal{G}'$ such that $\vdash_{PA^{Rfn}} \mathcal{G}' \leftrightarrow \neg Theor_{PA^{Rfn}}(\ulcorner \mathcal{G}' \urcorner)$ and $\mathcal{G}' \neq \mathcal{G}$. Now there is surely something pretty dubious about the relation between Tennant's deflationary proposal and the inexhaustibility phenomenon. The soundness extension, of course, is a recipe for the incompleteness of any formal

---

[2]It is worth recalling here that the notion of 'deductive inexhaustibility' can be fully characterised in plain recursion theoretic terms by means of both the well-known notions of *creative set* and *productive set* (Rogers 1987). Such a kind of characterisation turns out to be completely independent of classical model theory.

system; this means that as we decide $\mathcal{G}$ by expanding PA to include the soundness principle $Theor_{\mathsf{PA}}(\ulcorner\alpha\urcorner) \to \alpha$, we can in turn expand PA^{Rfn} with the principle $Theor_{\mathsf{PA}^{Rfn}}(\ulcorner\alpha\urcorner) \to \alpha$ in order to decide $\mathcal{G}'$ and so on. But now the question raises an epistemologically subtle point: the extension which allows us to fill the present deductive hole is the very *cause* of replication of the deductive hole itself. Thus, as soon as we decide $\mathcal{G}$, the achievement of its truth proves a mere fig leaf because *the process itself of deciding $\mathcal{G}$* launches the question of deflationarily achieving the truth of the new independent proposition $\mathcal{G}'$.

From an epistemological standpoint, the situation would be radically different if at each deductive extension step we decided one (or finitely many) of the infinitely many independent propositions $\mathcal{G}_1, \mathcal{G}_2, \ldots$, so as to avoid a new deductive hole. Although this would be another case of deductive inexhaustibility, the mechanism of filling the deductive holes step by step would not give real cause for concern. On the contrary, the relation between Gödelian incompleteness and Tennant's deflationary strategy re-enacts the sort of regress exposed in Zeno's paradox of Achilles and the tortoise: decidability will always remain behind with respect to the independent proposition which is under focus at each step.

To sum up, the real lesson of the case is that we must be careful not to take heuristic insights too seriously from a logical point of view—what the *Rfn* principle states is nothing but the fact that a proof of $\vdash_{\mathsf{PA}^{Rfn}} Theor_{\mathsf{PA}}(\ulcorner\alpha\urcorner)$ makes it possible a proof of $\vdash_{\mathsf{PA}^{Rfn}} \alpha$ by means of a *modus ponens* application between $\vdash_{\mathsf{PA}^{Rfn}} Theor_{\mathsf{PA}}(\ulcorner\alpha\urcorner)$ and *Rfn* instantiated with α. On this view, Tennant's strategy for achieving the truth of $\mathcal{G}$ happens to be surprisingly close to the naive argument outlined in Sect. 5.1 whereby the sentence $\mathcal{G}$ is true for it declares that 'I'm unprovable' and in fact it is unprovable. However both the soundness and the consistency extensions are at odds with the need to stem new independent propositions. This is to conclude that any reliable approach to the problem of deflationarily recognising $\mathcal{G}$ as a true statement has to focus attention on the main concern of Gödel's work, which is the inexhaustibility of formal arithmetic. This is what we will do in the next section.

5.3 An Alternative Deflationary Proposal: The Constructive ω-Rule

5.3.1 The Unrestricted ω-Rule

The shift from PA to the so-called ω-logic (henceforth indicated by PA^{ω}) gives us our starting point for proving $\mathcal{G}$ while avoiding the surfacing of another independent proposition. Let us recall that PA^{ω} is obtained from PA by adding the ω-rule in place of the rule of induction: we can infer that $\forall x \alpha(x)$, provided we can prove $\alpha(n)$ for *each* natural number n. Formally,

5 A Deflationary Account of the Truth of the Gödel Sentence $\mathcal{G}$

$$\frac{\vdash_{\mathsf{PA}^\omega} \alpha(\overline{0}) \quad \vdash_{\mathsf{PA}^\omega} \alpha(\overline{1}) \quad \vdash_{\mathsf{PA}^\omega} \alpha(\overline{2}) \quad \cdots}{\vdash_{\mathsf{PA}^\omega} \forall x \alpha(x)} \; \omega\text{-rule.}$$

The above rule was first described in a published work by David Hilbert in 1931.[3] PA^ω does not only decide both the Gödelian propositions $\mathcal{G}$ and $Cons_{\mathsf{PA}}$ but has the capacity for providing a *syntactically complete* characterisation of first-order arithmetic (see Corollary A.21).

The inclusion of a new inference rule has the advantage over the axiomatic extensions of allowing us to avoid any cumbersome philosophical commitment on the capacity of a certain axiom schema to syntactically reproduce a metatheorical property. From an epistemological point of view, moreover, the ω-rule needs no particular justification: the inferential device it expresses is largely supported by our intuition about the structure of natural numbers so as to be straightforwardly sound when one refers to the standard model.

Yet, of course, this route to $\mathcal{G}$ via the infinitely many premises of the ω-rule corresponds to the collapse of the Hilbertian notion of formal system, and this fact is just another way of stating the First Incompleteness Theorem. The best we can say is that PA^ω makes a virtue out of the very ineffectiveness of the notion of proof by circumventing the phenomenon of deductive inexhaustibility of formal arithmetic, given that the predicate $Theor_{\mathsf{PA}}(x)$ cannot be upgraded to $Theor_{\mathsf{PA}^\omega}(x)$. Anyway, we are driven back to a notion of truth which is *symmetric* with respect to the thin version sponsored by the advocate of deflationism: a thick absolute as the upshot of an infinitary nonconstructive reasoning. In effect, why not say that the movement from PA to PA^ω is anything but a *semantical* transition? The infinitary nature of ω-rule suggests that the rule intervenes as an external device to stretch syntactically the Tarskian definition of truth for the universal quantifier within arithmetical theories:

$\forall x \alpha(x)$ is true if, and only if, $\alpha(n)$ is true for all $n \in \mathbb{N}$.

This is why PA^ω succeeds in achieving completeness (see the proof of Theorem A.20), while the semantical completeness of PA with respect to the standard model $\mathcal{N}$ does not exceed the level of Σ_1-formulas.

The present situation is puzzling for a deflationary approach to incompleteness: on the one hand, the First Incompleteness Theorem pulls any epistemologically well-founded attempt to decide $\mathcal{G}$ away from the notion of formal system; on the other hand, any deductive strategy involving non-formalisable devices like the ω-rule is a *semantical* strategy in disguise. To be more precise, the question now before us is whether there is room between formal arithmetic and its classical semantics for a genuine syntactical manoeuvre able to achieve the provability of $\mathcal{G}$.

---

[3] For an accurate history of the ω-rule, see Isaacson (1991).

5.3.2 The Constructive ω-Rule

A path for an affirmative answer to this question is provided by the *constructive* ω-rule:

if $\alpha(\overline{n})$ admits a *prototype* proof w.r.t. $n \in \mathbb{N}$, then conclude $\vdash_{\mathsf{PA}} \forall x \alpha(x)$.

Following Michael Detlefsen (1979), the term 'prototype' assumes the meaning attached to it by Jacques Herbrand in 'Sur la non-contradiction de l'Arithmetique' (1931): 'when we say that a theorem is true for all x, we mean that for each x individually it is possible to iterate its proof, which may just be considered a prototype of each individual proof' (Herbrand 1931). In other words, a prototype proof provides 'a reasoning which uniformly holds for all arguments, and this uniformity allows (and it is guaranteed by) the use of a generic argument' (Longo 2011).

Alan Bundy and his co-workers regard the constructive ω-rule as a device for capturing the notion of *schematic* proof:

> The constructive ω-rule is a refinement of the ω-rule that can be used in practical proofs. It has the additional requirement that the $\varphi(n)$ premises be proved in a *uniform* way, i.e. that there exists a recursive program, `proof`$_\varphi$, which takes a natural number n as input and returns a proof of $\varphi(n)$ as output. [...] The recursive program `proof`$_\varphi$ formalises our notion of schematic proof (Bundy et al. 2005).[4]

Perhaps the most articulated mathematical approach to prototype proofs has been devised by Giuseppe Longo with regard to impredicative type theory (Longo 2000). As far as the epistemological nature of the constructive ω-rule is concerned, he observes:

> [...] the proof of a universally quantified statement is not understood by following the naif (Tarskian style) interpretation of "$\forall x \ldots$" as "for all $x \ldots$": in no way "$\forall x \ldots$" is used in a proof in the sense of the inspection of "all instances" in the intended model, yet its meaning and use refer to x as generic in a prototype proof (Longo 2000).

Henceforth, we indicate by $\omega_\downarrow$ and $\mathsf{PA}^{\omega\downarrow}$, respectively, the constructive version of the ω-rule and the deductive system obtained by weakening PA^ω through the replacement of the unrestricted ω-rule with $\omega_\downarrow$ and the consequent reintroduction

---

[4] Yet, this definition tends to blur the distinction between the *constructive* and the *recursive* versions of the ω-rule. Following Shoenfield (1959), Torkel Franzén writes: 'A proof of ϕ in a system incorporating the recursive ω-rule is either a pair $\langle \phi, 0 \rangle$ where ϕ is an axiom, or a sequence $\langle \phi, e_1, \ldots, e_n \rangle$ where e_i is a proof of ψ_i, and ϕ follows from $\psi_1, \ldots, \psi_n$ by some ordinary inference rule, or, if ϕ is $\forall x \psi$, a pair $\langle \phi, e \rangle$, where e is the index of a total recursive function such that $\{e\}(n)$ is a proof of $\psi(\overline{n})$ for every n' (Franzén 2004). Indeed, the constructive ω-rule turns out to be a particular kind of recursive ω-rule insofar as its implementation requires the specification of a recursive function able to return a proof of $\psi(\overline{n})$ for every $n \in \mathbb{N}$ in input. On the other hand, the specification of a recursive function able to return a proof for each one of the numerical instances does not necessarily induce uniformity in demonstrative reasoning.

5 A Deflationary Account of the Truth of the Gödel Sentence $\mathcal{G}$ 81

of the induction principle. Both proofs of $\vdash_{PA^\omega} \mathcal{G}$ and $\vdash_{PA^\omega} Cons_{PA}$ are based on prototype arguments. Let us produce, for instance, the proof of $\vdash_{PA^{\omega\downarrow}} \mathcal{G}$. Its crucial prototype juncture is displayed in detail below:

1. $\vdash_{PA} Dem_{PA}(\overline{n}, \ulcorner\mathcal{G}\urcorner)$ hypotheses by absurd
2. $\vdash_{PA} \exists x Dem_{PA}(x, \ulcorner\mathcal{G}\urcorner)$ $\exists$-introduction
3. $\vdash_{PA} Theor_{PA}(\ulcorner\mathcal{G}\urcorner)$ definition of $Theor_{PA}(x)$
4. $\vdash_{PA} \neg \mathcal{G}$ Diagonalisation Lemma
5. $\not\vdash_{PA} \neg \mathcal{G}$ First Incompleteness Theorem
6. $\not\vdash_{PA} Dem_{PA}(\overline{n}, \ulcorner\mathcal{G}\urcorner)$ absurd from 4,5
7. $\vdash_{PA} \neg Dem_{PA}(\overline{n}, \ulcorner\mathcal{G}\urcorner)$ Δ_0-decidability.

Clearly, for any $m \in \mathbb{N}$, this argument allows the generation of a proof of $\vdash_{PA} \neg Dem_{PA}(\overline{m}, \ulcorner\mathcal{G}\urcorner)$ just by replacing $\overline{n}$ with $\overline{m}$. As PA is a subsystem of $PA^{\omega\downarrow}$, we straightforwardly have, for any $m \in \mathbb{N}$, that $\vdash_{PA^{\omega\downarrow}} \neg Dem_{PA}(\overline{m}, \ulcorner\mathcal{G}\urcorner)$; finally, by a step of the $\omega_\downarrow$-rule, we can conclude that $\vdash_{PA^{\omega\downarrow}} \forall x \neg Dem_{PA}(x, \ulcorner\mathcal{G}\urcorner) \equiv \mathcal{G}$. Along similar lines, we can obtain a proof of $\vdash_{PA^{\omega\downarrow}} Cons_{PA}$ by the Second Incompleteness Theorem.[5]

The epistemological cleavage produced through this process of 'constructivisation' is remarkable. The epistemological dividend that this process can pay may be fully appreciated when it is realised that the unrestricted ω-rule can be viewed as a sort of general pattern from which one can specify some different constructive versions. In this respect, the induction principle may be conceived of as a *specific* constructive instance of the ω-rule, where the infinitely many premises for the universal quantification are generated by a well-defined recursive function which consists in the proof by induction itself. On the other hand, the infinite premises of $\omega_\downarrow$ are generated by a certain prototype proof through successive replacements.[6]

Both these constructive versions succeed in capturing a widespread pattern of reasoning in mainstream number theory, even in the most radically constructive contexts. As regards the $\omega_\downarrow$-rule, it is worth mentioning a family of diagrammatic proofs of basic arithmetical facts (Jamnik 2001). But, whereas the inductive mechanism can be compressed into a single axiom of a formal system like PA, the notion of 'prototypicality' seems to present intrinsic intensional features far from being formally reproducible (Longo 2011).

---

[5]In general, the provability of the Gödelian propositions is due to the fact that the enriched theory $PA^{\omega\downarrow}$ enjoys the following additional derivability condition:

$$\mathcal{D}_\omega : \text{ for any formula } \alpha, \not\vdash_{PA^{\omega\downarrow}} \alpha \Rightarrow \vdash_{PA^{\omega\downarrow}} \neg Theor_{PA}(\ulcorner\alpha\urcorner).$$

Clearly, $\mathcal{D}_\omega$ does not hold true in PA, otherwise $\mathcal{G}$ would be provable and unprovable at the same time. As far as the validity of $\mathcal{D}_\omega$ in $PA^{\omega\downarrow}$ is concerned, the reader can find all the technical details by looking at Theorem A.22 and Corollary A.23.

[6]Specific implementations of constructive $\omega_\downarrow$-rule, especially in view of automatic deduction treatments, are afforded in Baker et al. (1992) and Bundy et al. (2005).

The $\omega_\downarrow$-rule radically diverges from its unrestricted ancestor in the very logical way it introduces the universal quantifier. When the unrestricted ω-rule is applied, the inferential step leading from the infinitely many premises $\alpha(0), \alpha(1), \alpha(2), \ldots$ to the conclusion $\forall x \alpha(x)$ actually expresses, we might say, a *synthetic* inference inasmuch it abridges the infinity of its premises into a finite syntactical expression. On the contrary, the constructive ω-rule introduces the universal quantifier in an *analytical* way, for the infinitary information about premises is already finitarily encompassed into the logical structure of the prototype argument. This means that it is the *whole* prototype argument that accomplishes the synthetic task of enclosing the infinite into a finite demonstrative device. This is the reason why $\omega_\downarrow$-rule has a double logical nature: finitary and non-formalisable at the same time.

An immediate consequence of such an epistemological reversal is that, unlike the unrestricted pattern of inference conveyed by the ω-rule, $\omega_\downarrow$ does not lie within the province of classical semantics. In fact, in the process of constructive specification leading to $\omega_\downarrow$, the ω-rule loses its unconditional generality, so that it can no longer faithfully reproduce the Tarskian definition for the universal quantifier. In epistemological terms, the constructive requirement divorces the $\omega_\downarrow$-rule from the nonconstructive semantics grounded on bivalence.

The notion of uniformity underlying the constructive ω-rule clearly recalls the first-order logical principle of universal generalisation (GU) which licenses the inference from $\varphi(t)$ to $\forall x \varphi(x)$ provided the absolute *genericity* of the term t (technically, x does not appear as a free variable in $A(t)$ and $A(x)$ is the result of replacing all occurrences of t in $A(t)$ by x) (Cellucci 2009). Of course GU and $\omega_\downarrow$ cannot coincide; otherwise $\mathcal{G}$ would be provable against the First Incompleteness Theorem. Since their difference lies in the fact that $\omega_\downarrow$ is restricted to natural numbers, one might object that the adoption of $\omega_\downarrow$ entails a strong semantical commitment to them. Our reply is that a prototype proof do *not* presuppose the set of natural numbers, but *characterises* it as the set of all the numerical entities to which the schematic argument at issue applies. In other words, when a prototype argument is turned into a universal quantification, the quantifier is meant to range over all the numerical objects capable of instantiating the argumentative schema. So, we may say that any prototype argument implicitly *defines* a set of numbers. This aspect highlights another important difference with the unrestricted ω-rule which actually enjoins us to assume a numerical ontology in order to achieve to whole set of its premises. Once that the relation between semantics and syntax has been reversed, the constructive ω-rule can save its deflationary skill.[7]

---

[7]For similar reasons, this rule cannot be expressed by means of the *Uniform Reflection Scheme*:

$$Urs: \forall x Theor_{PA}(\overline{\ulcorner \alpha(x) \urcorner}) \to \forall x \alpha(x).$$

Indeed, PA^{Urs}—i.e. PA with Urs added as a new axiom—is still an incomplete formal system by the First Incompleteness Theorem, whereas $PA^{\omega_\downarrow}$ is syntactically complete. This fact, of course, cannot be avoided when our principle is formulated as a rule: *if* $\vdash_{PA} \forall x Theor_{PA}(\overline{\ulcorner \alpha(x) \urcorner})$, *then* $\vdash_{PA} \forall x \alpha(x)$.

5 A Deflationary Account of the Truth of the Gödel Sentence $\mathcal{G}$

5.3.3 Deconstructing Dummett's Argument

To test our deflationary proposal, let us accept the challenge raised by Tennant when he claims that 'in so much as stating the philosophical crux of Gödel's theorem, Dummett has furnished the kind of use of the truth-predicate that any deflationist would wish to deconstruct' (Tennant 2002, p.552).

In 1963, Dummett sketched a procedure for achieving the truth of $\mathcal{G}$:

> The argument for the truth of [$\mathcal{G}$] proceeds under the hypothesis that the formal system in question is consistent. The system is assumed, further, to be such that, for any decidable predicate $B(x)$ and any numeral $\bar{n}$, $B(\bar{n})$ is provable if it is true, — $\neg B(\bar{n})$ is provable if $B(\bar{n})$ is false (the notions of truth and falsity for such a statement being, of course, unproblematic). The particular predicate $A(x)$ [i.e., $\neg Dem(x, \ulcorner\mathcal{G}\urcorner)$] is such that, if $A(\bar{n})$ is false for some numeral $\bar{n}$, then we can construct a proof in the system of $\forall x A(x)$. From this it follows — on the hypothesis that the system is consistent — that each of $A(0)$, $A(1)$, $A(2)$, ... is true'. (Dummett 1963, p. 192)

As is so often the case, the fortune of a certain idea profits from a certain vagueness in its formulation, so that many authors have tried to make Dummett's hint more specific. However, it is rather surprising that it has never been detected here the offices of a prototype argument. It is because Dummett's argument has a prototypical nature, indeed, that the shift from PA to $\mathsf{PA}^{\omega\downarrow}$ yields a deflationary reduction of it. Specifically, two logical circumstances make this reduction possible. The first refers to the 'analytical' way in which the $\omega_\downarrow$-rule introduces the universal quantifier. This epistemological feature matches Dummett's claim that '[...] the transition from saying that all the statements are true to saying that $\forall x \alpha(x)$ is true is trivial' (Dummett 1963, p. 192). The second circumstance concerns the fact that 'the argument for the truth of [$\mathcal{G}$] proceeds under the hypothesis that the formal system in question is consistent' (Dummett 1963, p. 192). Indeed, the $\omega_\downarrow$-rule allows us to cut the Gordian knot of the consistency hypothesis, for $Cons_{\mathsf{PA}}$ turns out to be (deflationarily) *provable* within $\mathsf{PA}^{\omega\downarrow}$. In this way, we can get rid of the most cumbersome inflationary commitment, namely, the resort to the soundness of PA.

5.4 Concluding Remarks

The problem addressed in this chapter is that of showing that a deflationary view of incompleteness (incompletability, indeed) is possible, so that the Gödel phenomena are not disastrous for deflationism about truth. We do not mean to argue in favour of a reappraisal of Hilbert's foundational programme through the constructive ω-logic as, for instance, Detlefsen does (Detlefsen 1979). What we claim is that the introduction of the constructive ω-rule for achieving $\mathcal{G}$ is in tune with a deflationary point of view: the path leading to $\mathcal{G}$ consists of a proof of $\mathcal{G}$ which drops any reference to a genuine property of 'truth'. The fact that the proofs of $\mathcal{G}$ and $Cons_{\mathsf{PA}}$ within $\mathsf{PA}^{\omega\downarrow}$ are not relevant to a foundational point of view need not worry us, as these proofs exploit the First and Second Incompleteness Theorems which *assume* the consistency of PA. But such a foundational irrelevance pinpoints the deflationary nature of the present proposal, by spelling out the irrelevance of the truth value of $\mathcal{G}$ to the grasp of incompleteness phenomena.

Moreover, the deflationary character of our proposal can also be stressed from a broader perspective. If we look at the historical development of number theory—and so assuming the point of view of the mathematical practice—the prototypical reasoning turns out to be a very weak arithmetical demonstrative strategy. It seems indeed that only trivial arithmetical statements can be proved through pure prototype arguments.[8]

This is the case, to pick one example, of the proof of the transitivity of the divisibility property: *for all $a, b, c \in \mathbb{N}$, if $a|b$ and $b|c$, then $a|c$*. On this view, the weakness of the $\omega_\downarrow$-rule has to be read in opposition to the stronger number theoretical methods of induction and infinite descent employed for proving relevant arithmetical properties (Weil 1984). This aspect can be fully grasped by observing that both mathematical induction and infinite descent presuppose a prototype argument. Consider, for instance, the demonstrative method of mathematical induction. As remarked by Longo, an implicit prototypical passage is silently at work when one proves that the inductive step $\alpha(n) \to \alpha(n + 1)$ holds for all $n \in \mathbb{N}$ (Longo 2011). Similarly, any proof by descent is built over prototypical assumptions. On the one hand, this leads to the conclusion that one of the favourite inflationary tenets—the idea that any demonstrative method non-formalisable within PA has to be epistemologically stronger than the methods encompassed by PA—is shown to be flawed. On the other hand, we get a strong epistemological reason for deflationarily accepting the $\omega_\downarrow$-rule: its refusal would imply the refusal of the induction principle itself.

In conclusion, what emerges from our proposal is a notion of deflationism which clearly sponges on the epistemological authority of the mathematical practice, specifically that of number theory. Indeed, the deflationary licitness of the constructive ω-rule has been supported by stressing its undeniable *status* of universally accepted method in the practice of number theory: *explicitly* used through pure applications of prototype arguments, *implicitly* at work as a hidden basic subprinciple in proofs by induction or descent. The same epistemological move provides one further reason for discarding the unrestricted ω-rule to the extent that it expresses a purely semantical principle, too much abstractly shaped for exhibiting any kind of paradigmatic application *in corpore vili*.

---

[8]The situation seems to be radically different in geometry. Indeed, the very recognition that a deduction about *particular* constructions produces knowledge of *general* validity is at the heart of the emergence of Greek deductive mathematics. For example, the proof of the statement that in every triangle the sum of the three angles is equal to 180° considers a generic triangle while holding uniformly for all triangles. There is considerable plausibility in Reviel Netz's idea that the feeling of generality that Greek mathematicians gain at the end of a proof arises from the conviction that the proof concerned with a particular object is *repeatable* for any similar object (Netz 2003, p. 256, 269). This explanation makes Greek proofs prototype proofs *avant la lettre*.

Appendix: Technical Backgrounds and Proofs

A.1 Peano Arithmetic: Theory and Models

Definition A.1 (Peano Arithmetic). The language of PA is given by the language of first-order logic with identity enriched with the individual constant $\bar{0}$, the unary functional symbol $Succ(\_)$ (the successor) and the two binary functional symbols $+$ and $\cdot$. Moreover, the specific deductive apparatus of PA is defined by the following nine proper axioms:

(1) $x = y \rightarrow (x = z \rightarrow y = z)$
(2) $x = y \rightarrow Succ(x) = Succ(y)$
(3) $\bar{0} \neq Succ(x)$
(4) $Succ(x) = Succ(y) \rightarrow x = y$
(5) $x + \bar{0} = x$
(6) $x + Succ(y) = Succ(x + y)$
(7) $x \cdot \bar{0} = \bar{0}$
(8) $x \cdot Succ(y) = (x \cdot y) + x$
(9) For every formula $\alpha(x)$ of PA such that x occurs free in α,
$\vdash_{\mathsf{PA}} \alpha(\bar{0}) \rightarrow (\forall x(\alpha(x) \rightarrow \alpha(Succ(x))) \rightarrow \forall x \alpha(x))$.

We abridge with $\bar{n}$ the numeral $Succ(Succ \ldots Succ(\bar{0}) \ldots)$ resulting from n applications of the successor function to the constant $\bar{0}$. For any pair of terms t and s, $t \neq s$ is intended to be defined as $\neg(t = s)$.

Definition A.2 (Structure $\mathcal{N}$). The structure $\mathcal{N} = (\mathbb{N}, 0, +^{\mathcal{N}}, \cdot^{\mathcal{N}}, Succ^{\mathcal{N}})$ is formed by the set of non-negative integers $\mathbb{N} = \{0, 1, 2, \ldots\}$, the distinguished number $0 \in \mathbb{N}$ (which interprets the constant $\bar{0}$), the functional symbols $+^{\mathcal{N}}$ and $\cdot^{\mathcal{N}}$, respectively, corresponding to the familiar sum and product and the successor function $Succ^{\mathcal{N}}(x) \stackrel{def.}{=} x +^{\mathcal{N}} 1$.

Theorem A.3 (Soundness). $\mathcal{N}$ *is a model of* PA *and, moreover,* PA *is sound w.r.t.* $\mathcal{N}$.

Proof. For establishing that $\mathcal{N}$ is a model of PA (in symbols $\mathcal{N} \models \mathsf{PA}$), we have to show that each of the PA axioms is interpreted in $\mathcal{N}$ as a true statement. It is immediate to check that $\mathcal{N}$ satisfies axioms 1–8. As far as the induction principle is concerned (axiom 9), since the domain of $\mathcal{N}$ exactly coincides with the set of naturals $\mathbb{N}$, the inductive mechanism is indeed able to cover the totality of the elements of $\mathcal{N}$, so as to justify the introduction of the universal quantifier.

As far as the soundness property is concerned, the proof consists in showing that, for any formula α, if $\vdash_{\mathsf{PA}} \alpha$, then $\mathcal{N} \models \alpha$. We proceed by induction on the length of the PA proof ending with $\vdash_{\mathsf{PA}} \alpha$. The base is clearly provided by the fact that $\mathcal{N} \models \mathsf{PA}$. Then, it is easy to see that the logical inference rules transmit the truth from premisses to conclusions. □

Remark A.4 (Soundness and Consistency). Due to the strict bivalence of classical semantics, if a theory is sound w.r.t. a certain model, it is consistent (otherwise the theory would prove a false statement).

Remark A.5 (Standard Model). The structure $\mathcal{N}$ is said to be the *standard model* for PA.

A.2 The Incompleteness Theorems

Definition A.6 (Deductive Independence). A formula α is said to be independent of PA if $\nvdash_{\mathsf{PA}} \alpha$ and $\nvdash_{\mathsf{PA}} \neg\alpha$.

Definition A.7 (ω-Consistency). A certain arithmetical theory T is said to be ω-consistent if the following two conditions are mutually excluding:

- For all $n \in \mathbb{N}$, $\vdash_\mathsf{T} \alpha(\overline{n})$,
- $\vdash_\mathsf{T} \exists x \neg \alpha(x)$.

Remark A.8 ω-consistency is stronger than consistency so like any ω-consistent theory is also consistent.

The proofs of the incompleteness theorems are here merely sketched; for the technical details the reader is referred to Rautenberg (2000).

Theorem A.9 (First Incompleteness Theorem). *There exists a formula $\mathcal{G}$ such that if PA is ω-consistent, then $\mathcal{G}$ is independent of PA.*

Proof. The proof is developed through the following five points.

(1) There exists a 1–1 assignment of natural numbers to formulas and demonstrations of PA. $\ulcorner \alpha \urcorner$ and $\overline{\ulcorner \alpha \urcorner}$, respectively, indicate the number associated with α (its Gödelian code) and its corresponding numeral: if $\ulcorner \alpha \urcorner = n$, then $\overline{\ulcorner \alpha \urcorner} = \overline{n}$. In the same way, $\ulcorner \pi \urcorner$ and $\overline{\ulcorner \pi \urcorner}$, respectively, denote the Gödelian code of the proof π and the corresponding numeral.

(2) It is possible to define a Δ_0-formula $Dem_\mathsf{PA}(x, y)$ such that $\vdash_\mathsf{PA} Dem_\mathsf{PA}(\overline{n}, \overline{m})$ if, and only if, n encodes a PA demonstration of the formula α with $\ulcorner \alpha \urcorner = m$.

(3) Consider the predicate $Theor_\mathsf{PA}(y) \stackrel{def.}{=} \exists x Dem_\mathsf{PA}(x, y)$. Its negation admits a formula $\mathcal{G}$ as a fixed point, i.e.

$$\vdash_\mathsf{PA} \mathcal{G} \leftrightarrow \neg Theor_\mathsf{PA}(\overline{\ulcorner \mathcal{G} \urcorner}).$$

(4) $\vdash_\mathsf{PA} \mathcal{G}$ implies $\vdash_\mathsf{PA} \neg\mathcal{G}$ and so, if PA is consistent, $\nvdash_\mathsf{PA} \mathcal{G}$.
(5) If PA is ω-consistent, then $\nvdash_\mathsf{PA} \mathcal{G}$ implies $\nvdash_\mathsf{PA} \neg\mathcal{G}$.

Finally, $\mathcal{G}$ is independent of PA. □

Theorem A.10 (Second Incompleteness Theorem). *Consider the formula*

$$Cons_{PA} \equiv \neg Theor_{PA}(\ulcorner 0 = 1 \urcorner)$$

asserting the consistency of PA*: it is independent from* PA *as well as* $\mathcal{G}$.

Proof. The proof consists in showing that $Cons_{PA}$ is provably equivalent to $\mathcal{G}$, i.e. $\vdash_{PA} Cons_{PA} \leftrightarrow \mathcal{G}$. In such a way, $\vdash_{PA} Cons_{PA}$ and $\vdash_{PA} \neg Cons_{PA}$ would, respectively, imply $\vdash_{PA} \mathcal{G}$ and $\vdash_{PA} \neg \mathcal{G}$, against the First Incompleteness Theorem. □

A.3 Σ_1-Completeness and Related Results

Definition A.11 (Logical Complexity).

- A formula α belongs to the set Δ_0 if it is equivalent to a closed formula α' in which all the quantifiers, if any, are bounded.
- A formula α belongs to Σ_1 (resp. Π_1) if it is equivalent to a closed formula $\alpha' \equiv \exists x \beta(x)$ (resp. $\alpha' \equiv \forall x \beta(x)$) such that $\beta[t/x] \in \Delta_0$.
- A formula α belongs to Σ_{n+1} (resp. Π_{n+1}) if it is equivalent to a closed formula $\alpha' \equiv \exists x \beta(x)$ (resp. $\alpha' \equiv \forall x \beta(x)$) such that $\beta[t/x] \in \Pi_n$ (resp. $\beta[t/x] \in \Sigma_n$).

Example A.12 Both the Gödelian propositions $\mathcal{G}$ and $Cons_{PA}$ are Π_1-statement.

Remark A.13 Whereas $\alpha \in \Sigma_n$ if, and only if, $\neg \alpha \in \Pi_n$, the set of Δ_0-formulas is closed under negation.

Proposition A.14 *Let t, s be two closed arithmetical terms:*

(1) *If $\mathcal{N} \vDash t = s$, then $\vdash_{PA} t = s$,*
(2) *If $\mathcal{N} \vDash t \neq s$, then $\vdash_{PA} t \neq s$,*
(3) $\vdash_{PA} \bar{n} \geqslant \bar{m} \rightarrow (\bar{n} = \bar{0} \vee \bar{n} = \bar{1} \vee \ldots \vee \bar{n} = \bar{m})$.

Proof. The reader can find all the proofs in Rautenberg (2000). □

Theorem A.15 (Δ_0-Decidability). *If α is a closed Δ_0-formula, then either $\vdash_{PA} \alpha$ or $\vdash_{PA} \neg \alpha$.*

Proof. Let $\alpha \in \Delta_0$; we proceed by induction on the number of logical connectives occurring in α.

Base. If no logical connective occurs in α, then $\alpha \equiv t = s$ with t, s closed terms. It is either $\mathcal{N} \vDash t = s$ or $\mathcal{N} \vDash t \neq s$ and so Proposition A.14 gives us the basis.
Step. Proposition A.14 enables us to stress the following conversions

$$\exists x \leqslant k \alpha(x) \Leftrightarrow \alpha(0) \vee \alpha(1) \vee \ldots \vee \alpha(k)$$
$$\forall x \leqslant k \alpha(x) \Leftrightarrow \alpha(0) \wedge \alpha(1) \wedge \ldots \wedge \alpha(k),$$

for turning any quantified Δ_0-formula into an equivalent one without quantifiers. Then it is easy to see that any Boolean composition of decidable propositions is, in turn, decidable. □

Corollary A.16 (Δ_0-**Completeness**). *For any closed* $\alpha \in \Delta_0$, *if* $\mathcal{N} \vDash \alpha$, *then* $\vdash_{\mathsf{PA}} \alpha$.

Proof. Let $\mathcal{N} \vDash \alpha$, but $\nvdash_{\mathsf{PA}} \alpha$. For $\alpha \in \Delta_0$, by Theorem A.15, it would be $\vdash_{\mathsf{PA}} \neg\alpha$ against the soundness of PA w.r.t. $\mathcal{N}$. □

Theorem A.17 PA *is* Σ_1-*complete w.r.t.* $\mathcal{N}$ *if, and only if, it is* Δ_0-*decidable.*

Proof. ($\Rightarrow$) Let $\nvdash_{\mathsf{PA}} \alpha$, with α closed and in Δ_0. By the Σ_1-completeness, we obtain $\mathcal{N} \nvDash \alpha$ and so $\mathcal{N} \vDash \neg\alpha$. Since $\neg\alpha \in \Delta_0$, we perform a further step of Σ_1-completeness so as to obtain $\vdash_{\mathsf{PA}} \alpha$.

($\Leftarrow$) We proceed by absurd: let $\exists x \alpha(x)$ be a closed Σ_1-formula such that $\mathcal{N} \vDash \exists x \alpha(x)$, but $\nvdash_{\mathsf{PA}} \exists x \alpha(x)$. For $\mathcal{N} \vDash \exists x \alpha(x)$, there is an $n \in \mathbb{N}$ such that $\mathcal{N} \vDash \alpha(n)$. Since $\alpha(n) \in \Delta_0$, we can apply the just proved Δ_0-completeness and obtain $\vdash_{\mathsf{PA}} \alpha(\overline{n})$. As a matter of logic, we finally obtain $\vdash_{\mathsf{PA}} \exists x \alpha(x)$ which contradicts our assumption that $\nvdash_{\mathsf{PA}} \exists x \alpha(x)$. □

Corollary A.18 (Σ_1-**Completeness**). PA *is* Σ_1-*complete w.r.t.* $\mathcal{N}$.

Proof. Straightforwardly by Theorems A.15 and A.17. □

Corollary A.19 *If* $\alpha \in \Pi_1$ *is independent of* PA, *then* $\mathcal{N} \vDash \alpha$. *In particular, we have that* $\mathcal{N} \vDash \mathcal{G}$ *and* $\mathcal{N} \vDash \mathit{Cons}_{\mathsf{PA}}$.

Proof. By the Σ_1-completeness, we obtain $\mathcal{N} \nvDash \neg\alpha$ from $\nvdash_{\mathsf{PA}} \neg\alpha$, and so $\mathcal{N} \vDash \alpha$. Both the Gödelian propositions $\mathcal{G}$ and $\mathit{Cons}_{\mathsf{PA}}$ instantiate the case just explained so that $\mathcal{N} \vDash \mathcal{G}$ and $\mathcal{N} \vDash \mathit{Cons}_{\mathsf{PA}}$. □

A.4 ω-Logic, Constructive ω-Logic and Some Related Results

Theorem A.20 *For any formula* α, $\mathcal{N} \vDash \alpha$ *if, and only if,* $\vdash_{\mathsf{PA}^\omega} \alpha$.

Proof. (*Soundness*) It is a matter of extending the proof of Theorem A.3 so as to include the ω-rule. In order to show that any instance of the ω-rule transmits the truth from premisses to the conclusion, it is sufficient to remark that the ω-rule just provides a syntactical rendition of the Tarskian definition of the universal quantifier: if $\mathcal{N} \vDash \alpha(0)$, $\mathcal{N} \vDash \alpha(1)$, $\mathcal{N} \vDash \alpha(2)$ and so on, then $\mathcal{N} \vDash \forall x \alpha(x)$.

(*Completeness*) We proceed by induction on the logical complexity of α. The Δ_0-completeness provides the base of our induction. Then, we distinguish two cases:

- Let $\alpha \equiv \exists x \beta(x) \in \Sigma_{n+1}$. $\mathcal{N} \vDash \exists x \beta(x)$ means that there is an $n \in \mathbb{N}$ such that $\mathcal{N} \vDash \beta(n)$ with $\beta(n) \in \Pi_n$. By inductive hypothesis $\vdash_{\mathsf{PA}^\omega} \beta(\overline{n})$ and so we can introduce the existential quantifier for finally achieving $\vdash_{\mathsf{PA}^\omega} \exists x \beta(x)$.

- Let $\alpha \equiv \forall x \beta(x) \in \Pi_{n+1}$. $\mathcal{N} \vDash \forall x \beta(x)$ means that for all $n \in \mathbb{N}$, $\mathcal{N} \vDash \beta(n)$ with $\beta(n) \in \Sigma_n$. By inductive hypothesis we have that for all $n \in \mathbb{N}$, $\vdash_{\mathsf{PA}^\omega} \beta(\overline{n})$. Finally, the ω-rule enables us to introduce the universal quantifier so as to obtain $\vdash_{\mathsf{PA}^\omega} \forall x \beta(x)$. □

Corollary A.21 PA^ω *is syntactically complete, namely, for any formula α, either* $\vdash_{\mathsf{PA}^\omega} \alpha$ *or* $\vdash_{\mathsf{PA}^\omega} \neg \alpha$.

Proof. We show that $\nvdash_{\mathsf{PA}^\omega} \alpha$ implies $\vdash_{\mathsf{PA}^\omega} \neg \alpha$. Let $\nvdash_{\mathsf{PA}^\omega} \alpha$; by Theorem A.20 it is $\mathcal{N} \nvDash \alpha$ and so $\mathcal{N} \vDash \neg \alpha$. Then another application of Theorem A.20 allows us to conclude that $\vdash_{\mathsf{PA}^\omega} \neg \alpha$. □

Theorem A.22 *For any formula α, if $\nvdash_{\mathsf{PA}^{\omega\downarrow}} \alpha$, then $\vdash_{\mathsf{PA}^{\omega\downarrow}} \neg Theor_{\mathsf{PA}}(\ulcorner \alpha \urcorner)$.*

Proof. Suppose by absurd that there is an $n \in \mathbb{N}$ such that $\vdash_{\mathsf{PA}} Dem_{\mathsf{PA}}(\overline{n}, \ulcorner \alpha \urcorner)$. This latter would imply the existence of a PA proof π of α such that $\ulcorner \alpha \urcorner = n$. This is in contrast with our hypothesis that $\nvdash_{\mathsf{PA}^{\omega\downarrow}} \alpha$ and so we conclude $\nvdash_{\mathsf{PA}} Dem_{\mathsf{PA}}(\overline{n}, \ulcorner \alpha \urcorner)$. Then, the Δ_0–decidability allows us to turn $\nvdash_{\mathsf{PA}} Dem_{\mathsf{PA}}(\overline{n}, \ulcorner \alpha \urcorner)$ into $\vdash_{\mathsf{PA}} \neg Dem_{\mathsf{PA}}(\overline{n}, \ulcorner \alpha \urcorner)$. The argument just explained is clearly prototypical w.r.t. n (being, in turn, the proof of Theorem A.15 prototypical w.r.t. the formula α) so as, by a step of $\omega_\downarrow$-rule, we can conclude $\vdash_{\mathsf{PA}^{\omega\downarrow}} \forall x \neg Dem_{\mathsf{PA}}(x, \ulcorner \alpha \urcorner)$, that is, $\vdash_{\mathsf{PA}^{\omega\downarrow}} \neg Theor_{\mathsf{PA}}(\ulcorner \alpha \urcorner)$. □

Corollary A.23 $\mathsf{PA}^{\omega\downarrow}$ *decides both the Gödelian propositions $\mathcal{G}$ and $Cons_{\mathsf{PA}}$.*

Proof. Suppose by absurd that $\mathcal{G}$ is not provable in $\mathsf{PA}^{\omega\downarrow}$. By Theorem A.22, we would obtain $\vdash_{\mathsf{PA}^{\omega\downarrow}} \neg Theor_{\mathsf{PA}}(\ulcorner \mathcal{G} \urcorner)$ from $\nvdash_{\mathsf{PA}^{\omega\downarrow}} \mathcal{G}$. Now, we know that $\vdash_{\mathsf{PA}^{\omega\downarrow}} \mathcal{G} \leftrightarrow \neg Theor_{\mathsf{PA}}(\ulcorner \mathcal{G} \urcorner)$ and so we would be able to deduce $\vdash_{\mathsf{PA}^{\omega\downarrow}} \mathcal{G}$ against the fact that we assumed $\nvdash_{\mathsf{PA}^{\omega\downarrow}} \mathcal{G}$. Such an argument leads us to reject $\nvdash_{\mathsf{PA}^{\omega\downarrow}} \mathcal{G}$, that is, to affirm $\vdash_{\mathsf{PA}^{\omega\downarrow}} \mathcal{G}$.

The proof of $\vdash_{\mathsf{PA}^{\omega\downarrow}} Cons_{\mathsf{PA}}$ proceeds in an analogous way. □

Acknowledgements The second author acknowledges support from FAPESP Post-Doc Grant 2013/22371-0, São Paulo State, Brazil.

References

Baker, S., A. Ireland, and A. Smaill. 1992. On the use of the constructive omega-rule within automated deduction. In *LPAR*, 214–225. Berlin/Heidelberg: Springer.
Bundy, A., M. Jamnik, and A. Fugard. 2005. The nature of mathematical proof. *Philosophical Transactions of Royal Society* 363(1835): 2377–2391.
Cellucci, C. 2009. The universal generalization problem. *Logique & Analyse* 52: 3–20.
Cieśliński, C. 2010. Truth, conservativeness, and provability. *Mind* 119: 409–422.
Detlefsen, M. 1979. On interpreting Gödel's second theorem. *Journal of Philosophical Logic* 8(1): 297–313.
Dummett, M. (1963) The philosophical significance of Gödel's theorem. Reprinted in *Truth and other enigmas*. London: Duckworth, 1978.

Feferman, S. 1962. Transfinite recursive progressions of axiomatic theories. *Journal of Symbolic Logic* 27: 259–316.
Feferman, S. 1991. Reflecting on incompleteness. *Journal of Symbolic Logic* 56: 1–49.
Field, H. 1999. Deflating the conservativeness argument. *The Journal of Philosophy* 96: 533–540.
Franzén, T. 2004. Transfinite progressions: A second look at completeness. *Bulletin of Symbolic Logic* 10(3): 367–389.
Gödel, K. 1965. On formally undecidable propositions of principia mathematica and related systems. In *The undecidable*, ed. M. Davis. New York: Raven Press.
Halbach, V. 2001. How innocent is deflationism? *Synthese* 126: 167–194.
Halbach, V. 2011. *Axiomatic theories of truth*. Cambridge/New York: Cambridge University Press.
Herbrand, J. 1931. Sur la non-contradiction de l'Arithmetique. *Journal fur die reine und angewandte Mathematik* 166: 1–8.
Hilbert, D. 1931. Beweis des Tertium non datur – *Nachrichten von der Gesellschaft der Wissenschaften zu Göttingen. Mathematisch-Physikalische Klasse*, 120–125.
Isaacson, D. 1991. Some considerations on arithmetical truth and the ω-rule. In *Proof, logic and formalization*, ed. Michael Detlefsen, 94–138. London: Routledge.
Isaacson, D. 2011. Necessary and sufficient conditions for undecidabillity of the Gödel sentence and its truth. In *Vintage Enthusiasms: Essays in Honour of John Bell*, ed. Peter Clark, David DeVidi, and Michael Hallett, 135–152. University of Western Ontario Series in the Philosophy of Science. Heidelberg/New York: Springer.
Jamnik, M. 2001. *Mathematical reasoning with diagrams*. Stanford: CSLI.
Ketland, J. 1999. Deflationism and Tarski's paradise. *Mind* 108: 69–94.
Ketland, J. 2005. Deflationism and the Gödel phenomena: Reply to Tennant. *Mind* 114: 75–88.
Longo, G. 2000. Prototype proofs in type theory. *Mathematical Logic Quaterly* 46(3): 257–266.
Longo, G. 2011. Reflections on (concrete) incompleteness. *Philosophia Mathematica* 19(3): 255–280.
Netz, R. 2003. *The shaping of deduction*. Cambridge/UK: Cambridge University Press.
Piazza, M., and G. Pulcini. 2013. Strange case of Dr. Soundness and Mr. Consistency. In *The Logica Yearbook* 2013, 161–172. College Publications.
Rautenberg, W. 2000. *A concise introduction to mathematical logic*. Berlin: Springer.
Rogers, H. 1987. *Theory of recursive functions and effective computability*. Cambridge: MIT.
Shapiro, S. 1998. Proof and truth: Through thick and thin. *Journal of Philosophy* 95: 493–521.
Shoenfield, J.R. 1959. On a restricted ω-rule. *Bulletin de l'Académie Polonaise des Sciences. Série des sciences, mathématiques, astronomiques et physiques*, VII(7): 405–407.
Tennant. N, 2002. Deflationism and the Gödel phenomena. *Mind* 111: 551–582.
Tennant, N. 2005. Deflationism and the Gödel phenomena: Reply to Ketland. *Mind* 114: 89–96.
Tennant, N. 2010. Deflationism and the Gödel-phenomena: Reply to Cieslinski. *Mind* 119(474): 437–450.
Weil, A. 1984. *Number theory*. Boston: Birkhäuser.
Wright, C. 1994. About "The philosophical significance of Gödel's theorem": Some issues. In *The philosophy of Michael Dummett*. Dordrecht/The Netherlands: Kluwer.

Chapter 6
Rule-Following and the Limits of Formalization: Wittgenstein's Considerations Through the Lens of Logic

Paolo Pistone

6.1 Introduction

The justification of logical rules stumbles upon a celebrated remark by Wittgenstein: it takes rules to justify the application of rules. This inner circularity affecting logic makes it difficult to explain what by following a rule (typically, by endorsing the alleged compulsion of deduction) is inexorably left implicit.

Loosely following Searle's revision (see Searle 1969) of Kant's vocabulary, I'll style "constitutive" those conditions (if any) on which the existence of the activity of (correctly) applying a given rule depends. Then, a series of well-known arguments (that I briefly recall in Sect. 6.2) seem to undermine the possibility to provide logical rules with such constitutive conditions, since the description of the latter would involve the use of the same rules they should be constitutive of; it would be indeed patently circular to explain the existence of a successful practice of correctly applying the rule of, say, *modus ponens* by relying on the existence of the practice of *modus ponens*!

On the other hand, in Sect. 6.3, I'll try to show that by exploiting the advances made by proof theory on the dynamics of logic, a different approach to rules is possible: my thesis is indeed that the focus on the concrete syntactic manipulation which underlies the application of logical rules within formal systems can represent a fertile direction where to look for a way out from the "blind spot" of circular explanations.

The formalization of logical inference, as it is clear from the developments of mathematical logic from Frege's *Begrisschrift* to most recent programming languages, may carry some arbitrariness: for instance, we know that Frege's formalism,

---

P. Pistone (✉)
Department of Philosophy, Communication and Visual Arts, Università Roma Tre,
via Ostiense 234, 00144, Rome, Italy
e-mail: heighteight@gmail.com

natural deduction and sequent calculus actually code exactly the same notion of classical provability, though in quite different ways. Yet, the apparently superficial differences between these formal systems hide relevant structural information: for instance, Gentzen was able to show that, among those three, only sequent calculus exposes in a clear way the symmetries underlying classical inference, leading to his celebrated *Hauptsatz*.

It seems quite tempting to call those aspects regarding logical formatting the *subjective* side of logic, for at least two reasons: first, since they are generally considered as inessential to the characterization of "objective" logic (in our example, classical provability) and, second, since they are directly related to the way logic is actually written, that is, to the most significative observable form of access to logical inference. In particular, since writing is an activity concretely happening in space and time, it is quite natural to expect this syntactic dimension to be subject to the results and perspectives coming from computational complexity theory. Just to give an idea, taking λ-calculus or Turing machines as our favourite formalism for calculus will produce exactly the same "objective" class of computable functions, but with a very remarkable difference of grain: the single, atomic step of computation in λ-calculus (called β-reduction) requires a polynomial time to be simulated on a Turing machine; so to say, *what appears on one side as the immediate application of a rule takes quite a long time to be executed on the other*.

Finally, in the last section, one will find a technical sketch of the mathematical perspective of Girard's *transcendental syntax*, which reconstructs logic in the algebra and geometry of linear operators, building on proof-theoretic results coming from linear logic: such a reconstruction provides mathematical content to the philosophical reflections motivating this article, so as a concrete connection with computational complexity and its theoretical open questions (concerning the space and time in which formal devices evolve).

6.2 The Blind Spot of Rules

In Wittgenstein's *Tractatus*, it is stated that matters about logical formatting, i.e. about the choice of rules and notations for logic, cannot even be meaningfully formulated: these matters would indeed *say* something on the *logical form*, that is, on the way in which logical languages are involved in referential practices, i.e. somehow attached to the world; since the logical form plays a normative role, the one of assuring the indisputability of deduction, then it cannot be the object of discussion and challenge, since any challenge should presuppose it as a constitutive condition.

Logical conventionalism, with Carnap's distinction between *internal* and *external* questions (see Carnap 1950), was intended to save the theory of the logical form, eliminating at the same time all reference to "the mystic": "what one should be silent of" (Wittgenstein 2001), with Wittgenstein's words, became in Carnap's work the object of purely pragmatical considerations.

On the other hand, the celebrated criticism of conventionalism contained in Quine (1984), where it is stated that

[...] if logic is to proceed mediately from conventions, logic is needed for inferring logic from the conventions. Quine (1984) seems to show that the theory contained in the *Tractatus* cannot be reduced to a conventionalist one[1]; that criticism was indeed grounded on the remark that, in the description of logical rules, the explicitation of what is required in order to correctly apply those rules inexorably eludes the description: as Wittgenstein would say, it is *shown* but cannot be *said*; Quine's argument is just a variation on a theme published by Lewis Carroll forty one years before (see Carroll 1895). To make the long story short, Carroll's (and *a fortiori* Quine's) argument is based on the remark that in order to describe how to follow the rule of *modus ponens*

$$\frac{A \quad A \Rightarrow B}{B} \; (MP) \tag{6.1}$$

one is naturally led to produce an inflated argument which is nothing but another form (rather, a sequence of instances) of the rule of *modus ponens*, whose major premise is a formula expressing the content of the schema (6.1):

$$\frac{A \Rightarrow B \quad \dfrac{A \quad A \Rightarrow ((A \Rightarrow B) \Rightarrow B)}{(A \Rightarrow B) \Rightarrow B} \; (MP)}{B} \; (MP) \tag{6.2}$$

which leads naturally to a more inflated argument and so on... the attempt at justifying our tentative to follow the logical convention expressed by the *modus ponens* schema, in definitive, ends up in an infinite regress.

A similar argument is notoriously part of Wittgenstein's discussion on rule-following in the *Philosophical Investigations*, where he observes that in order to act in accordance with what a given rule, say *modus ponens*, requires him to do (i.e. in order to apply the rule), one has to *interpret* a certain schema (e.g. the schema (6.1)); on the other hand, if a person were asked to justify, or simply to explain, his course of actions as an interpretation of the schema or, as one would say, as somehow *compelled* by the associated rule, then he would end up picking up another schema (think of (6.2)) to be interpreted, giving rise to an infinite regress. Wittgenstein claims, however, that in practice mutual comprehension takes place without any regress of that kind:

> §201 It can be seen that there is a misunderstanding here from the mere fact that in the course of our argument we give one interpretation after another; as if each one contented us at least for a moment, until we thought yet of another standing behind it. [...]

---

[1]This way of reading the *Tractatus* presupposes the thesis that there is a theoretical unity tying together the theory of the logical form and what Wittgenstein writes later on the following rules: in both cases, one indeed finds that constitutive conditions of "saying" cannot be expressed linguistically without falling into circularity or regress.

> §217 If I have exhausted the justifications I have reached bedrock, and my spade is turned. Then I am inclined to say: 'This is simply what I do'. Wittgenstein (2009)

Again, in the infinite regress, one would find himself trying to *say*, continuously missing the point, what is implicit in the adoption of the rule, that is, what he should content himself to *show* in his linguistic practice.

As we have already remarked, the appeal to the distinction between internal and external questions (so as to the more familiar distinction between *object language* and *metalanguage*) does not constitute a viable way out of the presented arguments for the regress: to assert that a rule is justified at the meta-level by an argument which employs the alter ego of the rule itself cannot indeed constitute a proper justification of the rule, since practically every rule could be justified in that way: choose a rule R, and then write a semantics for R, by exploiting a rule *meta-R*; with few technical work, soundness and completeness (for instance, if R enjoys a kind of subformula property, enabling thus an inductive argument) will be proved. There would be thus no difference in principle between the rule of *modus ponens*, the most unassailable of deductive rules, and the worst rule one can imagine!

On one side, as a consequence, rules, meta-rules and meta-languages dramatically fail to provide constitutive conditions for deduction; on the other, it seems quite natural to expect that logically correct rules must be somehow explained, since otherwise they would simply be *arbitrary* ones.[2]

The alternative conclusion that logical deduction is fundamentally ungrounded, since we cannot discursively express what the validity of its rules consists in, is discussed in epistemological terms in *On certainty*: Wittgenstein maintains that grammatical rules are in charge of establishing what is constitutively exempt from doubt, in the sense that one could not make sense of what a doubt whatsoever about it could mean, since he could not display nor imagine any use of it:

> §314. Imagine that the schoolboy really did ask: "And is there a table there even when I turn around, and even when no one is there to see it?"
>
> §315 [...] The teacher would feel that this was only holding them up, that this way the pupil would only get stuck and make no progress.[...] this pupil has not learned how to ask questions. He has not learned the game that we are trying to teach him. Wittgenstein (1969)

The idea conveyed by these examples is that the activity of rule-following, and thus, in the case of logic, the alleged normativity of deduction, is one whose constitutive conditions, when we try to grasp them, inexorably slip out of our hands, hiding themselves behind the mirror effects of interpretations (semantics and meta-languages) obsessively reproducing the same questions at higher levels: we cannot *say* what underlies any form of saying, so as we cannot see the blind spot which enables any form of seeing.

The point at the heart of the work of the late Wittgenstein is that if we cannot say anything about conditions of possibility of rules, then we cannot even raise

---

[2]This expectation clearly goes well beyond Wittgenstein's view on rules and logic, since he believed that logical rules are truly arbitrary ones, though enjoying a special position in language.

meaningful doubts concerning them: in the course of our infinite interpretative regress, "doubt gradually loses its sense" Wittgenstein (1969).

> §307. And here the strange thing is that when I am quite certain of how the words are used, have no doubt about it, I can still give no grounds for my way of going on. If I tried I could give a thousand, but none as certain as the very thing they were supposed to be grounds for. Wittgenstein (1969)

This is to say that whereas the adoption of a given system of rules (a linguistic game, so to say) commits to a structured arrangement of referential claims, its "semantic ascent", with Quine's words, nothing can be said about the legitimacy to impose such a structure on experience. This remark revives the "mystical" question: why just these rules? To make an example, the adoption of a certain system of rules for modal logic (say $S5$) naturally leads to the acceptance of truth claims regarding propositions on possible worlds ("Is water the same as H_2O in every possible world?" "Is gold the same as goodness only knows what in some possible world?"), but what is to assure us, not about the truth-value of those sentences but rather about the right by which we are authorized to take such sentences as demanding for a definite truth-value? In a word, to quote a celebrated passage from Kant, *quid iuris*?

The moral to be drawn from the problem of the logical form would be thus that no serious questioning of logical formatting can be raised: entitlement to consider "logic" this or that system of rules would thus reduce to a matter of mere convention or to Wittgenstein's Promethean attitude (the "mystic"). In definitive, logic would be thus simply incapable to answer to a matter of right.

6.3 Proofs as Programs, Rules as Behaviours

The aim of this section is to discuss the significance of logical formatting ("why just these rules?") from a more specifically logical (loosely technical) viewpoint: this will be achieved by privileging an approach directed to the *dynamics* of logic, heritage of the so-called Curry-Howard correspondence, which establishes a connection (indeed an isomorphism) between logical proofs and programs written in a functional language like λ-calculus: a proof, as isomorphic with a program, is considered not only as a mathematical object to be *constructed* but also as a program to be *executed*; for instance, a proof of $\forall x \in A B(x)$ corresponds to a program which, when taken as input an element $t \in A$, produces, once executed, a proof of $B(t)$.

Through the mathematical analysis of cut-elimination algorithms, directly corresponding to the normalization of terms in λ-calculus, the rules of logic, as we'll sketch, are characterized by the symmetries which discipline the use (i.e. the dynamics) of proofs/programs. Particular attention is given to the problem of the termination of executions, i.e. of the computations that represent the interaction between proofs and programs: a program can indeed be of some use only when its execution achieves a certain goal within a finite amount of time and, on the other side, that logicalness constitutes an answer to the question "does this program actually do what it is designed for?" is well known by computer scientists.

That logical syntax is in charge to tame the dynamics of programs by forcing termination,[3] is best seen from the viewpoint of the so-called *formulas-as-types* correspondence: by this, formulas become sets of proofs/programs satisfying the norms associated by logical rules to formulas. For instance, a program π is typed $A \Rightarrow B$ (and it is indeed a proof of $A \Rightarrow B$) when, for every proof/program σ-typed A, the execution of the application $(\pi)\sigma$ of π to σ terminates producing a proof-/program-typed B.

In this way, the *static* viewpoint which individuates rules by their *schemas*, for instance, the *modus ponens* schema, is replaced by a *dynamic* one, in which proofs manifesting the correct application of a rule are those whose execution satisfies opportune *behavioural* (i.e. observable through interaction) norms: rules are then not imposed *a priori* on proofs, but rather different proofs manifesting the same behaviour can be "typed" in the same way. The input of computer science is crucial here: a program is a source code written in a given language (λ-calculus, for instance); it is thus a "subjective" (in the sense explained) device; yet, one is generally only interested in what the program does when it is run. The behaviour of the program can in fact be largely independent from its code, and programs written in different languages can behave exactly in the same way. These considerations prompt the search for a behavioural explanation of rules, that is, one in which the "objective" content of a rule is expressed as a problem of *interface* between proofs (more detailed, though informal, descriptions of this bridge between logic and computer science can be found in Joinet (2009, 2011)).

From the viewpoint of sequent calculus, resolving a problem of interface corresponds to eliminate cuts: let's see, through an easy example, how properties of rule-schemas can be derived from the behaviour of proofs through cut-elimination. Let's suppose to cut an arbitrary proof Π^4 of conclusion $A \wedge B$ with a *cut-free* proof Λ:

$$\cfrac{\vdash A \wedge B \quad \cfrac{\vdots \Lambda \\ \Delta, A, B \vdash \Gamma}{\Delta, A \wedge B \vdash \Gamma} (L\wedge)}{\Delta \vdash \Gamma} \; cut \tag{6.3}$$

---

[3] And by that logical correctness, since the eliminability of cuts (which implies the termination of all reductions) has as a corollary the logical coherence of the calculus: since no introduction rule is given for the absurdity $\bot$, if a proof of $\bot$ existed, then it would reduce to a proof whose last rule is an introduction for $\bot$, thus no such proof can exist.

[4] In this section, I'll adopt the convention to name sequent calculus proofs with capital Greek letters $\Pi, \Lambda, \ldots$ and with small letters $\pi, \lambda, \ldots$ the programs, respectively, associated.

6 Rule-Following and the Limits of Formalization: Wittgenstein's...

In order not to let interaction fail, Π must be reduced to a proof that we can split into a proof Π_1 of conclusion A and a proof Π_2 of conclusion B:

$$\cfrac{\cfrac{\begin{array}{c}\vdots\ \Pi_1\\ \vdash A\end{array} \quad \begin{array}{c}\vdots\ \Lambda\\ \Delta, A, B \vdash \Gamma\end{array}}{\Delta, B \vdash \Gamma}\ cut \quad \begin{array}{c}\vdots\ \Pi_2\\ \vdash B\end{array}}{\Delta \vdash \Gamma}\ cut \qquad (6.4)$$

that is, by cut-elimination, the request that Π interfaces with Λ has two consequences: first, the rule-schema for $\wedge$ must have the form one might expect[5]

$$\frac{\vdash A \quad \vdash B}{\vdash A \wedge B}\ (L\wedge) \qquad (6.5)$$

Second, it must be possible to transform a proof of conclusion $A \wedge B$ into one whose *last rule* is expressed by the schema (6.5). Behavioural norms impose thus, through cut-elimination, syntactical constraints on the form of rules and proofs.

A crucial remark in the argument above is that the rule-schema for $\wedge$ has been derived by taking into consideration an arbitrary cut-free proof *dual* to an arbitrary proof of $A \wedge B$; more generally, let's say that two proof-programs are *polar* each other when they can be cut together giving rise to a terminating execution (i.e. cut-elimination converges); given a set T of proofs-programs, we can define the set $T^\perp$ as the set of all proof-programs polar to every element of T. A set of this form will be called an "abstract type" (remark that these definitions make no use of rule-schemas). A simple argument shows that an abstract type T satisfies the equation

$$T = T^{\perp\perp} \qquad (6.6)$$

this equation asserts that T is *complete* in the sense that an arbitrary proof-program polar to an arbitrary proof-program in $T^\perp$ is already in T.

Let now S be the set of proofs/programs of a given formula A built following the rules of a certain logical syntax. By embedding the programs in S in a more general functional language, we can ask the nontrivial question: is S an abstract type? If we recognize the abstract type $S^\perp$ as a set of tests for S and thus $S^{\perp\perp}$ as the smallest abstract type containing S, our question can be rephrased as

$$S^{\perp\perp} \subseteq S? \qquad (6.7)$$

This way of embedding syntactic proofs into abstract types defined by their bipolar is reminiscent of the notion of *completion* in topology and functional analysis. It can easily be acknowledged that what (6.7) expresses is a form of completeness

---

[5] We neglect for simplicity matters regarding the discipline of contexts, which would require a more sophisticated argument to reach exactly the same conclusion.

(usually styled *internal completeness*; see Girard 2001) which makes no reference to semantics, that is, radically internal to syntax: it is a query on the internal properties of the logical format, that is, on the symmetries between rules: is anything missing? Is there anything syntax cannot see?

It is to be observed that a test for A isn't but a proof/program of $A \vdash \Gamma$, what, in the case in which $\Gamma = \emptyset$, means a possibly *incorrect* proof of $A \vdash$, that is, a possibly incorrect proof of $\neg A$: the interaction between a proof and a test can thus be interpreted as a *dispute* between two players (this perspective is made explicit in *ludics*; see Girard 2001). The focus on the dynamics of logic, so as the interest in wrong proofs, after all, is just a way to bring *dialectics* back at the heart of logic.

In definitive, what the view briefly sketched tries to achieve is an exposition of logic in which rules and their schemata are not primitive notions, but are derived from the recognition of the normative requirements necessary to avoid divergence, i.e. lack of mutual communication: if you want to sustain $A \wedge B$ and you want your proof to be accepted by others (i.e. your program to interface successfully), then (6.3) tells you that the principal rule of your proof *ought* to be a splitting between A and B.

A similar deconstruction of logical rules can be found in the tradition of proof-theoretical semantics based on Gentzen's *Hauptsatz* (see Prawitz 1973; Dummett 1991). For instance, Dummett describes the dynamics between proofs as the interaction between the proponent of an introduction rule (that he calls the "verificationist") and the proponent of an elimination rule (that he calls the "pragmatist"); the search for an equilibrium between the expectations of both leads to his notion of "logical harmony", corresponding to the requirement of eliminability of cuts.

In Dummett's and Prawitz's theories, though, the building blocks of logical proofs are rule-schemata (for instance, introduction rules); as a consequence, the recognition of something as being obtained through the correct application of a rule (be it harmonious or not) remains a precondition for their interactive explanation of logic, so that Wittgenstein's rule-following infinite regress still applies (see Cozzo (2004) for a brief discussion): Dummett himself, in Dummett (1978), declares that a satisfying justification of logical rules imposes to reject Wittgenstein's point about rule-following (e.g. by endorsing his proposal of a molecularist theory of meaning).

In the approach here sketched (which builds on Girard's *transcendental syntax* program, as it will be made more precise in the next section), on the contrary, rule-schemata do not constitute primitive notions, since proofs are represented by "pure" programs: this means that proofs, like programs in λ-calculus, can be written independently of logical rules; behavioural constraints are then used to retrieve rule-application through the notion of "last rule" (remark that the example (6.3) discussed above clearly recalls Dummett's "fundamental assumption" (Dummett 1991)). Explaining rules through the notion of polarity with respect to a given set of tests, i.e., again, pure programs, yields then a completely behavioural description of logic.

This strategy appears *prima facie* as providing a way out of the mirror effects, i.e. the infinite regress arguments; let's see how, for instance, in the case of Carroll's argument on the *modus ponens*, the step leading to the regress is due to the possibility to pass from the rule-schema of *modus ponens*

$$\frac{A \quad A \Rightarrow B}{B} \quad (MP) \tag{6.8}$$

to the logical formula expressing it:

$$A \Rightarrow ((A \Rightarrow B) \Rightarrow B) \tag{6.9}$$

which is used as premise of the inflated argument. Notably, the conclusion that Carroll himself and many others (Dummett included) have drawn from the argument is that this passage is somehow to be forbidden: a rule should not be expressed by a logical sentence. The typical argument (a good recognition can be found in Engel (2005)) is that a rule represents a *dynamic* pattern of linguistic practice, whereas a logical sentence expresses a *static* content. Dummett (1973) makes appeal to Frege's notion of "assertoric force" to distinguish the assertion (a piece of concrete linguistic practice) of the sentence A as premise of a rule of inference from its mere occurrence in the formula $A \Rightarrow ((A \Rightarrow B) \Rightarrow B)$:

> [...] it cannot be the same, because then "*Peter is a Jew; if Peter is a Jew, Andrew is a Jew; therefore Andrew is a Jew*" would be the same as "*If both Peter is a Jew and if Peter is a Jew, then Andrew is a Jew, then Andrew is a Jew*" and it was precisely Lewis Carroll's discovery (in 'What the Tortoise said to Achilles') that it was not. Dummett (1973)

In contrast with such a diagnosis, in what follows I'll accept to pass from the rule-schema (6.8) to the inflated schema, but with the remark that, in the latter, the formula expressing the schema must occur as conclusion of a proof/program μ; here is a sequent calculus proof that could be associated to μ:

$$\frac{\dfrac{\dfrac{A \vdash A \quad B \vdash B}{A, A \Rightarrow B \vdash B}}{A \vdash (A \Rightarrow B) \Rightarrow B}}{\vdash A \Rightarrow ((A \Rightarrow B) \Rightarrow B)} \tag{6.10}$$

My thesis is that one is not compelled to go on inflating the argument: an explanation of *modus ponens* is given by showing that, given arbitrary *correct* proofs/programs σ, τ of conclusions, respectively, A and $A \Rightarrow B$, the interaction $(\mu)\sigma\tau$ is convergent, what amounts, essentially, to a (strong) normalization argument. Remark that, by representing a rule by a proof-program of a logical sentence, its dynamic content (so as the assertoric force of its premises and conclusions) is not lost, since the *application* of the rule (6.8) is translated into the *execution* of the program μ. Questioning (6.8) amounts then to ask for an argument which asserts the compatibility of μ with the interface induced by the introduction rules for A and $A \Rightarrow B$: in other words, Dummett's argument for logical harmony is retrieved without presupposing that proofs are built following rules-schemata.

Furthermore, one could well imagine to invert the process, namely, to accept μ as representing a correct rule and to adopt it as a test for proofs of A and $A \Rightarrow B$: one will say that σ and τ are correct when $(\mu)\sigma\tau$ is convergent. This isn't but a way to pose the issue of internal completeness: do all correct σ, τ actually come from your favourite syntax?

Summing up, the strategies just presented have the following form:

1. Start by μ and prove that correct proofs of A and $A \Rightarrow B$ are *polar* to μ: σ and τ are thus tests for μ, i.e. the rule is tested.
2. Take arbitrary σ, τ polar to μ and prove that they come, respectively, from proofs of A and $A \Rightarrow B$: μ becomes here a test for σ and τ, i.e. the rule is the tester.

> If a blind man were to ask me "Have you got two hands?" I should not make sure by looking. If I were to have any doubt of it, then I don't know why I should trust my eyes. For why shouldn't I test my *eyes* by looking to find out whether I see my two hands? *What* is to be tested by *what*? Wittgenstein (1969)

Regress is then translated into the balance between introduction and elimination rules, that is, in the acknowledgement of the inner symmetries of (good) syntax.

A fundamental objection to the argument just presented is that it is at least implausible that an argument for normalization can be carried over without appeal to some instance of *modus ponens*, so that circularity would not be eliminated. This is tantamount to saying that it is at least implausible that a serious argument for justifying logic be carried over without using logic somewhere (this remark was indeed the starting point of our discussion). It must be conceded then that such a form of circularity impedes any form of definitive foundation for logic.

By the way, this indirect form of circularity is not the same as the plain circularity exposed by Carroll's argument, since, as we are going to see, it puts on the foreground the syntactical complexity of normalization arguments: these are indeed usually very complex logical arguments and may heavily depend on the formalism chosen. Thus, if one were somehow sceptic about the normalization argument (for defendable reasons, as we'll point out), then he would find himself in the situation hypothesized by Carroll in his article, namely, that of accepting the premises of the *modus ponens*, accepting the schema of the rule and yet remaining perplexed about the conclusion.

> If A and B and C are true, Z *must* be true," the Tortoise thoughtfully repeated. That's another *Hypothetical*, isn't? And, if I failed to see its truth, I might accept A and B and C, and *still* not accept Z, mightn't I?" Carroll (1895)

Suppose indeed to have laid down a normalization proof for $(\mu)\sigma\tau$, where Σ, T vary among all the proofs of A and $A \Rightarrow B$ in a given formalism, which establishes a hyper-exponential run time for normalization (this is actually the case for classical and intuitionistic propositional logic): to accept such a proof would mean to accept that by composing σ and τ, a program representing a proof of B can be obtained by execution. By the way, can we reasonably expect to *see* the reduction of $(\mu)\sigma\tau$ terminate? What the normalization proof asserts is indeed the existence of a normal form, without providing a viable procedure to produce one (in a reasonable time). In the case of (6.3), to make another example, the theorem asserts the existence of a tower of exponentials bounding the time necessary to split the proof Π of a conjunction into a conjunction of proofs: our universe could not plausibly assist to such an event!

One could thus well imagine a "strict finitistic" tortoise accepting the schema expressed by μ but questioning the application of the rule, since the latter depends on an argument which provides no concrete mean to show that whose existence it asserts. Again, why not to imagine a "Turing-tortoise" refusing β-reduction since it takes too much time to be implemented on the tape of a Turing machine?

These examples are not meant to motivate the endorsement of a more or less strictly finitistic view on mathematics, but only to show how a (more or less) legitimate doubt could be raised on *formats* disciplining the use of deductive rules, a doubt that could not be raised from a purely rule-theoretic standpoint. A somehow Kantian issue on legitimation is recovered by observing that by adopting sequent calculus or λ-calculus, one is led to accept as valid something he cannot actually *see* as valid. Again: by what right?

Gödel's incompleteness enables us to push the matter even farther: it is well known that since cut-elimination implies coherence, normalization proofs for systems enabling inductive/recursive constructions (and more generally, impredicative second-order constructions) cannot be performed within those systems themselves. This can be read in terms of ordinal run times for normalization: the so-called proof-theoretic ordinal number is associated to every proof as a measure of the computational cost of its normalization. A worst-case run time, which embodies the general normalization proof, is then obtained as the least not proof-theoretic ordinal (usually, a limit ordinal): such a proof cannot be a proof in the system.

What this argument shows is that by incompleteness, in no way one can be certain of the logical correctness of a given format: in no way, one will satisfy once for all the demands by a sceptic tortoise, since normalization arguments must employ principles whose logical correctness is even harder to justify.

Here we stumble again on the already discussed circle of logic presupposing logic, but with a significant quality/price advantage: in his remarks, Wittgenstein clearly stated that all questioning must stop somewhere, namely, when the "bedrock" is reached, i.e. the practice of a given "grammar" and all that is implicit in it. Now, the proof-theoretical and/or computational complexity of strong normalization proofs fills the gap between what can be directly seen and what must be left implicit in the use of formal proofs.

In this sense, the reformulation of Kant's *quid iuris* question in the preceding section reduces to a question on the implicit complexity of syntactic manipulation: can we actually write this? Can we do that in a reasonable time and space?

Summing up, if the argument presented is right (or at least, not drastically wrong, given its quite informal and sketchy description), then the infinite regress of logical rules interpreting themselves must stop at the level of formats, namely, at the level of syntactical prescriptions. As a consequence, a rigorous mathematical exposition of these preconditions of logic (which demands for more sophisticated techniques than the one sketched here; see the next section) might replace the Wittgensteinian "thereof we must be silent" and open the way for an exploration of what lies behind the apparently unquestionable evidence of the practice of logic and its rules.

6.4 Linearity and Logical Syntax

The search for an internal explanation of logical rules, through a profound mathematical investigation of formal systems, represents the core of the *transcendental syntax* program (see Girard 2011b), launched in 2011 by Jean-Yves Girard as the foundational counterpart to the refinement of proof-theoretical tools obtained through linear logic and its developments (for instance, *ludics* and *geometry of interaction*). In what follows, I'll try to give a hint at how, through the mathematical treatment of linearity in logic, it is possible to give ground to the discussions of the previous sections and to develop a promising framework to reconstruct logic.

From the viewpoint of syntax, linearity is obtained when the structural rules of *weakening* and *contraction*

$$\frac{\vdash \Gamma}{\vdash \Gamma, A} \; (W) \qquad \frac{\vdash \Gamma, A, A}{\vdash \Gamma, A} \; (C) \tag{6.11}$$

are eliminated; logically speaking, this amounts to reject the logical principles $\bot \Rightarrow A$ and $A \Rightarrow A \wedge A$. This choice, which could seem at first rather arbitrary, is explained by observing that, in sequent calculus, the structural rules (W) and (C)[6] form an independent group with respect to identity rules (axiom and cut) and logical rules (introduction and elimination of connectives).

Historically, the first reason for styling "linear" the fragment of logic without weakening and contraction[7] was a semantical one: linear logic proofs could indeed be interpreted as *linear maps* between coherent spaces, that is, stable maps on cliques (see Girard 1987a) satisfying

$$f\left(\sum_i^k a_i\right) = \sum_i^k f(a_i) \tag{6.12}$$

where the sum is intended as a disjoint union of sets. However, it was only with the development of *proofnet* theory and *geometry of interaction* (*GoI* in the following) that it was realized that linear proofs could be represented as linear operators in the traditional, matricial, sense. These operators act on the (usually complex) linear space generated by the occurrences of atomic formulas in the formulas to be proved: more precisely, to every occurrence of an atomic formula A, a finite linear space

---

[6] Along with the rule (Ex) of *exchange* which allows to permute the order of appearance of occurrences of formulas in a sequent and whose rejection leads to *non-commutative logic* NL, see Abrusci and Ruet (2000).

[7] Which happens to be very well organized, with classical connectives splitting into a *multiplicative* and an *additive* version, and with modalities (known as *exponentials*) reintroducing, in a linear setting, controlled versions of weakening and contraction.

$|A|$ is associated; to the general formula $F(A_1,\ldots,A_n)$, where the A_i denote the occurrences of atomic formulas in F, the linear space $|F| = \bigoplus_i^n |A_i|$ is associated. Remark the *locativity* of this approach: every occurrence of formula is assigned a distinct *location* (i.e. an independent linear space); for instance, to construct the occurrence space of $(A \Rightarrow A) \Rightarrow (A \Rightarrow A)$, one first has to distinguish the four distinct occurrences of A as $(A_1 \Rightarrow A_2) \Rightarrow (A_3 \Rightarrow A_4)$ and then to associate to the formula the linear space $|A_1| \oplus |A_2| \oplus |A_3| \oplus |A_4|$.

Now, if we have a look at how structural rules are reflected in cut-elimination, we notice that they cause the dimension of the "occurrence space" not to be preserved during the computation: for instance, in the case of contraction, this dimension is, in the worst case, multiplied by the number of contracted occurrences:

$$\dfrac{\dfrac{\vdots\ \Pi}{\vdash \Gamma, A, A}\ (C)\quad \vdots\ \Lambda}{\vdash \Gamma, \Delta}\ cut \quad \rightsquigarrow \quad \dfrac{\dfrac{\dfrac{\vdots\ \Pi}{\vdash \Gamma, A, A}\quad \dfrac{\vdots\ \Lambda}{\vdash \Delta, \neg A}}{\vdash \Gamma, \Delta, A}\ cut \quad \dfrac{\vdots\ \Lambda}{\vdash \Delta, \neg A}}{\dfrac{\vdash \Gamma, \Delta, \Delta}{\vdash \Gamma, \Delta}\ (C)}\ cut \qquad (6.13)$$

In case of nested contractions, this phenomenon is responsible for the hyper-exponential growth rates of reductions in classical logic.

On the other hand, in the absence of those principles, the dimension of the occurrence space is preserved (the space of all the occurrences of formulas strictly shrinking during reduction), as shown in the following example: selectfont

$$\dfrac{\dfrac{\vdots\ \Pi_1 \quad \vdots\ \Pi_2}{\vdash \Gamma_1, A \quad \vdash \Gamma_2, B}}{\vdash \Gamma, A \wedge B}\quad \dfrac{\vdots\ \Lambda}{\vdash \neg A, \neg B, \Delta}\ cut \quad \rightsquigarrow \quad \dfrac{\vdots\ \Pi_2}{\vdash \Gamma_2, B}\quad \dfrac{\dfrac{\vdots\ \Pi_1}{\vdash \Gamma_1, A}\quad \dfrac{\vdots\ \Lambda}{\vdash \neg A, \neg B, \Delta}}{\vdash \neg B, \Gamma_1, \Delta}\ cut$$
$$\vdash \Gamma, \Delta \qquad\qquad\qquad\qquad \vdash \Gamma, \Delta \qquad (6.14)$$

The matricial representation of a linear proof in *GoI* is given by the symmetry (called *wire*) on the occurrence space representing the exchange between the dual occurrences of formulas in the axioms of the proof; for instance, to the following proof Π^8

$$\dfrac{\dfrac{\vdash A, \sim A \quad \vdash B, \sim B}{\vdash A \otimes B, \sim A, \sim B}}{\vdash A \otimes B, \sim A \,\mathfrak{B} \sim B} \qquad (6.15)$$

---

[8] Where $\sim$, $\otimes$, $\mathfrak{B}$ denote, respectively, linear negation, multiplicative conjunction (*tensor*) and multiplicative disjunction (*par*).

one straightforwardly associates the symmetry below, acting on the space generated by $A_1, B_1, \sim A_2, \sim B_2$ (isomorphic to $\mathbb{C}^4$),[9] which exchanges, respectively, $|A_1|$ with $|\sim A_2|$ and $|B_1|$ with $|\sim B_2|$:

$$\pi = \begin{pmatrix} 0 & 0 & 1 & 0 \\ 0 & 0 & 0 & 1 \\ 1 & 0 & 0 & 0 \\ 0 & 1 & 0 & 0 \end{pmatrix} \tag{6.16}$$

Not every symmetry on the occurrence space comes from a proof of the formula associated; this is because the representation of proofs as wires radically forgets about rules (except axiom rules): it erases all sequentiality (i.e. rule-application) from the proofs.

In order to characterize the wires which represent correct sequent calculus proofs, the elegant solution, coming from proofnet theory, is to represent rules as *tests* to be performed on the wires: every logical rule is interpreted as a set of permutation matrices on the occurrence space, so that inductive composition of rules produces a set of general permutations τ_i (which can still be seen as wires) on the occurrence space, that can interact with the wire. In *GoI*, interaction (cut-elimination) can be directly implemented on wires: the intuition behind that is that the interaction of two wires is obtained by "plugging" them together to produce another wire:

$$\tag{6.17}$$

In geometrical terms, this means to go through all the paths produced alternating the two wires. Let's see a little more in detail how to use this idea to interpret cut-elimination in sequent calculus: the cut rule (in locative form)

$$\frac{\begin{array}{c} \vdots \Pi \\ \vdash \Gamma, A_1 \end{array} \quad \begin{array}{c} \vdots \Lambda \\ \vdash \sim A_2, \Delta \end{array}}{\vdash \Gamma, \Delta} \; cut \tag{6.18}$$

is represented (as every rule) as a test given by the following symmetry on the occurrence space which is the direct sum of the occurrence spaces of π and λ:

$$\sigma = \begin{pmatrix} I & 0 & 0 & 0 \\ 0 & 0 & I & 0 \\ 0 & I & 0 & 0 \\ 0 & 0 & 0 & I \end{pmatrix} \tag{6.19}$$

---

[9]In this section, I'll adopt the convention to name sequent calculus proofs with capital greek letters $\Pi, \Lambda, \ldots$ and with small letters $\pi, \lambda, \ldots$ the wires, respectively, associated.

σ acts as the identity on the subspace $|\Gamma| \oplus |\Delta|$ and as a symmetry on the space $|A_1| \oplus | \sim A_2|$ exchanging every atomic occurrence in A_1 with its dual one in $\sim A_2$.

We interpret the cut-elimination algorithm as follows: suppose to start in a point in $|\Gamma|$ and to reach, by the symmetry π, a point in $|A_1|$; then, by the symmetry σ, you are sent to a point in $|A_2|$, and two distinct possibilities arise:

- The symmetry λ sends you into $|\Delta|$, and you're done, since you've exited the "combat zone".
- λ sends you to another point in $|A_2|$; by σ, you are sent back into $|A_1|$. Again, two distinct possibilities arise....

Strong normalization (i.e. logical correctness) warrants that you will always exit $|A_1| \oplus |A_2|$ (towards $|\Gamma| \oplus |\Delta|$) in a finite amount of time, that is, that you won't get into a cycle in $|A_1| \oplus |A_2|$.

To simplify things, let's suppose $\Delta = \emptyset$; then, given the block decomposition,

$$\pi = \begin{pmatrix} \pi_{\Gamma\Gamma} & \pi_{\Gamma A} \\ \pi_{A\Gamma} & \pi_{AA} \end{pmatrix} \quad (6.20)$$

the general algorithm can be translated into the expression

$$\pi_{\Gamma\Gamma} + \pi_{\Gamma A}\big(\sigma + \sigma\pi_{AA}\sigma + \sigma\pi_{AA}\sigma\pi_{AA}\sigma + \cdots + \sigma(\pi_{AA}\sigma)^n + \ldots\big)\pi_{A\Gamma} \quad (6.21)$$

which, if the series $\sum_n^\infty (\pi_{AA}\sigma)^n$ converges, corresponds to the algebraic expression[10]

$$\pi_{\Gamma\Gamma} + \pi_{\Gamma A}\sigma\big(I - \pi_{AA}\sigma\big)^{-1}\pi_{A\Gamma} \quad (6.22)$$

Strong normalization is then translated into the *nilpotency* (corresponding to acyclicity) of $\pi_{AA}\sigma$, namely, the existence of an integer N, the *order of nilpotency*, such that $(\pi_{AA}\sigma)^N = 0$, what trivially implies convergence: nilpotency reduces indeed the expression (6.22) to the finite one below

$$\pi_{\Gamma\Gamma} + \pi_{\Gamma A}\big(\sigma + \sigma\pi_{AA}\sigma + \sigma\pi_{AA}\sigma\pi_{AA}\sigma + \cdots + \sigma(\pi_{AA}\sigma)^N\big)\pi_{A\Gamma} \quad (6.23)$$

corresponding to a finite computation.

It is now possible to implement typing through tests: let's say that two wires σ, τ on the same occurrence space are *polar* (notation, $\sigma \perp \tau$) when $\sigma\tau$ is nilpotent, and let $S^\perp := \{\tau \in \mathcal{M}(\mathcal{X}) | \sigma \perp \tau, \sigma \in S\}$ for every set S of wires on a given occurrence space $\mathcal{X}$. We recover thus logical types as the "abstract types" of the previous section: they are polar to sets of tests induced by logical rules, that is, as

---

[10] A simple calculation shows that, for u a linear operator, $(I - u)^{-1} = \sum_n^\infty u^n$, given the convergence of the series on the right or, equivalently, the invertibility of $I - u$.

the sets of *all* the wires following a given sequence of rules. The concrete translation of the rules of logic into tests follows the ideas in Girard (1987b), but I won't enter here into the details.

Notice that this framework enables the issue of *internal completeness*: do all wires in a given logical type come from actual sequent calculus proofs? A fundamental result, giving substance to the whole theory at the end of the eighties, was the theorem establishing completeness in the case of multiplicative linear logic (see Girard 1987b).

The justification for *modus ponens* we gave in the previous section can now be translated in the vocabulary of *GoI* as follows: given a wire σ on the occurrence space $|A|$ representing a proof of A and a wire ϕ on the occurrence space $|A| + |B|$ representing a proof of $A \Rightarrow B$, we can obtain a wire $[\phi]\sigma$ (intuitively, the application of ϕ to σ) on the occurrence space $|B|$ representing a proof of B; in fact ϕ can be pictured as the block matrix

$$\phi = \begin{pmatrix} \phi_{AA} & \phi_{AB} \\ \phi_{BA} & \phi_{BB} \end{pmatrix} \tag{6.24}$$

where ϕ_{XY} represents a linear operator from $|X|$ to $|Y|$. Following (6.22), $[\phi]\sigma$ is defined as follows:

$$[\phi]\sigma = \phi_{BB} + \phi_{BA}\sigma(I - \phi_{AA}\sigma)^{-1}\phi_{AB} \tag{6.25}$$

what can be done only in case the series $\sum_i^\infty (\phi_{AA}\sigma)^n$ converges (i.e. $I - \phi_{AA}\sigma$ is invertible). As expected, for wires coming from logical syntax, convergence is assured by the proof that $\phi_{AA}\sigma$ is nilpotent (this constituted in fact the main theorem of *GoI* – see Girard (1989)).

Remark that the nilpotency order N can be arbitrarily big, and it can be read as a measure of the complexity involved in the computation: it provides indeed a bound on the number of iterations necessary to reach convergence. Consider the following (linear) proof

$$\frac{\dfrac{A_3 \vdash A_1 \quad A_2 \vdash A_4}{A_1 \Rightarrow A_2, A_3 \vdash A_4}}{A_1 \Rightarrow A_2 \vdash A_3 \Rightarrow A_4} \tag{6.26}$$

represented by the wire

$$\phi = \begin{pmatrix} 0 & 0 & 1 & 0 \\ 0 & 0 & 0 & 1 \\ 1 & 0 & 0 & 0 \\ 0 & 1 & 0 & 0 \end{pmatrix} \tag{6.27}$$

Let's call σ the simple wire $\begin{pmatrix} 0 & 1 \\ 1 & 0 \end{pmatrix}$ on the occurrence space corresponding to the first two occurrences of A in the previous proof, representing the axiom $A \Rightarrow A$; we can easily compute $[\phi]\sigma$ and find that the nilpotency order is bounded by 2, the dimension of the occurrence space in which ϕ and σ actually "meet". Again, no room for doubt, no room for the Tortoise.

The wire ϕ corresponds to the λ-calculus term $\lambda f.\lambda x.(f)x$, usually taken as the *Church numeral* $\underline{1}$. On the other hand, the general Church numeral $\underline{n} = \lambda f.\lambda x.(f)^n x$ requires the encoding in *GoI* of n contractions. This means that we must transform, in a linear way, a space of dimension $k + n$ into a space of dimension n. This is done by means of a *linear isomorphism* $\Phi : \mathcal{X}^2 \to \mathcal{X}$. The existence of these isomorphisms requires the adoption of infinitary operator algebras, like C^*-algebras. To give an example, if we consider an operator algebra acting over the Hilbert space $\ell^2(\mathbb{N})$, we can take

$$\Phi\left(\sum_n (\alpha_n e_n \oplus \beta_n e_n)\right) = \sum_{2n} \alpha_{2n} e_{2n} + \sum_{2n+1} \beta_{2n+1} e_{2n+1} \qquad (6.28)$$

which follows the coding of pairs of integers into integers known as "the Hilbert Hotel". Concretely, the Hilbert Hotel is what is required in order to double a wire τ into two distinct isomorphic copies τ', τ'': here is a typical feature of *GoI*, namely, the exposition of the procedures implicit in the manipulation of formal systems (introducing superscripts may indeed require to "make space on the blackboard").

The subjectivity of notations is thus eventually recovered: the Hilbert Hotel by no means constitutes a canonical choice for the isomorphism Φ, and different choices are not warranted to agree, i.e. to interact without producing interferences; Φ can be written as $\Phi = P + Q$ with the property that for every $u, v \in \mathcal{X}$,

$$P^*uPQ^*vQ = Q^*vQP^*uP$$
$$P^*Q = Q^*P = 0 \qquad (6.29)$$

it can thus be used to produce commutations between arbitrary operators, by sending them into disjoint copies of the same universe. Remark that, when interpreting logic, this doubling of the universe has the effect to enlarge the dimension of the occurrence space (without violating linearity). On the other hand, any P and Q satisfying (6.29) will do the job, so that we can imagine infinite many different possibilities to implement an apparently simple task. By the way, the nilpotency order of computations (and thus the complexity involved in normalization proofs), as I'm going to make a bit more clear, will be highly sensible to the choice of P and Q, since the latter discipline the growth of the occurrence space during computation.

We can represent, for instance, the numeral $\underline{2}$, associated to the following proof (where $A_{f(2,4)}, A_{g(3,5)}$ denote fresh locations, respectively, depending on $2, 4$ and $3, 5$),

$$\cfrac{\cfrac{\cfrac{A_1 \vdash A_2 \quad A_3 \vdash A_4 \quad A_5 \vdash A_6}{A_2 \Rightarrow A_3, A_4 \Rightarrow A_5, A_1 \vdash A_6}}{A_{f(2,4)} \Rightarrow A_{g(3,5)}, A_1 \vdash A_6}\,(C)}{A_{f(2,4)} \Rightarrow A_{g(3,5)} \vdash A_1 \Rightarrow A_6} \qquad (6.30)$$

as follows: from the 6×6 wire,[11] representing the axioms,

$$\begin{pmatrix} 0 & 0 & 0 & 0 & 1 & 0 \\ 0 & 0 & 1 & 0 & 0 & 0 \\ 0 & 1 & 0 & 0 & 0 & 0 \\ 0 & 0 & 0 & 0 & 0 & 1 \\ 1 & 0 & 0 & 0 & 0 & 0 \\ 0 & 0 & 0 & 1 & 0 & 0 \end{pmatrix} \qquad (6.31)$$

we pass, through two linear isomorphisms $P : |A_2| \oplus |A_4| \to |A_{f(2,4)}|$ and $Q : |A_3| \oplus |A_5| \to |A_{g(3,5)}|$ satisfying (6.29), to the 4×4 contracted wire

$$\phi(2) = \begin{pmatrix} 0 & PQ^* & P^* & 0 \\ QP^* & 0 & 0 & Q^* \\ P & 0 & 0 & 0 \\ 0 & Q & 0 & 0 \end{pmatrix} \qquad (6.32)$$

We can verify now that $[\phi(2)]\sigma$ still verifies nilpotency, but its order is now bounded by 4, meaning that the occurrence space of σ has been doubled by P and Q during interaction (or, equivalently, that σ has been doubled in $P\sigma P^* + Q\sigma Q^*$, that is, in what we would naturally write σ', σ''): in order to use $\phi(2)$, one has "made space on the blackboard" to create new copies.

On the other hand, suppose one makes the two distinct copies of integer representations interact, i.e. to represent the computation $(\underline{m})\underline{n}$, which in λ-calculus leads to an exponential growth:

$$\begin{aligned} (\underline{n})\underline{m} &\equiv (\lambda f.\lambda x.(f)^n x)\lambda g.\lambda y.(g)^m y \\ &\leadsto_\beta \lambda x.(\lambda g.\lambda y.(g)^m y)^{n-k} \lambda y.(x)^{m^k} y \\ &= \lambda x. \underbrace{(\lambda g.\lambda y. \underbrace{(g) \ldots (g)}_{m \text{ times}} y) \ldots (\lambda g.\lambda y. \underbrace{(g) \ldots (g)}_{m \text{ times}} y)}_{n-k \text{ times}} \lambda x. \underbrace{(x) \ldots (x)}_{m^k \text{ times}} y \\ &\leadsto_\beta \lambda x.\lambda y. \underbrace{(x) \ldots (x)}_{m^n \text{ times}} y \equiv \underline{m^n} \end{aligned} \qquad (6.33)$$

From the viewpoint of formal systems, this poses strictly no problem (since variables and terms can be copied as many times as desired), but, in *GoI*, where every occurrence must receive a distinct location, one needs isomorphisms

---

[11] To be precise, the wire corresponds to a linear operator u whose *support* is a projection π of dimension 6, i.e. such that $u = \pi u \pi$, and which can thus be thought of as a 6×6 matrix. In the following, I'll freely use matrices to represent linear operators of finite support.

whose composition (inducing a group structure) provides enough space to implement the computation (6.33). In this manner, time and space requirements are translated into the algebraic properties of the group of isomorphisms chosen (see Girard 2007 for a discussion).

Now, we can well imagine a tortoise accepting *modus ponens* and demanding for more instructions for its application: since lacking the relevant – though logically arbitrary – isomorphisms, she simply does not know how to write something which would conform to the rule, and so she asks for further explanations. For instance, in (6.33) she could well question the convergence of reductions, since the possibility to write (6.33) undermines the validity of what (6.33) expresses: in order to show convergence of a notation for exponentials, one has to adopt an explicitly exponential notation (look at the underbraces)!

By inflating the argument, as relying on higher forms of the *modus ponens* schema (or some schema defining exponential functions), still the tortoise would not know how to accept the conclusion, since she would find herself in a completely isomorphic situation: the instructions she needs are inexorably left implicit in the description of logical rules; in order to investigate what it is to follow the rules of logic, then, one has to take into consideration the concrete manipulation of symbols in time and space.

To conclude, I add that, from a purely mathematical perspective, the purpose to characterize in purely algebraico-geometrical terms the subjectivity of formal systems and to describe logically their interactions has already led to some considerable results (concerning the implicit characterization of complexity classes like *ELEMENTARY*, *P* or *NL*, see for instance Girard (2012), Girard (2011a), and Aubert and Seiller (2012)) and constitutes one of the most promising technical aspects of research on *transcendental syntax*.

References

Abrusci, Vito Michele and Paul Ruet. 2000. Non-commutative logic I: the multiplicative fragment. *Annals of Pure and Applied Logic* 101(1): 29–64.
Aubert, Clément and Thomas Seiller. 2012. Characterizing co-NL by a group action. arxiv preprint abs/1209.3422.
Carroll, Lewis. 1895. What the Tortoise said to Achilles. *Mind* 104(416): 691–693.
Carnap, Rudolf. 1950. Empiricism, semantics and ontology. *Revue Internationale de Philosophie* 4: 20–40.
Cozzo, Cesare . 2004. Rule-following and the objectivity of proof. In *Wittgenstein today*, ed. A. Coliva and E. Picardi. Padova: Il poligrafo.
Dummett, Michael. 1973. *Frege's philosophy of language*. Cambridge: Harvard University Press.
Dummett, Michael. 1978. Wittgenstein's philosophy of mathematics (1959). In *Truth and other enigmas*. Cambridge: Harvard University Press.
Dummett, Michael. 1991. *The logical basis of metaphysics*. New York: Columbia University Press.
Engel, Pascal. 2005. Dummett, Achilles and the tortoise. HAL-SHS preprint, ijn_00000571.
Girard, Jean-Yves. 1987a. Linear logic. Theoretical Computer Science. 50: 1–102.
Girard, Jean-Yves. 1987b. Multiplicatives. In *Logic and computer science: new trends and applications*, ed. G. Lolli, 11–34. Rosenberg & Sellier.

Girard, Jean-Yves. 1989. Geometry of interaction 1: Interpretation of system F. *Studies in Logic and the Foundations of Mathematics* 127: 221–260.
Girard, Jean-Yves. 2001. Locus solum: from the rules of logic to the logic of rules. *Mathematical Structures in Computer Science* 11(3): 301–506.
Girard, Jean-Yves. 2007. *Le point aveugle Tome 2, Cours de logique, Vers l'imperfection*. Hermann: Vision des sciences.
Girard, Jean-Yves. 2011a. Geometry of interaction V: logic in the hyperfinite factor. *Theoretical Computer Science* 412(20): 1860–1883.
Girard, Jean-Yves. 2011b. La syntaxe transcendantale, manifeste. http://iml.univ-mrs.fr/~girard/syntran.pdf.
Girard, Jean-Yves. 2012. Normativity in logic. In *Epistemology versus ontology – essays on the philosophy and foundations of mathematics in honour of Per Martin-Löf*, 243–263. Dordrecht: Springer.
Joinet, Jean-Baptiste. 2009. Ouvrir la logique au mond. In *Ouvrir la logique au mond: Philosophie et Mathématique de l'interaction*. Hermann et CCI-Cerisy.
Joinet, Jean-Baptiste. 2011. Logique et métaphysique. In *O que è metafisica?* Editora da Universidade Federal do Rio Grande do Norte.
Prawitz, Dag. 1973. Towards a foundation of a general proof theory. In *Proceedings of the fourth international congress for logic, methodoly and philosophy of science*, Bucharest, 1971, vol. 74, 225–250.
Quine, Willard Van Orman. 1984. Truth by convention (1936). In *Philosophy of matemathics: selected readings*, ed. H. Putnam, P. Benacerraf. Cambridge: Cambridge University Press.
Searle, John. 1969. *Speech acts: an essay in the philosophy of language*. New York: Cambridge University Press.
Wittgenstein, Ludwig. 1969. *On certainty*, ed. by G.E.M, Anscombe and G.H von Wright. Basil Blackwell.
Wittgenstein, Ludwig. 2001. *Tractatus logico-philosophicus (1921)*. Routledge.
Wittgenstein, Ludwig. 2009. *Philosophical investigations (1953)*. Wiley-Blackwell. German text, with an English translation by G.E.M. Anscombe, P.M.S. Hacker and Joachim Schulte.

Chapter 7
Paradox and Inconsistency: Revising Tennant's Distinction Through Schroeder-Heister's Assumption Rules

Luca Tranchini

In natural deduction systems (Prawitz 1965), negation is usually governed by the following rules and reduction:

$$\dfrac{[A]}{\dfrac{\bot}{\neg A}}\neg I \qquad \dfrac{A \quad \neg A}{\bot}\neg E \qquad \dfrac{\dfrac{[A]}{\mathscr{D}_2} \quad \dfrac{\mathscr{D}_1}{A} \quad \dfrac{\bot}{\neg A}\neg I}{\bot}\neg E \quad \overset{\neg\text{-RED}}{\triangleright} \quad \begin{array}{c}\mathscr{D}_1 \\ A \\ \mathscr{D}_2 \\ \bot\end{array}$$

A sentence λ equivalent to its negation—a sort of proof-theoretic liar—can be characterized as follows (Prawitz 1965; Tennant 1982; Schroeder-Heister 2012):

$$\dfrac{\neg\lambda}{\lambda}\lambda I \qquad \dfrac{\lambda}{\neg\lambda}\lambda E \qquad \dfrac{\dfrac{\mathscr{D}}{\neg\lambda}}{\dfrac{\lambda}{\neg\lambda}} \quad \overset{\lambda\text{-RED}}{\triangleright} \quad \dfrac{\mathscr{D}}{\neg\lambda}$$

Using the rules of λ, it is possible to produce a closed derivation $\mathbf{\Lambda}$ of $\bot$:

$$\dfrac{\dfrac{\dfrac{\lambda^{(1)}}{\neg\lambda}}{\dfrac{\bot}{\neg\lambda}\neg I\,(1)}\lambda I}{\bot} \qquad \dfrac{\dfrac{\lambda^{(2)} \quad \dfrac{\lambda^{(2)}}{\neg\lambda}\lambda E}{\dfrac{\bot}{\neg\lambda}\neg I\,(2)}\neg E}{\neg E} \qquad (\mathbf{\Lambda})$$

L. Tranchini (✉)
Computer Science Department, Tübingen University, Sand 13 72076, Tübingen, Germany
e-mail: luca.tranchini@gmail.com

The derivation Λ is not normal because the major premise of $\neg$E that yields the conclusion of the derivation is obtained by $\neg$I. The result of applying the negation reduction ($\neg$-RED) to Λ is the derivation Λ':

$$
\begin{array}{c}
\dfrac{\dfrac{\dfrac{(1)}{\lambda} \quad \dfrac{\overset{(1)}{\lambda}}{\neg\lambda}}{\dfrac{\bot}{\neg\lambda}\,(1)}}{\lambda} \quad \dfrac{\dfrac{\overset{(2)}{\lambda} \quad \dfrac{\overset{(2)}{\lambda}}{\neg\lambda}}{\dfrac{\bot}{\neg\lambda}\,\neg\text{I}\,(2)}}{\dfrac{\lambda}{\neg\lambda}\,\lambda\text{I}}\,\lambda\text{E} \\[2pt] \hline \bot
\end{array} \qquad (\Lambda')
$$

The major premise of the last application of $\neg$E is the conclusion of two consecutive applications of λI and λE. Hence, the derivation Λ' is not normal either. By applying λ's reduction (λ-RED) to Λ', we obtain Λ again. That is, the reduction sequence enters what Tennant (1982) called an 'oscillating loop'.

Prawitz (1965, appendix B) and in a more systematic way Tennant (1982) proposed to take this feature as the proof-theoretic characterization of paradoxes: namely, that they give rise to derivations of $\bot$ all reduction sequences of which enter a loop.

More precisely, what deserves the attribute 'paradoxical' is, for Tennant, the set of sentences figuring in the special inference rules—called by Tennant *id est* inferences—which make it possible to derive $\bot$.

Tennant's id est inferences for the liar (and for other paradoxes as well) are formulated using the truth predicate:

$$\dfrac{\neg T\ulcorner\lambda\urcorner}{\lambda} \qquad \dfrac{\lambda}{\neg T\ulcorner\lambda\urcorner}$$

Consequently, the inference rules governing the truth predicate

$$\dfrac{A}{T\ulcorner A\urcorner} \qquad \dfrac{T\ulcorner A\urcorner}{A}$$

are also needed in order to derive $\bot$. Here I am not interested in the conditions at which a paradoxical sentence like λ is definable in a language (for instance, that the language contains a truth predicate satisfying the above rules and some quotation device to construct in the language names for its expressions). For this reason, the use of the truth predicate is here avoided and λ (which can just be viewed as a nullary operator) is directly characterized in terms of rules which encode its negative self-reference. Rather than speaking of id est inference, I will speak of 'special' inference rules, with which I simply mean any rule not belonging to those of the natural deduction system for intuitionistic logic NI.

In Λ these are the rules of λ. The loop in the reduction sequence of Λ thus shows that λ is a paradoxical sentence.

7.1 Paradox, Inconsistency, and Open Assumptions

Tennant (1982) distinguishes between Λ, the derivation of $\bot$ generated with the rules of λ, and other, far more innocuous, derivations of $\bot$, such as the following (we call it **I**):

$$\frac{\dfrac{A \wedge \neg A}{A} \qquad \dfrac{A \wedge \neg A}{\neg A}}{\bot} \tag{I}$$

Let us compare Λ and **I**. The derivation Λ is not normal and cannot be normalized because of the loop arising in its reduction sequence. On the other hand, the derivation **I** is already normal.

For Tennant, a derivation of $\bot$ with looping reduction sequences shows that the sentences involved in its id est inferences are paradoxical. On the other hand, a normalizable open derivation of $\bot$ shows the inconsistency of its assumptions. Thus **I** simply shows the inconsistency of $A \wedge \neg A$.

Whereas in Λ all assumptions are discharged, **I** is an open derivation. However, being closed is not an essential feature of paradoxical derivations of $\bot$. It is the distinctive trait only of what Tennant calls *pure* paradoxes.

One of Kripke's (1975) lessons is that in particularly unfavourable circumstances virtually any sentence that expresses semantic facts can turn out to be paradoxical. That is, in general 'paradoxicality [...] is relative to the empirical facts' (Tennant 1982, p. 282).

For Tennant, this is reflected in Kripke's approach to paradoxes, in which a sentence is paradoxical if and only if it receives no truth value at any fixed point of his semantic construction. Since this construction starts from a fully interpreted language, being paradoxical is relative to the interpretation with which one started.

To recast this idea in proof-theoretic terms is not a trivial matter.

Tennant's proposal is roughly the following. Among the derivations of $\bot$ whose reduction sequences loop, he distinguishes those in which the conclusion depends on no assumptions (the pure paradoxes), from those in which some assumptions are undischarged. In the latter case, the set of sentences involved in id est inferences is said to be paradoxical *relative* to the class of models in which the assumptions are true.

So Tennant (1982, p. 283): 'A set of sentences is paradoxical relative to M if and only if there is some proof of $\bot$ from $\Theta(M)$, involving those sentences in id est inferences, that has a looping reduction sequence', where a proof from $\Theta(M)$ is 'a proof from assumptions that are truths in (every member of) M, and by means of rules for the logical operators and the truth predicate, as well as the id est rules of inference that are legitimate by every member of M'.

This generalization leaves untouched what 'paradoxical' is ascribed to: it is the set of sentences figuring in the id est inferences which is paradoxical and not the open assumptions on which the derivation depends.

On the other hand, the role played by open assumptions in a derivation of $\bot$ depends on whether the derivation normalizes. If all reduction sequences of the derivation enter a loop, the open assumptions constitute an empirical basis relative to which the sentences involved in the id est inferences are paradoxical. If the derivation normalizes, they constitute a set of propositions which is shown to be inconsistent. As an aside, for Tennant the assumptions of a normal derivation of $\bot$ are beliefs which the derivation shows to be false. In this way, Tennant most probably wants to avoid the need of negative facts and the like. These would otherwise be called into questions, were assumptions taken as expressions of states of affairs, facts, or some other ontologically loaded correlate of propositions.

Summing up, it looks as if Tennant is conceiving three kinds of cases.

1. Non-normalizing derivations of $\bot$ from no undischarged assumptions. An example is Λ, which uses special rules of inference in which the sentence λ figures. The derivation shows the (pure) paradoxical nature of λ.
2. Normal derivations of $\bot$. Tennant presupposes that in these derivations, such as **I**, the conclusion $\bot$ depends always on one or more open assumption. These may be viewed as beliefs about empirical facts, and the derivation shows them to be (collectively) false.
3. Non-normalizing derivations of $\bot$ depending on a set of undischarged assumptions Δ. Tennant presupposes that special rules are used in the derivation which involve a set of sentences, say Γ. The derivation shows that the set of sentences Γ is paradoxical relative to the class of models in which the assumptions Δ are true.

I will show that the presuppositions made by Tennant in the second and third case are not correct. In particular:

(i) Thanks to special inference rules, we can generate normal derivations of $\bot$ from no open assumptions;
(ii) Under a very natural generalization of the notion of reduction, we have non-normalizing derivations of $\bot$ which do not use special inference rules.

Thereby, Tennant's distinction between the roles played, respectively, by open assumptions and sentences figuring in id est sentences faces two distinct problems.

In case (i), there are no assumptions to which inconsistency is to be ascribed. In case (ii), there are no special rules which allow the identification of a set of sentences as paradoxical.

Hence, a sharp distinction between (possibly inconsistent) assumptions and (possibly paradoxical) sentences which figure in id est inferences turns out not to be always appropriate.

I propose to discuss the issue in terms of the extension of natural deduction proposed by Schroeder-Heister (1981). In this setting, the notion of assumption is enriched so that rules are admitted as a special kind of assumptions alongside with sentences. As a result, all derivations of $\bot$ will depend on some assumptions. According to whether the derivation is normalizable or not, the assumptions involved in the derivations will be said to be either inconsistent or paradoxical, independently of their being sentences or rules.

7.2 From `tonk` to $\bot$

Let us consider Prior's `tonk`:

$$\frac{A}{A \text{ tonk } B} \text{ tonkI} \qquad \frac{A \text{ tonk } B}{B} \text{ tonkE}$$

By means of the rules of `tonk`, one can easily produce a closed derivation of $\bot$. Given any closed derivation of any logically valid sentence, such as $A \supset A$, we extend it by `tonkI` to a closed derivation of $(A \supset A)$ `tonk` $\bot$ and then by `tonkE` to a closed derivation Π for $\bot$:

$$\frac{\dfrac{\dfrac{\overset{(1)}{A}}{A \supset A} \supset\!\text{I}\,(1)}{(A \supset A)\text{tonk}\bot} \text{tonkI}}{\bot} \text{tonkE} \qquad (\Pi)$$

What is special about `tonk` is the impossibility of specifying a reduction procedure to get rid of consecutive applications of `tonkI` and `tonkE` rules (as it is often said, the rules of `tonk` are not in harmony). Hence, the derivation Π is normal simply because there is no reduction procedure that can be applied to it. Thus, given Tennant's analysis, Π does not display the distinguishing feature of being paradoxical.

Incidentally, one could argue that a normal derivation should not be defined as one to which no reduction can be further applied, but as one which does not contain consecutive application of introduction and elimination rules. It is true that, on this definition, Π would not count as normal.[1] However, this is irrelevant as to whether Π passes Tennant's test for being paradoxical. For Tennant, the characteristic of paradoxes is their giving rise to derivations with looping reduction sequence. Thus, irrespective of its counting as normal or not, Π does not count as paradoxical: Its reduction sequence *does* terminate, since it does not even start!

Not being paradoxical, the derivation should show the inconsistency of its open assumptions. However, in Π there are no open assumptions at all to which inconsistency can be ascribed.

One way of responding to this problem could simply be that of denying that Prior's connective has nothing whatsoever to do with the phenomenon of semantic closure and hence with Tennant's analysis thereof.

However, this is not a viable answer for an inferentialist whose aim is that of characterizing the meaning of logical constants in terms of the rules governing their behaviour in deduction.

Since its very tentative formulations, the challenge for proof-theoretic semantics was that of giving criteria for excluding `tonk`-like connectives from the realm

---

[1] The two alternative definitions of 'normal' are discussed in detail in Tranchini (n.d.).

of meaningful expressions. Furthermore, a solution to (or at least a diagnosis of) paradoxes is in the agenda of whoever pursues any semantic program.

Thus, the development of a theoretical framework in which both paradoxical connectives (such as λ) and non-harmonious ones (such as `tonk`) could be uniformly investigated is of the greatest interest for proof-theoretic semantics.[2]

As a final remark in this section, I want to point out that Tennant himself discusses a normal derivation of $\bot$ from no assumption. This derivation results by formalizing Chihara's Sec Lib setting:

> in which many clubs have hired secretaries but have established rules excluding such secretaries from membership. Suppose that these secretaries form their own club, Secretary Liberation (or "Sec Lib" for short), the rules of which state: "A person is eligible to join this club if, and only if, he (she) is secretary of a club which he (she) is not eligible to join." All goes well for the club until it hires itself a secretary, a certain Ms Fineline, who has the misfortune of being a secretary of no other club. (Chihara 1979, p. 593–594)

The assumptions are elicited by Tennant in terms of inference rules which allow to produce a normal derivation of $\bot$ from no assumptions. As the derivation of $\bot$ is normal, Tennant's opinion is that the Sec Lib setting should be considered as merely inconsistent and not (as in Chihara's opinion) paradoxical. The derivation Π is thus analogous to the one resulting by the Sec Lib case.

Also in this case, the derivation contains no open assumptions, and hence one may ask to what should inconsistency be ascribed. In Sect. 7.5, Tennant's distinction between paradox and inconsistency will be reformulated in a way that provides a natural answer to this question.

7.3 Ekman's Paradox

Take $A \leftrightarrow B =_{def} (A \supset B) \wedge (B \supset A)$. Now replace each application of λI and λE in Λ with two consecutive applications of $\wedge$E and $\supset$E as follows:

$$\dfrac{\dfrac{\mathcal{D}}{\neg\lambda}}{\lambda}\lambda\text{I} \quad \rightarrow \quad \dfrac{\dfrac{\mathcal{D}}{\neg\lambda} \quad \dfrac{\lambda \leftrightarrow \neg\lambda}{\neg\lambda \supset \lambda}\wedge\text{E}}{\lambda}\supset\text{E}$$

$$\dfrac{\dfrac{\mathcal{D}}{\lambda}}{\neg\lambda}\lambda\text{E} \quad \rightarrow \quad \dfrac{\dfrac{\mathcal{D}}{\lambda} \quad \dfrac{\lambda \leftrightarrow \neg\lambda}{\lambda \supset \neg\lambda}\wedge\text{E}}{\neg\lambda}\supset\text{E}$$

---

[2]In Tranchini (2014), I suggest a way in which Tennant's ideas can be used to extend Dummett-Prawitz-style proof-theoretic semantics so that it can be applied to paradoxical phenomena as well.

7 Paradox and Inconsistency: Revising Tennant's Distinction Through...

Furthermore, replace all occurrences of λ with occurrences of a propositional variable A. Like the original Λ, also this derivation, call it $\Lambda_{\leftrightarrow}$, is not normal, since the major premise of the concluding application of $\neg$E is the conclusion of a $\neg$I:

$$
\cfrac{\cfrac{(1)}{A} \quad \cfrac{\cfrac{\cfrac{A \leftrightarrow \neg A}{A \supset \neg A} \wedge\text{E}}{\neg A} \supset\text{E}}{\cfrac{\bot}{\neg A} \neg\text{I}(1)} \quad \cfrac{\cfrac{A \leftrightarrow \neg A}{\neg A \supset A} \wedge\text{E}}{A} \supset\text{E}}{\bot} \quad \cfrac{\cfrac{(2)}{A} \quad \cfrac{\cfrac{A \leftrightarrow \neg A}{A \supset \neg A} \wedge\text{E}}{\neg A} \supset\text{E}}{\cfrac{\bot}{\neg A} \neg\text{I}(2)} \neg\text{E}}{\neg\text{E}}
$$
$(\Lambda_{\leftrightarrow})$

We call $\Lambda'_{\leftrightarrow}$ the derivation resulting by applying $\neg$-RED to $\Lambda_{\leftrightarrow}$:

$(\Lambda'_{\leftrightarrow})$

As first observed by Ekman (1998), although $\Lambda'_{\leftrightarrow}$ is formally speaking normal, it still contains a redundant chunk.

This corresponds to the one constituted by the two consecutive applications of λI and λE that was contained in Λ'. Replacing the applications of λ-rules with applications of $\wedge$E and $\supset$E does not help in getting rid of the redundancy:

$$
\cfrac{\cfrac{\neg\lambda}{\lambda} \lambda\text{I}}{\neg\lambda} \lambda\text{E} \quad \dashrightarrow \quad \cfrac{\neg A \quad \cfrac{\cfrac{A \leftrightarrow \neg A}{\neg A \supset A} \wedge\text{E}}{A} \supset\text{E}}{\neg A} \cfrac{A \leftrightarrow \neg A}{A \supset \neg A} \wedge\text{E} \supset\text{E}
$$

The redundancy constituted by the consecutive applications of λI and λE can be get rid of using λ-RED. It is thus natural to enrich the set of reduction procedures, by introducing a new one to get rid of the redundancy resulting by going back and forth between the antecedent and the consequent of a bi-implication:[3]

$$
\cfrac{\neg A \quad \cfrac{\mathcal{D} \quad \cfrac{A \leftrightarrow \neg A}{\neg A \supset A} \wedge\text{E}}{A} \supset\text{E}}{\neg A} \cfrac{A \leftrightarrow \neg A}{A \supset \neg A} \wedge\text{E} \supset\text{E} \quad \overset{\leftrightarrow\text{-RED}}{\triangleright} \quad \cfrac{\mathcal{D}}{\neg A}
$$

---

[3] 'I shall always assume that eliminating detours such as the one mentioned, that lead from a sentence occurrence to another occurrence of the same sentence, is an acceptable part of any procedure of normalising proofs'. (Tennant 1982, p. 270)

Once this reduction procedure is introduced, the reduction sequence of the derivation $\Lambda_\leftrightarrow$ behaves exactly like that of Λ, namely, it enters a loop. (By applying $\leftrightarrow$-RED to $\Lambda'_\leftrightarrow$, one obtains $\Lambda_\leftrightarrow$ again.)

However, there are two differences between Λ and $\Lambda_\leftrightarrow$. Whereas Λ is closed, $\Lambda_\leftrightarrow$ depends on the undischarged assumption $A \leftrightarrow \neg A$. Furthermore (and crucially for showing the limits of Tennant's analysis), $\Lambda_\leftrightarrow$ does not contain any id est inference.

Hence, whereas Tennant's analysis tells us that Λ shows λ to be a pure paradox, it is not clear what it has to tell about $\Lambda_\leftrightarrow$.

The reduction sequence of $\Lambda_\leftrightarrow$ does not terminate. Thus, according to Tennant's analysis, some paradoxical sentence is involved in the derivation. In spite of this, there is no way of identifying the paradoxical sentences. For Tennant, the paradoxical sentences are those involved in the id est inferences used in the derivation. But in $\Lambda_\leftrightarrow$ there are no such inferences.

Furthermore, since the conclusion of $\Lambda_\leftrightarrow$ depends on an open assumption, the paradoxical sentences supposedly involved in the derivation would not be *pure* paradoxes, but paradoxical relative to the models in which the open assumption of the derivation is true. Since the open assumption $A \leftrightarrow \neg A$ fails to be true in any model, following Tennant one would reach the rather awkward conclusion that the supposedly paradoxical sentence contained in $\Lambda_\leftrightarrow$ would be paradoxical relative to no model at all.[4]

7.4 Rules as Assumptions

In natural deduction, the set of rules that can be applied in derivations is usually taken to be fixed in advance. A derivation begins by assuming some sentence and then it proceeds by applying one of the available rules first to the assumptions and then to the conclusions of the previously applied rules. A generalization of natural deduction is achieved by allowing one to assume not only sentences but also rules. To assume a rule simply means to suppose that the rule in question is available, that is, to suppose that one is allowed to pass over from the premises of the rule to its conclusion. Following Schroeder-Heister (1981), we use $\Gamma \Rightarrow A$ to refer to the rule allowing to pass over from the set of sentences Γ to A.

Given the possibility of assuming rules, the conclusion of a derivation will not only depend on the undischarged assumptions but also on the rules assumed in the course of the derivation. These are just the rules that do not belong to the basic reserve of rules (in our case, the rules of the intuitionistic natural deduction system).

---

[4]For the problems that Tennant's analysis of paradox has to face when confronted with the phenomenon observed by Ekman, see also Schroeder-Heister and Tranchini (n.d.).

7 Paradox and Inconsistency: Revising Tennant's Distinction Through...

For example, in the following derivation, the conclusion $C \vee D$ depends on two assumptions, the formulas A and B and the rule $A, B \Rightarrow C$:

$$\cfrac{A, B \Rightarrow C \quad \cfrac{A \qquad B}{C}}{C \vee D} \vee I_1$$

Although a rule $A \Rightarrow B$ and the corresponding implication $A \supset B$ are formally distinct, they are strongly related. In particular, to assume the implication is essentially the same as to assume the rule (Schroeder-Heister 2011). Like an application of the assumption rule, an application of modus ponens entitles one to pass over from A to B:

$$\cfrac{A \supset B \quad A}{B} \supset E \qquad A \Rightarrow B \; \cfrac{A}{B}$$

Furthermore, we can always derive the implication $A \supset B$ from the sole assumption of the rule $A \Rightarrow B$:

$$\cfrac{A \Rightarrow B \; \cfrac{\overset{(1)}{A}}{B}}{A \supset B} \supset I \, (1)$$

Once the notion of assumption is enriched so to admit also rules and not only formulas to be assumed, the notion of rule can be enriched as well, by allowing rules to discharge not only formulas but also assumption rules. In this way, a hierarchy of rules of increasing complexity is achieved. Formulas are taken as rules of level 0 and constitute the basis of the hierarchy. A rule of level $n + 2$ is one discharging rules of level n. The rule $\supset I$ is thus of level 2, since it discharges formulas, i.e. rules of level 0.

Usual natural deduction systems are constituted by rules whose level is not greater than 2. One of the reasons for introducing rules of higher level (i.e. of level 3 or higher) is that they allow to formulate introduction and elimination rules for propositional operators in a uniform manner (see Schroeder-Heister 1984).

We will not tackle here this issue. Rather, we suggest that a general framework in which rules as much as formulas are treated as assumptions allows to clarify the two problematic cases for Tennant's analysis of paradoxes.

7.5 Assumption Rules, Inconsistency, and Paradox

According to Tennant, a paradox is identified by looking at the inference rules used in derivations of $\bot$ with looping reduction sequences. Assumptions provide always empirical information: in normalizable derivations of $\bot$, this information is shown to be inconsistent; in derivations of $\bot$ with looping reduction sequences, this information selects the models relative to which the sentences figuring in id est inferences are paradoxical.

The adoption of the rule assumption setting suggests to revise the way in which the distinction of roles between assumptions and rules is formulated by Tennant. I propose that the whole set of assumptions (sentences and rules) of a derivation of $\bot$ should be viewed as either paradoxical or inconsistent depending on whether the reduction sequences of the derivation terminate.

The revision of Tennant's picture provides a better account of the two problematic cases discussed in the previous sections.

In Sect. 7.2, using the rules of tonk, we produced the derivation Π that has $\bot$ as conclusion. The derivation Π is normal and thus bears the mark of inconsistency rather than that of paradox. However, it contains no undischarged assumption to which, in Tennant's analysis, inconsistency can be ascribed.

In the revised picture suggested, the problem disappears since the conclusion of Π does depend on assumptions: the rules for tonk. The derivation thus shows the inconsistency of the rules of tonk. (The notion of inconsistency here discussed is clearly not to be confused with those introduced by Belnap (1962) and Read (2010) in the context of criteria of admissibility of the rules governing a connective. Contrary to the notion of inconsistency here discussed, both are meant to apply as much to tonk as to λ.)

For the inferentialist, the meaning of a connective is given by laying down the inference rules governing it. In this perspective, to assume the rules of tonk is to assume that there is an expression whose meaning is characterized by these rules. This assumption is reduced ad absurdum. That is, the rules of tonk fail to endow tonk with meaning.

Analogously, the Sec Lib case shows the inconsistency of the Sec Lib club's eligibility rule together with the club's hiring of Ms Fineline. Therefore, one can conclude either that the eligibility rule is not valid (as Tennant does in his own analysis) or that at least one of the assumptions (possibly viewed as beliefs) about Ms Fineline is false.

Whereas the derivation of $\bot$ using tonk forces one to ascribe inconsistency to rules, Ekman's derivation of $\bot$ forces one to look for the paradox not in the rules but in the assumptions. According to Tennant, those sentences are paradoxical that are used in the rules by means of which a non-normalizable derivation of $\bot$ can be produced. No rules other than those of intuitionistic logic are used in the derivation of Ekman's paradox. If we do not want to view them as the source of paradox, we must reject Tennant's view that the role of assumptions in paradoxes is at best to provide only a material base for paradoxes. That is, we have to admit that the source of paradox lays in the assumptions as well as in the rules.

In the revised picture suggested, rules are assumptions no more and no less than sentences. In this view it is very natural to view Ekman's derivation as paradoxical. In the setting allowing rule assumptions, to assume an implication is the same as assuming a rule to pass over from the antecedent to the consequent. Hence, the assumption $A \leftrightarrow \neg A$ in the proof of Ekman's paradox $\Lambda_\leftrightarrow$ is the same as the assumption of λ-rules in Λ. It is therefore desirable that both assumptions are treated as paradoxical, since the two assumptions are essentially the same.[5]

Beside providing a smooth analysis of the problematic cases, the alternative formulation of the distinction between paradox and inconsistency has also the advantage (at least for the inferentialist) of not mentioning any notion of model whatsoever.

Thus it can be viewed as a welcomed improvement of Tennant's analysis of paradoxes in proof-theoretic terms.

Acknowledgements I thank Alberto Naibo and Peter Schroeder-Heister for having pointed to Ekman's paper. His 'paradox' has been also independently observed by Matteo Plebani, whom I thank for his brilliant work as discussant of my talk 'Proof-theoretic semantics, paradoxes and the distinction between sense and denotation' at the V Latin Meeting of Analytic Philosophy, Lisbon, November 2011. Thanks also to an anonymous referee for several suggestions which helped improving the readability of the paper.

References

Belnap, N.D. 1962. Tonk, plonk and plink. *Analysis* 22(6): 130–134.
Chihara, C. 1979. The semantic paradoxes: a diagnostic investigation. *The Philosophical Review* 88(4): 590–618.
Ekman, J. 1998. Propositions in propositional logic provable only by indirect proofs. *Mathematical Logic Quarterly* 44: 69–91.
Kripke, S. 1975. Outline of a theory of truth. *Journal of Philosophy* 72(19): 690–716.
Prawitz, D. 1965. *Natural deduction. A proof-theoretical study*. Stockholm: Almqvist & Wiksell.
Read, S. 2010. General-elimination harmony and the meaning of the logical constants. *Journal of Philosophical Logic* 39: 557–576.
Schroeder-Heister, P. 1981. *Untersuchungen zur regellogischen Deutung von Aussagenverknüpfungen*, PhD thesis, Bonn University.
Schroeder-Heister, P. 1984. A natural extension of natural deduction. *The Journal of Symbolic Logic* 49(4): 1284–1300.
Schroeder-Heister, P.: 2011. Implications-as-rules vs. implications-as-links: an alternative implication-left schema for the sequent calculus. *Journal of Philosophical Logic* 40(1): 95–101.
Schroeder-Heister, P. 2012. Proof-theoretic semantics, self-contradiction, and the format of deductive reasoning. *Topoi* 31(1): 77–85.
Schroeder-Heister, P. and L. Tranchini. n.d. Ekman's paradox. Under review. Available at http://ls.inf.uni-tuebingen.de/psh
Tennant, N. 1982. Proof and paradox. *Dialectica* 36: 265–296.
Tranchini, L. 2014. Proof-theoretic semantics, paradoxes, and the distinction between sense and denotation. *Journal of Logic and Computation*. doi: 10.1093/logcom/exu028.
Tranchini, L. n.d. Harmonizing harmony. To appear in the *Review of Symbolic Logic*.
Tranchini, L. and P. Schroeder-Heister. n.d. Ekman's paradox and general natural deduction rules. Under review. Available at https://sites.google.com/site/lucatranchini/
von Plato, J. 2000. A problem of normal form in natural deduction. *Mathematical Logic Quarterly* 46(1): 121–124.

---

[5] According to von Plato (2000), this holds only when the elimination rule of implication is modus ponens, but not in a natural deduction system with general elimination rules. See Tranchini and Schroeder-Heister (n.d.) for a thorough criticism of von Plato.

Chapter 8
Constructibility and Geometry

Alberto Naibo

8.1 Introduction

The purpose of this work is to investigate the possibility of capturing a somehow intuitive idea of *geometrical constructivity* by means of logical notions of constructivity. The motivation does not simply consist in providing a conceptual analysis of an allegedly loose notion by making appeal to others which are supposed to be clearer and better understood. What we are interested in is also suggesting an account of geometry that respects some epistemic constraints and that can eventually be compatible with a mechanization of geometrical practice.

Two more preliminary comments should be made before entering into a discussion of these topics. The first one is that, for simplicity, we will restrict our analysis exclusively to plane geometry. The second one is just to acknowledge our intellectual debts: fundamental sources of inspiration for this work have been Vesley (2000), von Plato (1995, 2010a), and Beeson (2010).

8.1.1 Constructivity in Geometry

In the traditional Euclidean sense, geometry is usually reckoned to be a science of constructions, as it studies the production of certain objects, i.e., figures, or configurations of objects, once a given set of rules or instruments has been fixed (e.g., the possibility of appealing only to the ruler and the compass). In general,

---

A. Naibo (✉)
IHPST (UMR 8590), Université Paris 1 Panthéon-Sorbonne, CNRS, ENS.
13 rue du Four, 75006 Paris, France
e-mail: alberto.naibo@univ-paris1.fr

we could say that the aim of geometry is to study those actions performable by an abstract human geometer once certain constraints have been previously imposed.

It is not difficult to see how this situation shares many similarities with Turing's analysis of the so-called *human computor* (Gandy 1988; Sieg 1994).[1] In this case, the definition of *computable function* is based on the analysis of those possible actions that a human agent can perform during the calculation of the values of a function, supposing that certain limitations have been previously imposed, such as the locality of the moves or the finitude of the number of symbols and internal states of the computor (Turing 1937, pp. 231–232). But, as this behaviouristic and quasi-physical analysis of the human computor finally leads to the abstract and linguistic formal definition of Turing machine, or, equivalently, of recursive function or λ-definable function, we may wonder in the same way whether the analysis of the human geometer can be fixed at some linguistic formal level. In other words, we would like to understand if there exists a syntactic formal manner with which to capture the idea of geometrical constructivity. And because syntactic formal considerations usually fall under the scope of logic, it is quite natural to look at logical constructivity as a reasonable candidate for playing that role.

8.1.2 Constructivity in Logic

Turning now to logic, we can see that in this domain, the notion of constructivity is essentially polymorphic, in the sense that it can be declined in many different ways (cf. Troelstra 1991; Troelstra and van Dalen 1988, §1.1; Rathjen 2005, §3). In particular, the term "constructive" is basically used as synonymous for:

C1. Intuitionistically admissible (Brouwer, Heyting, Dummett).
C2. Provable by finitistic, direct, or effective means (Hilbert, Brouwer, Bishop, Martin-Löf).
C3. Algorithmically executable or executable by a program (Markov, Engeler, Bishop, Martin-Löf).

These features are not independent and definitely separated from each other. On the contrary, they frequently overlap, and keeping them distinguished is neither easy nor very sensible in the final analysis. In what follows we will try to understand which of these possible overlaps are needed in order to capture the notion of geometrical constructivity.

It is worth noting that the previous list is not exhaustive. There are in fact other important features that will not be considered in this work, e.g.:

C4. Predicativity (Poincaré, Weyl).
C5. Resource sensitivity (Lemoine, Lambek, Girard).

---

[1] A similar analogy is drawn in Beeson (2010, §1).

The reason is that, on the one hand, we will not be confronted with issues requiring quantification over sets (e.g., the continuity axiom), while on the other hand, we will limit here our analysis of the notion of constructivity to the abstract point of view. In order to clarify this latter point, a further comparison with the notion of computable function becomes fruitful. The notion of computable function is based on the analysis of a human computor, whose actions abstract from contingent limitations like a lack of attention or a lack of paper on which to perform the calculation. In the same way, the notion of logical constructivity only concerns what can be constructed in principle by an ideal mathematician and not what can be practically done by some particular agent in certain specific situations. In fact, according to Brouwer, ideal mathematicians are not agents totally freed from any kind of contingent constraints; on the contrary, they have the very same epistemic capacities that any other concrete human beings possess, the only difference being that their capacities are perfect. More precisely, like every concrete human being, they can deal with only a finite amount of resources and information, and their actions can be performed only in a finite amount of time and space; however, unlike concrete human beings, their finite capacities are not subject to any fixed bound.[2]

8.1.3 The Geometrical Space

The considerations we made on the ideal mathematician are essential also for understanding an implicit assumption standing behind the presentation of Euclidean geometry that we adopt in this work, and that would be difficult to justify if our analysis of Euclidean geometry simply reduced to a study of mathematical (algebraic) structures. Our idea is in fact to take seriously the parallelism we have sketched between Turing's human computor and the abstract human geometer. As the former works on a finite but indefinitely extensible string along its length, the latter works with a finite but indefinitely extensible sheet of paper along its length and height. If both of them worked instead with an infinite string or paper, then we would not be considering only an idealization of a concrete situation, but we would be considering a situation in which from the very beginning, agents are supposed to possess features going beyond human epistemic capacities: since human agents are by definition finite beings, they cannot have a full access to what is infinite.

In geometry, this shift from a finite potentially extensible space to an actual infinite space is far from trivial. A finite sheet of paper corresponds to a normed space and thus to a space equipped with a metric. The idea is that if a point is fixed with respect to a set of coordinates, given in this case by the borders of the sheet,

---

[2]The fact that we do not consider actions performed within a limited amount of resources should not be construed as deeming this aspect irrelevant for the discussion of the relation between geometry and logic. For further information about the connections between geometrical proofs and resource-sensitive logics, see Pambuccian (2004).

then the point will possess specific coordinates as well, determined by its distance from the borders of the sheet. This means that whenever the human geometer puts the tip of a pencil on the sheet, a specific point of that sheet has been identified. This entails two consequences. First, it becomes possible to ask which coordinates correspond to the point determined by the acts of the geometer (cf. Weyl 1949, p. 75), which means that the identity problem between two points can be reduced to the problem of knowing if the two points have the same coordinates. Secondly, if every position in space can possibly correspond to a point, then it becomes reasonable to think that we are working with a continuous space (cf. Poincaré 1902, p. 78), and the most natural mathematical representation of this kind of space is the field of real numbers.[3] More precisely, as Mumma (2012, p. 118) remarks, the activity of abstract human geometers is performed with a special medium: the spatial continuum. Hence, geometrical constructions involve parameters that can vary continuously, and the analysis of geometrical constructivity corresponds to isolating those properties and those constructions that are invariant with respect to these variations in the continuum. In Sect. 8.5.3 and in the conclusion, we will see that this assumption concerning the continuity of the space will play a fundamental role in our analysis. In particular, we will try to understand in which sense this assumption is compatible with the constructivist approach adopted here.

It should be noted that if we had considered the sheet of paper as actually infinite, there would be no norm on it. In this case, when a point is fixed on that paper, it would be aimless to ask which are the coordinates corresponding to that point, since there is no preconceived grid of coordinates. Hence, the system of points is not something that is already there, and to which we can refer, but instead it can be generated from a freely fixed point, around which to construct a system of coordinates from scratch. More precisely, starting from this point, and using only ruler and compass operations, it is possible to obtain a set of points forming what is called a Euclidean field, that is, a field where every nonnegative element can be represented as a square. But a Euclidean field does not necessarily coincide with the field of real numbers. In particular, it can have a countable cardinality, which could allow its elements to be generated in an algorithmic way and decide the equality relation between two elements of the field (cf. Rybowicz 2003).

Do we have then to conclude that this second manner of conceiving geometrical space is more constructive than the first one?[4] We do not think so. The problem

---

[3]Note that starting from the first French translation of the *Grundlagen der Geometrie*, in 1900, Hilbert adds the axiom of completeness in order to force the models of Euclidean geometry to be *maximal*, that is, isomorphic to the field of real numbers, thus obtaining categorical results (cf. Awodey and Reck 2002, §§3.2, 3.3). For further details, see Hallett (2008, §§8.2, 8.3) and Venturi (2011).

[4]The two ways of conceiving the geometrical space presented here reflect the two ways in which, according to Kreisel (1981), mathematical concepts are traditionally analyzed. Borrowing Pascal's terminology, these two ways can be called the *esprit de géométrie*, "where we think of arbitrary points, not only those constructed by use of Euclid's operations, a ruler and a pair of compasses" (1981, p. 70), and the *esprit de finesse*, where, as in algebra, the set of objects

with the second conception is that it assumes from the beginning that the space is infinite, and this risks obliterating any reference to a human agent supposed to realize the geometrical constructions. In this manner, Euclidean geometry would be considered just one algorithmic theory among many others. Of course, it can be objected that in this case the axioms of Euclidean geometry would identify a set of structures which would not necessarily exceed our epistemic capacities, since they could be generated in an algorithmic way, as is done, for example, by Seeland (1978).[5] However, this would not yet explain how we have been able to single out these structures. The problem is that they are stipulated in an axiomatic way, and what counts is only the alethic relation linking them to the axioms, that is, the characterization of these structures by referring to those sets of sentences which they render true and from which other true sentences can be inferred by purely logical reasoning. But, in this way, the analysis of Euclidean geometry becomes an abstract model-theoretical analysis, where the attention is focused only on the logical relation between the axioms and theorems by reference to some abstract structures, and not to the concrete observable activity of the geometer. On the other hand, the analysis of space that we propose takes into consideration this activity, but, as we have seen, in order to do this, an assumption regarding the continuity of geometrical space has to be made, and this assumption could be incompatible with the above constructivist approach. Our analysis will show that this is not the case and that such an assumption is not incompatible with a formal reconstruction of geometrical constructive activity capable of taking into account the epistemic constraints which go with a constructive interpretation of mathematics.

8.1.4 What Is Constructed?

After this very first general glance at what is intended by constructivity in geometry and in logic, respectively, a question naturally arises: On which entities is the notion of constructivity predicated? By taking into consideration what has been said in the two previous sections, it seems natural to think that geometrical constructivity applies to objects, while logical constructivity mainly acts at the level of a formal language. But this is not compulsory and thus, by a pure combinatorial analysis, four possible answers should be considered:

O1. *Objects*. For example, the construction of a specific triangle ABC possessing certain well-determined properties and a particular position in the plane; the algorithmic identification of the values of a specific given function.

---

considered is any set which is closed under a given list of operations, and in particular the smallest of these sets (1981, p. 71; cf. the difference from what has been said in Note 3).

[5]For a survey of this kind of algorithmic study of Euclidean structures, see Pambuccian (2008).

O2. *Types of objects.* For example, the construction of lines given distinct points; the identification of those computable functions that are definable exclusively by means of a given set of primitive functions.

O3. *Sentences.* For example, the verification that a certain sentence A is provable in a direct way or by using only intuitionistically acceptable inference steps.

O4. *Types of sentences.* For example, the verification that sentences of the form Π_n, for any given n, are decidable or provable with direct means.

It is quite evident that these distinctions are at the moment still excessively general and imprecise; the examples that we will present later will play an essential role in clarifying them.

8.2 Axiomatic Presentations

As already mentioned, the principal aim of this work is to understand whether the previous characterizations of logical constructivity are enough to capture the constructive features proper to geometrical practice. In order to do that, we will take into account different syntactic formal presentations of geometrical theories. The first to be analyzed is an axiomatic one: Tarski's theory of elementary geometry, EG for short (Tarski 1959).

From a purely morphological point of view, EG is formulated in a first-order language (with identity) of signature $\Sigma = \{\beta, \delta\}$, where β is a ternary predicate expressing the betweenness relation and δ is a quaternary predicate expressing the equidistance relation. The formula $\beta(x, y, z)$ is read as "the point y lies between the point x and the point z," (the case where y coincides with x or z is not excluded), while $\delta(x, y, z, w)$ is read as "x is distant from y as z is from w." The logical deductive framework underlying EG is classical first-order logic.[6]

Apart from being a consistent theory, the main peculiarity of EG is to be *syntactically complete*: for every closed formula (i.e., sentence) A formulated in the language of EG, by starting from the axioms of EG, either A is derivable or $\neg A$ is derivable, i.e.,

$$\text{for all } A \in Sen_{\mathscr{L}(EG)}, \vdash_{EG} A \text{ or } \vdash_{EG} \neg A.$$

The proof is given using the method of quantifier elimination, where sentences are syntactically reduced to Boolean combinations of atomic (open) formulas. In the case of EG, these atomic formulas are then reduced to tautologies ($\top$) or contradictions ($\bot$). What is relevant is that this type of reduction can be performed algorithmically (see Quaife 1989) so that the following result holds.

---

[6]An axiomatic presentation of Euclidean geometry based instead on an intuitionistic framework can be found in Lombard and Vesley (1998). This system is to EG as Heyting arithmetic is to Peano arithmetic: when its axioms are interpreted classically, one gets a theory equivalent to classical Euclidean geometry.

Proposition 1 *EG is decidable: For every sentence $A \in \mathscr{L}(EG)$, it is possible to establish algorithmically whether A is a theorem or not.*

An immediate consequence is that EG can be considered as constructive in the sense of C3. The validity of a sentence can be verified in every situation and in every moment within a finite amount of time and without any appeal to ingenuity (Vesley 2000, p. 291). In this sense, constructivity coincides with *mechanizability*: a procedure is constructive when it can be executed following always the same rule and without being affected by variations in the context of application. On the contrary, a non-constructive procedure lacks this universal applicability and requires some ingenuity in order to be adapted to each specific situation.

Another interesting aspect of EG is that its constructive character seems to be compatible with a realist position, at least if, following Dummett (1963), realism is understood in a semantical rather than ontological sense; each sentence of $\mathscr{L}(EG)$ is derivable or not and thus (by the soundness theorem) possesses a well-determined truth value independently of the strategies and methods that we can conceive in order to establish it. It is a property of the theory in itself to decide about the derivability of all its sentences, and this property is exclusively due to its syntactic form, so that in principle no appeal to the capacities of a human subject is needed in order to explain the attribution of a specific truth value to each sentence of the theory. This form of constructivism is thus quite distant from the Brouwerian and, more generally, intuitionistic one. The reason is that its attention is focused only on an algorithm acting on a rather "superficial" level, that is, the level of the (preorder) derivability relation between sentences (cf. O3), while no analysis is made of the epistemic properties of the objects guaranteeing this relation to hold (e.g., the proofs). Moreover, the decidability of the derivability relation is not a constructive feature peculiar to geometry; there are plenty of other theories that are decidable. Thus, if we want to make our investigations more precise, we must focus on two aspects that have been omitted from the previous analysis of the constructivity of EG:

P1. *Nothing is said about the objects of the theory.* No constructions are performed on the objects the theory speaks about. In particular, no constructions on them are needed in order to prove the syntactic completeness of the theory. The decision algorithm is "blind" with respect to the objects of the theory, in the sense that there is no relation between what is done at the level of the sentences and what is done at the level of the objects.

P2. *Nothing is said about the derivational structure of the proofs of the theorems.* It is possible to mechanically decide that a sentence is derivable without knowing anything about how its derivation is made. It is not necessary to effectively construct a proof, thus making its inferential structure explicit, in order to determine whether a sentence is derivable or not. In other words, the decision algorithm does not provide any explicit information concerning the inference steps linking the axioms to the theorems.

What follows will be essentially devoted to examining which strategies can be established in order to recover these two aspects and to understand whether they are sufficient in order to capture the notion of geometrical constructivity.

8.3 Constructivity and Objects

Let us start with the analysis of P1. Focusing on the language of the theory seems to us a reasonable way to understand which are the objects this theory commits to and, eventually, to introduce constructive features at the level of these objects.

8.3.1 A Purely Relational Theory

According to the language of EG, there is a unique universe of objects, and this universe contains only one sort of object: points. Moreover, as we already mentioned, this language is completely relational: there are no functions acting on points. Therefore, what the axioms of EG do is only to describe how these points are mutually organized. More precisely, the existence of a nonempty set of objects is implicitly assumed, and the axioms fix, at the linguistic level, the relations holding between them. This is exactly the picture coming out of the axiomatic method as it is conceived in the Hilbertian tradition:

> [...] contemporary axiomatics proceeds from the idea of a *system of objects* fixed from the outset. In geometry, for example, one considers the points, straight lines, and planes in their totality as such a system of things. Within this system one thinks of the relationship of incidence (a point lies on a straight line, or on a plane), of betweenness (a point lies between two others), and of congruence as being determined from the outset (Bernays 1930, p. 236).[7]

Hence, the role played by axioms consists exclusively in identifying "what might be called a relational structure" (Bernays 1967, p. 497). In this respect, EG is not different from other axiomatic theories that can be presented in a purely relational language, such as the theory of linear orders or lattice theory. But then, what is lost with this kind of presentation of Euclidean geometry is the very natural idea of the possibility of constructing new types of objects starting from other previously given ones. Let us give an example. In EG, lines, or similarly, segments, do not belong to the initially given system of objects, nor are they obtained as the result of an operation performed on points. In fact, their construction is replaced by the addition of an axiom establishing the existence of a point satisfying certain properties with respect to other given points, namely:

$$\forall x \forall y \forall u \forall v \exists z (\beta(x, y, z) \wedge \delta(y, z, u, v)).$$

---

[7]In the same vein, according to Mueller (1981, p. 14), "For Hilbert geometric axioms characterize an existent system of points, straight lines, etc. At no time in the *Grundlagen* is an object brought into existence, constructed. Rather its existence is inferred from the axioms."

More prosaically, we can say that lines are never really traced, but only ideally conceptualized. What the axiom does is to stipulate the existence of a particular point in order to prevent a concept from being empty. A segment l is nothing else but the concept under which two given points a and b fall, as well as every other point z satisfying one of the three relations $\beta(a,b,z)$, $\beta(a,z,b)$, $\beta(z,a,b)$ and being located at a certain (positive) distance from a and b (cf. Tarski and Givant 1999, p. 178). A consequence is that once a and b are given, the extension of (the concept) l contains at least one element different from a and b themselves. Therefore, in purely extensional terms, lines are nothing but sets of points satisfying certain given properties. This means that, in principle, Euclidean geometry can be considered as a branch of set theory: given the set of points, it is sufficient to have a separation axiom in order to define, or, better, to *reduce*, every geometrical object to a particular set of points.[8]

Which kinds of changes in the formal presentation of Euclidean geometry should be carried out in order to avoid this kind of set-theoretical reductionism and restore the intuitive idea that an essential character of Euclidean geometry is to carry out constructions on objects? Or even more generally, how is it possible to give a presentation of Euclidean geometry where the constructive aspects of the theory are detectable at the level of the objects?

8.3.2 *Functions as Constructions*

A first immediate reaction could be to change the signature of the theory by replacing relations with operations on objects. The underlying motivation is that geometrical constructions can be seen to be analogous to the calculation of the solutions of certain equations built up from a set of given primitive functions. This is the idea proposed by Moler and Suppes (1968).[9] Their theory is still one-sorted, but the relational predicates β and δ are replaced with the two functions:

$$S(x,y,u,v) = w \quad I(x,y,u,v) = w.$$

The first function corresponds to the laying out of lines, while the second one is used in order to construct the intersection point of two distinct lines. However, the two functions S and I can be explicitly defined in the language of EG.

---

[8]It is worth noting that what we said is not in contrast with Tarski's claim according to which his presentation of elementary geometry concerns "that part of Euclidean geometry which can be formulated and established without the help of any set theoretical notions" (Tarski 1959, p. 16). Actually, what Tarski wants to say here is simply that his formalization of geometry does not make any appeal to a higher-order language allowing quantification over sets (cf. Tarski 1959, p. 17, Szczerba, L.W. 1986, p. 908). But, evidently, this is a problem concerning the complexity of the language and it does not mean that there is no use at all of set-theoretical machinery as a background theory.

[9]Similar proposals have been developed by Engeler (1968, 1993) and the already mentioned Seeland (1978). For a survey of these positions, see Pambuccian (2008).

Let us consider the case of the first function. The following biconditional

$$S(x, y, u, v) = w \leftrightarrow (\neg(u = v) \vee (u = v \wedge x = y)) \rightarrow (\beta(u, w, v) \wedge \delta(u, w, x, y))$$

represents a correct definition of the function S, since the uniqueness of w can be proved, i.e., since we have that

$$\vdash_{EG} \forall x \forall y \forall u \forall v \exists! w ((\neg(u = v) \vee (u = v \wedge x = y)) \rightarrow (\beta(u, w, v) \wedge \delta(u, w, x, y))).$$

Now, thanks to the soundness and completeness theorem, for every model $\mathfrak{M}$ of EG,

$$\mathfrak{M} \vDash \forall x \forall y \forall u \forall v \exists! w ((\neg(u = v) \vee (u = v \wedge x = y)) \rightarrow (\beta(u, w, v) \wedge \delta(u, w, x, y))).$$

This means that the function S can be explicitly defined in the theory and behaves exactly like a set of 5-tuples of the form $\langle a, b, c, d, e \rangle$, that is, as a subset of the Cartesian product M^5, where M is the universe of the model $\mathfrak{M}$.

The function S is thus a completely extensional object, identifiable with its graph. Therefore, once again, what we are doing is just a mere stipulation of the manner in which the objects of the domain, i.e., the set of points, are mutually related; no new objects are constructed and no essential difference with the Tarskian approach can be established.

8.3.3 The Intuitionistic and Finitistic Proposal

A second proposal comes from a part of the intuitionistic tradition (Heyting 1925) which is strongly influenced by those types of requirements characterizing points C2 and C3. Roughly speaking, the guiding principle standing behind this position consists in the substitution of ideal or infinitely precise concepts with effective and finitely precise ones. In the case of geometry, this proposal becomes specific in the following sense.

A rather natural assumption is to consider two points on a line as two elements of the ordered set of real numbers: in other words, real numbers are generally considered as the elements on which the intended models of geometry are based. But this assumption is far from being harmless, because in order to speak of real numbers, a crucial passage from finitary to infinitary considerations has to be made: real numbers are defined indeed in terms of infinite sets of rational numbers satisfying certain specific properties (e.g., real numbers are represented as Cauchy sequences or Dedekind cuts of rational numbers). However, this step towards forms of infinitary reasoning is not the only problematic aspect. Another difficulty is that some of the properties used to characterize real numbers are proved by making appeal to non-predicative concepts (cf. C4). For instance, the standard proof of the least upper bound principle, i.e., the existence of a least upper bound for every subset S of $\mathbb{R}$, is based on the definition of a particular Dedekind cut X as the intersection of a given set $S^* = \{D_x \mid x \in S\}$ itself constituted by Dedekind cuts of the form $D_x = \{a \mid a \in \mathbb{Q} \land x < a\}$. By the continuity of $\mathbb{R}$, every Dedekind cut corresponds to a real number and, in particular, the real number corresponding to X is the least upper bound of S (Feferman 2005, pp. 597–598). The crucial point is that this purely set-theoretical definition leaves open the possibility for X to already belong to S^*, so that X would be defined in terms of itself, generating a circularity in the definition. In general, axiomatic definitions of sets (like the comprehension principle, the power set operation, or the axiom of choice), as well as impredicative definitions, stipulate in a single stroke the existence of infinite sets of objects without at the same time providing a method for specifying one by one the objects contained in these sets. But even if we change the definition of real numbers, by imposing a control on the generation of the infinite series of rational numbers (e.g., by generating real numbers using choice sequences in Brouwer's sense), or we restrict the operations on sets only to predicative ones (Weyl 1918; Feferman 1964), a problem still remains. Namely, when we assert that it is true that two real numbers are equal, we are asking that we be capable of identifying two infinite series of rational numbers, and, in principle, this task consists of an infinite verification. In particular, suppose a real number a to be defined as an infinite sequence of rational numbers $\{a_i\}_{i \in \mathbb{Z}^+}$ such that for all $m, n \in \mathbb{Z}^+$, $|a_m - a_n| \leq m^{-1} + n^{-1}$. In order to verify that $a = b$ is true, we have to verify that for all $k \in \mathbb{Z}^+$, $|a_k - b_k| \leq 2k^{-1}$ (Bishop 1967, p. 15). But this amounts to testing the property for an infinite number of elements and thus the verification remains purely ideal, never executable *in concreto*.

The intuitionistic solution is to abandon equality as a primitive relation and to replace it with the new primitive relation of *apartness* (Brouwer 1924), which can be defined in the following manner (obtained by rephrasing the definitions present in Bishop 1967, §2.2):

$$a \# b \text{ if and only if there exists } k \in \mathbb{Z}^+, \text{ such that } |a_{2k} - b_{2k}| > k^{-1}. \tag{8.1}$$

According to this definition,[10] if it is true that $a \# b$, then it is possible to recognize this in an algorithmic way: starting with $n = 1$, put $k = n$ and test if the property is satisfied for all $i \leq n$. If this is not the case, put $k = n + 1$ and repeat the procedure until an integer is found that satisfies the property, and then stop. Note that at every step, the number of positive integers for which the property

---

[10]Note that the definition (8.1) cannot be obtained simply via the negation of equality. On the one hand, the negation of a universal sentence cannot be in general intuitionistically transformed into an existential sentence. On the other hand, not even making appeal to classical negation would be sufficient: although the *definiens* of (8.1) entails the classical negation of equality, i.e., the existence of $k \in \mathbb{Z}^+$, such that $|a_k - b_k| > 2k^{-1}$, the converse does not hold (at least using the definitions present in Bishop 1967, §2.2).

has to be tested is a finite one and the operations of subtraction and comparison between rational numbers are both executable in a finite number of steps as well.[11]

Let us now take a closer look at the apartness relation. Axiomatically, it is characterized in the following way:

$$\forall x \neg (x \# x) \qquad \textbf{Irreflexivity}$$
$$\forall x \forall y \forall z (x \# y \rightarrow x \# z \lor y \# z) \qquad \textbf{Split}$$

The first axiom has an immediate reading. The second one is less intuitive. A possible explanation is the following. Suppose we fail to decide whether $y \# z$ (resp. $x \# z$)[12] but we have already verified that $x \# y$ holds, then we are also able to verify that $x \# z$ (resp. $y \# z$) holds.

From a theoretical point of view, the principal interest in the apartness relation results from the fact that an algorithm, acting directly on objects, is built inside one of the primitive relations of the theory. However, this is not yet sufficient in order to fully extend the satisfaction of decidability from the level of sentences to the level of objects. The reason is that apartness is just a *semidecidable* relation, not a decidable one. From the intuitionistic point of view, to recognize that $a \# b$ is false corresponds to possessing a procedure p such that, given any proof of $a \# b$, it entails a contradiction. By definition, this is equivalent to proving that for any $k \in \mathbb{Z}^+$, we possess a procedure p showing that it is impossible for $|a_{2k} - b_{2k}| > k^{-1}$ to hold. But this requires an infinite number of cases to be tested and thus we fall again into the same troubles we came across with equality considered as a primitive relation. For this reason it is quite natural to define equality as the negation of apartness:

$$a = b \equiv_{df} \neg (a \# b) \equiv_{df} a \# b \rightarrow \bot.$$

It is worth noting that the negation of this defined equality is weaker than apartness. The latter implies the former, but in general the converse does not hold: the proof that $a = b$ leads to a contradiction does not *a priori* give any lower bound for $|a - b|$. In particular, Brouwer showed that defining real numbers on the basis of

---

[11] It has to be mentioned that in Tarski (1959, p. 17), among the primitive relations of the language of EG, there appears a relation of "diversity." Unfortunately, Tarski does not explain how this relation should be intended; thus we cannot say if it corresponds to the apartness relation or not. Due to this lack of specification, as well as the fact that none of Tarski's collaborators paid any attention to this diversity relation, we have decided to present EG along the lines of Givant and Tarski (1999), where equality is a primitive relation and diversity is defined simply as its negation.

[12] This failure is not strictly due to some necessary epistemic limitations, as it could be the case that the two points are indeed the same: in this case an ideal precision would be required for the verification, and such capacity is not possessed by human beings. On the other hand, the failure could also be due to some contingent epistemic limitations. For instance, the computation could be so long that, after a certain amount of time, we become too tired to continue it and we thus decide to abandon it.

free choice sequences[13] allows the proof of $\neg \forall x (\neg (x = 0) \to x \# 0)$ and thus to conclude that negation of defined equality and apartness do not coincide (Heyting 1971, pp. 121–122). This result is not surprising because it agrees with the fact that in intuitionistic logic, double negation is not involutive. However, if some more restrictive definitions of real numbers are assumed, in particular those involving the Markov principle, then it can be proved that $\neg (a = b)$ and $a \# b$ are equivalent (Troelstra and van Dalen 1988, pp. 205–206).

Leaving aside these technical considerations, what should actually be retained for our discussion of geometrical constructivity is that the introduction of apartness entails an at least partial satisfaction of C3, and this satisfaction not only concerns the level of sentence (O3) but is also deeply rooted at the level of the objects of the theory (O1). Nonetheless, what is still missing is the possibility of creating new objects starting from already given ones. In other words, the presentation of geometry based on apartness is still a relational one, basically grounded on an axiomatic presentation. This relational character can be well detected in Heyting's presentation of intuitionistic geometry (Heyting 1959), where lines are taken exclusively as primitive objects, without any possibility of constructing them starting from points. In particular, the relations holding between points and lines are guaranteed by the constructive relation of *outsideness*, i.e., ω.[14] In analogy with apartness, a point x lies outside a line u, i.e., $x \omega u$, when there is a positive distance separating these two elements. The axiom describing the behavior of outsideness with respect to apartness is fixed by the constructive axiom of substitution for points:

$$\forall x \forall y \forall u (x \omega u \to x \# y \vee y \omega u) \quad \textbf{Csub}$$

Moreover, in analogy with equality, the incidence of a point on a line, denoted by $\in$, becomes a defined notion[15]:

$$a \in l \equiv_{df} \neg (a \omega l) \equiv_{df} a \omega l \to \bot.$$

---

[13]With the adjective "free," we want simply to point out that the subject generating a sequence can freely decide the degree of restrictions in the selection of each element of the sequence. At one extreme of the range each element is completely determined in advance by a law (see p. 154, *infra*), so that the choice of this element is temporally independent. On the other extreme, the choice is totally unrestricted, in the sense that there is no kind of *a priori* fixed rule guiding the generation of the sequence, so that each element is undetermined in advance but depends on the particular point in time at which the choice is made.

[14]The language used in Heyting's presentation of intuitionistic geometry is thus a two-sorted first-order language. In this work, we will not adopt any particular notation to distinguish between the two sorts of variables for points and lines, respectively; it will be the presence of relations that disambiguates the reading of formulas. However, in general, we will use the letters x, y, z to indicate (eigen)variables and a, b, c to indicate parameters for points, while for lines, u, v will be used to indicate (eigen)variables and l, m the parameters.

[15]The use of the membership symbol is due to the intuitive idea that a line is nothing else but a set of points. As we have seen, this idea is literally followed by the Tarskian approach, while in Heyting it is simply a *façon de parler*; it is in fact contradicted by taking points and lines to be two distinct sets, irreducible one to the other. A much more interesting aspect is that, historically,

To sum up, from the analysis of Tarski's and Heyting's presentations of geometry, we realize that if we do not want to reduce geometry to a mere relational theory, we have to introduce some kind of *constructors* operating on the objects of the theory itself. And since we have decided to work on a purely syntactic formal level, we need to find some syntactic entities that can play the role of such constructors. Now, according to the intuitionistic tradition, and in particular to the BHK explanation of logical connectives, a reasonable candidate for playing such a role are the proofs themselves. The possibility of finding a positive solution to problem P2 is thus a crucial point to analyze, and it constitutes the core of our investigation.

8.4 Constructivity and Proofs

A possible way to attack the problem stated in P2 is to answer the following question: Can EG be considered as constructive in the sense of C1, when evaluated with respect to the class of its theorems? At first sight, the answer seems to be absolutely trivial: having been formalized in a classical logic framework, EG cannot satisfy C1. Nonetheless, trying to point out which of the intuitionistic properties are concretely infringed by EG is far from trivial; moreover, as we will see, this kind of investigation sheds light also on the analysis of C2 and C3.

The first step to be done is to make explicit the conditions used for judging a theory T as intuitionistically acceptable. An intuitive idea could be to list all the proofs of every theorem of T, verify when an appeal to classical principles is made,

---

Heyting (1959) did not consider the outsideness relation as a primitive relation, but as defined from incidence and apartness, namely, $a \, \omega \, l \equiv_{df} \forall x (x \in l \to x \# a)$. However, as was noticed by von Plato (2010b), this definition is too weak from the intuitionistic point of view. If we understand it correctly, this means that when it is imposed to use the previous definition in conjunction with the axiom of substitution for points

$$\forall x \forall y \forall u (x \in u \wedge x = y \to y \in u) \quad \textbf{Sub}$$

then the formula $x \# a$ is no longer intuitionistically derivable from $x \in l$. Actually, what can be derived is $\neg(x = a)$, which is equivalent to $\neg\neg(x \# a)$, and, as we have seen, double negation elimination is not in general admissible in the intuitionistic theory of apartness (cf. pp. 134–135, supra):

$$\cfrac{\cfrac{\textbf{Sub}}{\cfrac{\forall x \forall y \forall u(x \in u \wedge x = y \to y \in u)}{\cfrac{x \in l \wedge x = a \to a \in l}{a \in l}} \forall_E} \quad \cfrac{x \in l \quad [x = a]^{1.}}{x \in l \wedge x = a} \wedge_I}{a \# a} \to_E \quad \cfrac{\cfrac{a \, \omega \, l}{\cfrac{\forall x(x \in l \to x \# a)}{a \in l \to a \# a} \forall_E}{df}}{\to_E} \quad \cfrac{\cfrac{\textbf{Irr}}{\cfrac{\forall x \neg (x \# x)}{\neg a \# a} \forall_E}}{\to_E}$$

$$\cfrac{\cfrac{\bot}{\neg(x=a)} \to_I 1.}{\neg\neg(x \# a)} df$$

and decide if it is eliminable or not. The problem of such a "generate and test" method is its never ending nature: not only is the number of theorems infinite, but also the number of possible proofs for each theorem is *a priori* unbounded. Usually, a more manageable way for testing whether a theory T is genuinely intuitionistic is to verify that T satisfies the two following properties:

Proposition 2 (Disjunction property). *Let $A, B \in Sen_{\mathscr{L}(T)}$. If $\vdash_T A \vee B$, then $\vdash_T A$ or $\vdash_T B$.*

Proposition 3 (Witness property). *Let $\exists x A(x) \in Sen_{\mathscr{L}(T)}$. If $\vdash_T \exists x A(x)$, then $\vdash_T A[t/x]$, for some term $t \in \mathscr{L}(T)$.*

It is quite evident that focusing on these two properties implies a shift of attention to the level of O4. Consider now EG. It is not difficult to see that, thanks to its syntactic completeness, it trivially satisfies the first property. The proof is straightforward. Suppose that $\vdash_{EG} A \vee B$ and $\nvdash_{EG} A$ and $\nvdash_{EG} B$. By syntactic completeness, we have that $\vdash_{EG} \neg A$ and $\vdash_{EG} \neg B$; this implies $\vdash_{EG} \neg A \wedge \neg B$, which is (intuitionistically) equivalent to $\vdash_{EG} \neg(A \vee B)$. Hence, we obtain a contradiction. But EG is a consistent theory (see Friedman 1999); therefore, we can classically conclude that $\vdash_{EG} A$ holds or $\vdash_{EG} B$ holds.[16]

Looking closer at Proposition 2, we can see that it has the form of an admissibility result: if $A \vee B$ is valid (in T), then A is valid (in T) or B is valid (in T). Nonetheless, it is not made explicit how the validity condition is defined. Now, when an intuitionistic point of view is adopted, proofs[17] are the objects that guarantee the validity of the sentences. Thus, an intuitionistically faithful reading of Proposition 2 should be the following: if we possess a proof π of $A \vee B$, then it is possible to extract from π a proof π' of A or a proof π'' of B; this extraction is usually obtained by a manipulation of the proof π, as it could be *detours* elimination (see p. 140, *infra*) or the permutation of the inference rules (see p. 141, *infra*). This situation seems not to be guaranteed by the proof that passes through syntactic completeness, because that proof does not establish any explicit link between the proof of $A \vee B$ and the proofs of A and of B.

---

[16]The last step of the proof corresponds to the *reductio ad absurdum*. This means that when we reason about the properties of the derivability relation in EG, we are reasoning classically. However, this does not yet preclude the possibility of a BHK reading of disjunction in EG. The reason is that some clauses of BHK semantics already induce forms of non-constructive meta-level reasoning (cf. von Plato 2013, p. 103). In particular, interpreting $\bot$ as a proposition which is never realized, i.e., for which no proof can be exhibited, already entails classical forms of meta-level reasoning on the properties of the derivability relation in a purely intuitionistic logical setting. For instance, consider the proof that the 0-ary rule $L\bot$ of **G3i** is valid, i.e., that $\overline{\bot, \Gamma \vdash \Delta}$ is valid. By assuming that $\bot$ is never realized, the antecedent is always false and therefore the sequent is vacuously valid. In other words, assuming that $\bot$ is never realized induces proving the validity of the previous sequent in the same way as the validity of a material implication is classically proved.

[17]By the term "proof," we here indicate a *closed* derivation, that is, a derivation having all the assumptions discharged. When we want to speak of derivations having non-discharged assumptions, we will call them *open* proofs (see p. 152, *infra*).

If we move now to the analysis of Proposition 3, we note that it is not possible to apply the same procedure we used for proving Proposition 2. The reason is that if we suppose $\vdash_{EG} \exists x A(x)$ and $\nvdash_{EG} A(t)$ for all terms t, then there are two possibilities: either t is a variable, let us say y, or it is a closed term. In the first case we cannot appeal to the completeness of EG in order to get $\vdash_{EG} \neg A(y)$ because completeness applies only to sentences, i.e., closed formulas. In the second case, even if we get $\vdash_{EG} \neg A(t)$, we cannot conclude $\vdash_{EG} \forall x \neg A(x)$, for the rule of $\forall$-introduction would be violated.[18]

The failure of the witness property seems to be confirmed by the fact that the proof of the existence of at least three non-collinear points is obtained simply by stating the correspondent axiom,[19] i.e.,

$$\exists x_1 \exists x_2 \exists x_3 \forall u \neg (x_1 \in u \land x_2 \in u \land x_3 \in u) \quad \mathbf{NC}$$

The proof is then a zero-step derivation: no inferential passages are needed, just the appeal to the axiom. A consequence is that there is no way to extract from this proof the witnesses of the three existential quantifiers. We have thus no guarantees that there is indeed (at least) an object that we can exhibit and that falls under the concept expressed by the axiom and stipulated as nonempty. In other terms, if the existence of certain specific objects is stipulated in a categorical way (i.e., without resting on any kind of assumptions), then there is no information that can be exploited in order to extract the desired witnesses. More generally, we are in a situation where what the axiom states is that certain objects are organized in a well-determined configuration, and where the coherence of the axiomatic systems guarantees that this configuration is logically possible and thus *a priori* not empty. But this is not yet sufficient for being guaranteed that this configuration can be effectively realized by some objects to which we have an authentic epistemic access.

---

[18] Some clarifications have to be made at this point. Actually, in the presence of some kind of infinitary rule, it could be possible to conclude $\vdash_{EG} \forall x \neg A(x)$ from the infinite number of premisses $\neg A(t_i)$, where every t_i denotes an object of the intended model of EG. But the use of an infinitary rule would be in conflict with C2. However, the problem de facto does not hold because the language of EG does not contain any kind of individual constants or function operators; therefore, it is not possible to form any closed term in EG. The second case analyzed is thus only a purely theoretical exercise, since it would be excluded from the very nature of the language of EG.

[19] Although a different language is used, namely, the one introduced in Sect. 8.3.3, our presentation of the axiom of non-collinearity closely follows that of the *lower-dimensional axiom* adopted by Tarski in EG, i.e.,

$$\exists x_1 \exists x_2 \exists x_3 (\neg \beta(x_1 x_2 x_3) \land \neg \beta(x_2 x_3 x_1) \land \neg \beta(x_2 x_1 x_2)).$$

It should be noted that sometimes this axiom is formulated by specifying that the three points are different, that is, by adding to the propositional matrix the conjunction $\bigwedge_{1 \leqslant i,j \leqslant 3} x_i \neq x_j$. However, this condition is not really necessary for our analysis. It is thus dropped in order to simplify the use of this axiom and **NC** as well.

More generally, we come up here against the distinction between proving that certain objects enjoy some specific properties and proving that this set of objects is actually not empty. More specifically, the problem is that even if a set of definitions for these objects is available, we still have no guarantee that they can be effectively reached. As Proclus states, geometry not only asks the question "What sort of thing is this object?" but also "Does the object exists as defined?" (Proclus 1970, p. 158). The answer to the first one is given by the proof of what Proclus calls a *theorem*, i.e., the proof that starting from the definition of this object, some of its intrinsic but non-evident properties can be inferred. The answer to the second one is given, instead, by the solution to what Proclus calls a *problem*, i.e., the setting of a procedure that allows to exhibit the object defined, once certain conditions are satisfied (Proclus 1970, p. 66). The solution of a problem is thus a way to guarantee that what is expressed in the consequences of a theory remains not only logically possible but also epistemically accessible.

8.4.1 Formalization of Geometrical Problems

The idea that geometrical problems correspond to the request for an effective existence proof of certain objects, once other objects are supposed to be given, has induced identifying their logical structure with that of Π_2 sentences (cf. Mäenpää and von Plato 1993; Avigad et al. 2009). More precisely, in the language of type theory, a problem is a sentence of the form[20]:

$$\forall x : S. \exists y : T. A(x, y).$$

The solution of a problem consists in the exhibition of an object of type T, once other objects of type S and a fixed set of construction procedures are given. This idea is in complete accordance with the intuitionistic reading of Π_2 sentences provided by the BHK explanation of quantifiers; in particular, it is very close to Kolmogorov's reading (Kolmogorov 1932, p. 328):

> A solution of $\forall x : S. \exists y : T. A(x, y)$ consists in the possession of an effective procedure p of type $S \to T$, such that for any object a of type S, it gives both an object $p(a)$ of type T and solution of $A(a, p(a))$.

A full comprehension of this given reading depends on the possibility of making explicit what the effective procedure p is. Only when this procedure is clearly identified can we constructivize the existential quantifier and extract the desired witness.

---

[20]For simplicity, we limit ourselves here to the case in which only one variable is universally quantified and no type dependencies are involved. We will see later some more specific examples which need an appeal to dependent types.

8.4.2 A Parallel with Arithmetic

In order to better understand what it means to constructivize the existential quantifiers of Π_2 sentences, let us concentrate on arithmetic. In particular, we will show how in this context the role of effective procedures is played by proofs, in conjunction with some algorithmic manipulations of their structure.

A first peculiar aspect of arithmetic is that a certain class of theorems, valid in a classical setting, is not altered when the logical framework is changed into an intuitionistic one. This class of theorems is exactly that of the Π_2 theorems.

Theorem 1 (Kreisel–Friedman). *Let A be a primitive recursive formula. If* $\vdash_{PA} \forall x \exists y A(x, y)$, *then* $\vdash_{HA} \forall x \exists y A(x, y)$.

This means that classical arithmetic, PA, is a conservative extension of intuitionistic arithmetic, HA, with respect to Π_2 sentences having a decidable propositional matrix A.[21]

A second fundamental aspect is that in HA, the validity of Proposition 3 is preserved.

Theorem 2. *Let* $\exists x A(x) \in Sen_{\mathcal{L}(HA)}$. *If* $\vdash_{HA} \exists x A(x)$, *then* $\vdash_{HA} A(n)$, *for some numeral n.*

What we are interested in is the way in which Theorem 2 is proved. One possibility is to proceed in three steps (Prawitz 1971, pp. 264–266; von Plato 2006):

(i) Transform the axioms of HA into an (provably) equivalent set of natural deduction inference rules, HA^R for short.
(ii) Prove the normalization theorem for HA^R.
(iii) Prove the last rule lemma[22] for HA^R.

For the presentation of a system of inference rules for arithmetic, we refer to von Plato (2006, p. 160). Actually, what is particularly relevant for our discussion is point ii. In a purely logical system, the normalization of a proof consists in the elimination of those formulas that are first introduced and subsequently eliminated. Two cases are possible. The first case is when the application of the elimination rule immediately follows the application of the introduction rule; we call this sequence

---

[21] The decidability of the propositional matrix A is a direct consequence of the fact that A is a primitive recursive formula, that is, a formula constituted by relations which are themselves primitive recursive. And if a relation is primitive recursive, then its characteristic function is a primitive recursive function, that is, a function computable in a mechanical way.

[22] The expression "last rule lemma" (or "last rule property") comes from the French word *lemme de la dernière règle* (or *propriété de la dernière règle*) that we have learned from G. Dowek's lectures on proof theory, but this terminology is not very standard in the proof-theoretical literature. An alternative formulation could be "introduction form property," as can be found in Schroeder-Heister (2014, §1.3). However, neither does this expression have a large diffusion. See below for the explanation of this property.

of rules a *detour*.²³ The second case is when the introduced formula is the minor premiss of a ∨-elimination ($\vee_E$) or of an ∃-elimination rule ($\exists_E$), then remains the same throughout a sequence of applications of these rules, and finally is eliminated (cf. Troelstra and Schwichtenberg 2000, p. 178). Actually, this latter case can be reduced to the first one modulo upward permutations of the elimination rule with respect to $\vee_E$ and $\exists_E$. But when we consider HA^R instead of a purely logical system, we also have to take into consideration new permutations that can be generated by arithmetical rules. For instance, the induction rule *Ind* may produce new types of permutations, namely,

$$
\dfrac{A[0/x] \quad \dfrac{[A[y/x]]^{1.}}{\vdots} \quad \dfrac{[A[t/x]]^{2.}}{\vdots}}{\dfrac{C}{E} \quad Ind\ 1.,2.} \quad \dfrac{\vdots}{D} \circ_E
$$

converts to

$$
\dfrac{A[0/x] \quad \dfrac{[A[y/x]]^{1.}}{\vdots} \quad \dfrac{\dfrac{[A[t/x]]^{2.}}{\vdots}\ \dfrac{\vdots}{D}}{E}\circ_E}{E}\ Ind\ 1.,2.
$$

where $\circ$ is an arbitrary logical connective, y is an eigenvariable not free in any assumptions of the derivation, and t is any term.

A fundamental aspect is that the operations of elimination of detours and of permutation conversion for HA^R can be performed algorithmically. Proving normalization consists then in providing an algorithm that converts every non-normal proof of HA^R into a normal one, that is, a proof where each major premiss of an elimination rule is either an assumption or a conclusion of an application of an elimination rule different from $\vee_E$, $\exists_E$ and *Ind*.

As to point iii, the last rule lemma can be proved as a corollary of the normalization theorem. This lemma affirms that if a certain non-atomic sentence A is provable, then there exists a derivation of A, the last rule of which is an instance of the introduction rule of the principal connective of A. This lemma is extremely powerful: when it applies, it allows transforming any proof of A into a direct proof of A (cf. C2).²⁴

---

[23] Here we are working in a natural deduction setting. When natural deduction proofs are translated into a sequent calculus setting, then the status of detours becomes clearer: the use of detours corresponds to cuts, and cuts correspond to the use of lemmas. A proof without detours is thus a proof where every inferential passage has been made completely explicit.

[24] The claim that every theory, in order to be meaningful, has to satisfy this property is known under the name of *Dummett's fundamental assumption*.

Suppose now that $\forall x \exists y A(x, y)$ is derivable in HA^R. By the normalization theorem, there exists a normal proof π of $\forall x \exists y A(x, y)$ such that, according to the last rule lemma, it terminates with a $\forall_I$ rule. This proof π can be seen as a program having $\forall x \exists y A(x, y)$ as specification and such that when an input a is given via an application of $\forall_E$, then the normalization algorithm provides a procedure for computing the value of the variable y and eventually producing an output b. Graphically, we have that

$$\begin{array}{c} \pi \\ \vdots \\ \dfrac{\forall x \exists y A(x, y)}{\exists y A(\underbrace{a}_{Input}, y)} \forall_E \end{array} \quad \xrightarrow{Normalization} \quad \begin{array}{c} \pi' \\ \vdots \\ \dfrac{A(a, \overbrace{b}^{Output})}{\exists y A(a, y)} \exists_I \end{array}$$

For instance, a proof of $\forall x \exists y (x = 2y \vee x = 2y + 1)$, when combined with the normalization algorithm, corresponds to a program that divides any given natural number by two.

To sum up, when Theorems 1 and 2 are put together, the following result holds: if a PA sentence of the form $\forall x \exists y A(x, y)$ with A primitive recursive has been proved classically, and the universal quantifier has been afterwards particularized, it is possible to algorithmically extract the value of y using only intuitionistic passages.[25] In other words, certain classical sentences can be constructivized in the sense of C3 as well as in that of C1 and C2, since the value of y is obtained by providing an algorithm that finds the direct intuitionistic proof of $\exists y A(a, y)$. Moreover, although this algorithm acts primarily at the level of O4 (being applied to Π_2 sentences), its execution concerns also the level of O2: given objects of a certain type as inputs, it transforms them into objects of another type as outputs. If it were possible to have such a result for geometrical theories, then we could have a syntactic formal manner of expressing the idea that geometry effectively constructs its objects.

---

[25]It is worth noting that if the propositional matrix A of a Π_2 sentence is not primitive recursive, the procedure we have just described cannot be carried out. Consider, for instance, the sentence $\forall x \exists y ((y = 0 \wedge B(x)) \vee (y = 1 \wedge \neg B(x)))$, where B is an undecidable predicate. It can be shown that this sentence could be derivable in PA, without being derivable in HA. In effect, if it was derivable in HA, we could instantiate the universal quantifier with a numeral n, eliminate all possible detours in the proof, and obtain then a witness t for the existential quantifier, so that

$$\vdash_{HA^R} (t = 0 \wedge B(n)) \vee (t = 1 \wedge \neg B(n)).$$

Since this last sentence is a closed one, and the proof does not contain open assumptions, we could then apply Proposition 2 and obtain one of the two disjoints. But this is impossible, because it amounts to rendering B decidable.

A crucial step in this direction seems to be the shift from the level of mere derivability to that of proofs. This shift entails also a move from a purely extensional setting to an intensional one: modulo normalization, i.e., modulo the algorithmic dynamics with which proofs are equipped, two different proofs of the same sentence represent two different manners of achieving the same result. More precisely, two non-normal proofs of the same sentence, not reducing to the same normal form, represent two programs taking the same inputs and giving back the same outputs but computing in different ways. This aspect is extremely important because it allows capturing, at a purely syntactic level, the intuitive idea that there exist different manners of constructing the same object starting from the same data and using the same instruments.

8.5 The Case of Geometry: A Troublesome Situation

For simplicity, let us abandon here the theory EG and restrict ourselves to a very elementary part of Euclidean geometry, namely, *incidence geometry*. In a nutshell, we will study the relationship between points and lines only with respect to the incidence relation, leaving aside the affine property of betweenness and the metrical one of equidistance (for more details, see Greenberg 1993, p. 50). The language considered will be composed of two sorts of variables, one for points and another for lines, and one binary multisorted relation of incidence between points and lines, denoted by $\in$. The logical framework is classical logic.

Consider the following Π_2 sentence:

$$\forall u \exists x \neg (x \in u). \tag{8.2}$$

For brevity, we will use the infix notation for the negation of the membership relation, i.e., write $\forall u \exists x (x \notin u)$. Its informal reading is that for every line, there exists a point lying outside it. According to what we said in the previous section, our expectation is that, given a certain line l, there exists an effective procedure for constructing, or at least exhibiting, a point c lying outside l. In particular, we expect that this procedure can be given by a mechanical manipulation of proofs.

Usually, the proof of (8.2) rests on the non-collinearity axiom we mentioned before. Consider the derivation

$$\cfrac{\cfrac{\cfrac{\text{NC}}{\exists x_1 \exists x_2 \exists x_3 \forall u \neg (x_1 \in u \land x_2 \in u \land x_3 \in u)}}{\exists x_1 \exists x_2 \exists x_3 \forall u (x_1 \notin u \lor x_2 \notin u \lor x_3 \notin u)} DM \quad \cfrac{\cfrac{[\forall u(x_1 \notin u \lor x_2 \notin u \lor x_3 \notin u)]^4 \quad [x_1 \notin l]^{1\cdot}}{x_1 \notin l \lor x_2 \notin l \lor x_3 \notin l} \quad \cfrac{[x_2 \notin l]^{2\cdot}}{\exists x(x \notin l)}^{\exists_I} \quad \cfrac{[x_3 \notin l]^{3\cdot}}{\exists x(x \notin l)}^{\exists_I}}{\exists x(x \notin l)}^{\exists_I}}{\exists x(x \notin l)}^{\exists_E 4.}$$

The terminating step is $\exists_E$ and not the desired rule $\exists_I$. Actually, $\exists_I$ is used in the proof, but there is no way to let it commute downward at the end of the proof, because the presence of axioms prevents the permutation of the order of application of the rules. Hence, it is not possible to have a direct proof of $\exists x(x \notin l)$. Once again, the presence of axioms seems to oppose a constructive account of geometrical practice. Nevertheless, we could still try to constructivize the existential quantifier by proceeding in an analogous way as we did for arithmetic. Two preliminary steps are then necessary.

1. The transformation of the classical theory into an intuitionistic one.[26]
2. The transformation of the axioms into inference rules.

It is worth noting that the first step is crucial if we want to obtain the same initial conditions used in the formulation of the Kreisel–Friedman theorem. In this theorem, the propositional matrix of Π_2 sentences is decidable. In geometry, instead, if the working space is that of the real numbers, it is unusual to work with decidable relations. The best we can do is to work with semidecidable relations, as intuitionistic relations are. Thus, the type of Π_2 sentences that we are going to consider are those whose propositional matrices do not contain negation. In this manner, when these sentences are provable, their propositional matrix can always be verified. Let us focus on **NC**. In order to transform it into an intuitionistically admissible axiom, we could follow what we said in Sect. 8.3.3 and replace the primitive incidence relation with that of outsideness, in order to obtain the following:

$$\exists x_1 \exists x_2 \exists x_3 \forall u (x_1 \, \omega \, u \vee x_2 \, \omega \, u \vee x_3 \, \omega \, u) \quad \textbf{CNC}$$

However, replacing **NC** with **CNC** does not induce any substantial change: with respect to the previous proof, nothing happens but the avoidance of the classical rule DM, which corresponds to one of De Morgan's laws, namely, $\neg(A \wedge B) \rightarrow (\neg A \vee \neg B)$.

Much more interesting is the attempt to transform **CNC** into an inference rule. A first idea could be to appeal to the method proposed by Negri and von Plato (2001, p. 209; 2011, §2.3) which allows transforming, in a deterministic and mechanical way, certain classes of axioms into sets of provably equivalent rules of inference. For example, if the propositional matrix of a universal axiom can be decomposed into a conjunction of sentences of the form

$$P_1 \wedge \ldots \wedge P_m \rightarrow Q$$

---

[26]For the sake of the argument, we assume, without actually proving it, that the transformation of the classical theory into the intuitionistic one is sound at least with respect to Π_2 sentences: if A is a Π_2 sentence formulated in the language of the classical theory T and provable in T, then its intuitionistic translation $\text{I}(A)$ is provable in $\text{I}(T)$ as well. For further details about the relation between classical theories and their corresponding intuitionistic ones, see Negri and von Plato (2005).

8 Constructibility and Geometry

where P_i and Q are atomic formulas, then each of the conjuncts can be transformed into the rule-scheme:

$$\frac{P_1 \ \ldots \ P_m}{Q} \ Reg$$

In this manner, only inference rules acting on atomic formulas are generated, so that their adjunction to the standard set of logical rules remains compatible with the normalization procedure defined for the latter. Moreover, properly speaking, these inference rules are neither introduction nor elimination rules,[27] thus no new detours are created (cf. Troelstra and Schwichtenberg 2000, pp. 197–198). More precisely, it is possible to show that when natural deduction rules are translated into the sequent calculus, no sequence of *Reg* rules is translated by making appeal to the Cut rule. The idea is that *Reg* rules are translated either as right rules of the form

$$\frac{\Gamma \vdash P_1 \ \ldots \ \Gamma \vdash P_m}{\Gamma \vdash Q}$$

or as left rules of the form

$$\frac{\Gamma, Q \vdash C}{\Gamma, P_1, \ldots, P_m \vdash C}$$

but not into both of them, so that no cuts between two *Reg* rules can be created.

Obviously, the problem is that **CNC** is not a universal axiom; hence, the transformation cannot be applied. Should we then abandon the idea of constructivizing geometrical problems along the lines of what we have done for arithmetic, or are there other methods for getting out of this impasse?

8.5.1 A Partial Solution: Skolemization

A first solution could be to perform a Skolemization on the axioms containing existential quantifiers. Suppose we do this for **CNC**. Roughly speaking, the idea is to substitute the existential quantifiers appearing in head position with individual constants. We thus obtain[28]:

$$\forall u (c_1 \ \omega \ u \lor c_2 \ \omega \ u \lor c_3 \ \omega \ u).$$

---

[27]Using a terminology borrowed from Skolem, we could say that these rules are just *principles of derivation* (*Erzeugungsprinzipien*; cf. von Plato 2007, p. 199) representing specific kinds of mathematical reasoning.

[28]Note that here we are reasoning in a purely hypothetical way. Skolemizing inside an intuitionistic setting is far from being trivial, in particular, because the derivability relation could be affected (as is the case for intuitionistic logic with equality; cf. Mints 2000). However, reasoning as if this

But the form of this new axiom is still not fit for the application of the aforementioned transformation, and thus it cannot be converted into an inference rule. Nonetheless, its propositional matrix is in disjunctive form, and a finite sequence of disjunctions is nothing else but an existential operator ranging over a finite domain. With an abuse of notation, we could rewrite the previous sentence as

$$\forall u \exists y \in \{c_1, c_2, c_3\}(y \, \omega \, u)$$

and, Skolemizing again, obtain

$$\forall u (f(u) \, \omega \, u)$$

where $f(u) \in \{c_1, c_2, c_3\}$. The propositional matrix of this new sentence is in the desired form

$$P_1 \wedge \ldots \wedge P_m \to Q$$

with $m = 0$ and $Q \equiv f(u) \, \omega \, u$. After the application of the transformation, we obtain the 0-ary rule

$$\frac{}{f(u) \, \omega \, u} \, Coll$$

and the proof of $\forall u \exists y (y \, \omega \, u)$ becomes

$$\cfrac{\cfrac{\cfrac{}{f(u) \, \omega \, u} \, Coll}{\exists y (y \, \omega \, u)} \, \exists_I}{\forall u \exists y (y \, \omega \, u)} \, \forall_I$$

Suppose we now proceed as in arithmetic, using this proof and the normalization procedure in order to extract the witness of the existential quantifier. Let a line l be the input for the algorithm. We thus obtain

$$\cfrac{\cfrac{\cfrac{\cfrac{}{f(u) \, \omega \, u} \, Coll}{\exists y (y \, \omega \, u)} \, \exists_I}{\forall u \exists y (y \, \omega \, u)} \, \forall_I}{\exists y (y \, \omega \, l)} \, \forall_E \quad \overset{Normalization}{\rightsquigarrow \ldots \rightsquigarrow} \quad \cfrac{\cfrac{}{f(u) \, \omega \, l} \, Coll}{\exists y (y \, \omega \, l)} \, \exists_I$$

---

kind of Skolemization was not only possible, but even free from pernicious consequences, is not problematic for our argumentative goals: what we want to show is just that even if it were possible to accomplish this kind of Skolemization, we would still not be able to reach the desired result, namely, to exploit normalization for constructing geometrical objects.

The existential sentence has been finally inferred using the introduction rule for the existential quantifier: its proof is a direct one. Nonetheless, we cannot consider ourselves fully satisfied by this proof, since it remains uninformative about the object which it is supposed to have constructed. Even if we can extract the witness for the existential quantifier, namely, $f(u)$, we still do not know which point, c_1, c_2, or c_3, lies outside the line l. In other words, the proof gives us the witness only in the form of a syntactic term, but it does not provide a way to identify the object corresponding to this term, because the Skolemization procedure is non-constructive: it gives us no information about the behavior of f, which remains instead a sort of black box. It is only when we provide a model of the theory that the function f takes a well-determined interpretation and a specific value for the argument u, eventually selecting one and only one of the three possible points. Hence, it is only in the presence of a semantical interpretation that the witness can be really exhibited.[29] However, the appeal to a model seems to be a solution vitiated by circularity: it is possible to exhibit a witness because, by possessing the model, we implicitly already possess it. The function f simply serves to denote this object in the model, but the problem of justifying how we got in possession of the model itself still remains open. An immediate consequence is that the possession of a direct proof is not a sufficient condition for the construction of certain objects.

8.5.2 An Alternative Solution: Primitive Object Constructors

Properly speaking, appealing to Skolem functions is a kind of solution based on the enrichment of the language in order to create a new theory which is "nearly the same" as the initial one, namely, a conservative extension of the initial one. As we have already seen, the problem is that when Skolem functions are introduced, we do not know *a priori* whether they correspond to computable functions or not. Thus, there are no guarantees that extracting the values of the function could be done in an algorithmic way. The appeal to a semantic interpretation, i.e., the exhibition of a model, is not a good solution either, because usually there is no way to generate a model in an effective way, unless we restrict the analysis to finitary models (cf. Van Bendegem 2010). We will discuss this point in Sect. 8.5.3. For the time being, if we want to avoid both Skolemization and the appeal to any semantic interpretations, we should find an alternative way to constructivize not only the existential quantifiers of **CNC**, but also the disjunctions present in its propositional matrix. In order to achieve this, it seems that the only viable way consists in the reformulation of the

---

[29]In fact, it could be objected that the witness can be obtained by executing the algorithm inscribed in the relation ω. This algorithm could give as a result a certain c_i ($1 \leqslant i \leqslant 3$) such that $c_i \, \omega \, l$, and thus, it would be possible to conclude $\exists y (y \, \omega \, l)$ by a $\exists_I$. The problem is that, taken as syntactic objects, individual constants do not possess any kind of structure; hence, the algorithm inscribed in ω cannot be used, since there is no data (structured) on which to operate. Once again it seems necessary to interpret constants on numerical values.

axiom **CNC**, so as to avoid any appeal to disjunction. Again, a way to do this is to change the language of the theory. In particular, we could enrich the language by adding a line constructor ln. More precisely, ln is a primitive object constructor operating in the following manner: given two distinct objects x and y of type "point" (Pt), it transforms them into a new object of type "line" ($Line$). Formally speaking, this means that the typing rule governing ln is

$$\frac{x : Pt, y : Pt \quad \pi : x \# y}{ln(x, y) : Line} \qquad (8.3)$$

where π is an object standing for a proof establishing that x and y are distinct. The other inference rules governing the behavior of ln have to assure the derivability of $a \in ln(a, b)$ and $b \in ln(a, b)$ for any a and b.

At first sight, ln behaves exactly like a function taking points as arguments and giving back lines as values. But in fact it is definitely not a function in the usual extensional and set-theoretical sense: ln cannot be represented as a set of ordered pairs of points. On the contrary, ln behaves like a program, in the sense that it is an intensional object. Let us explain what this means through an example. Suppose that four distinct points are given: let us say a, b, c, and d. Suppose we have established that $a \in ln(c, d)$ and $b \in ln(c, d)$. Then, thanks to the uniqueness axiom (von Plato 2010a, p. 146)

$$\forall x \forall y \forall u \forall v (x \in u \land x \in v \land y \in u \land y \in v \to x = y \lor u = v)$$

we can conclude that $ln(a, b) = ln(c, d)$. But this does not mean that $ln(a, b)$ and $ln(c, d)$ are identical in the sense that they are the same line. On the contrary, it just means that they lie in the same place. In other words, even if they occupy the same position on the plane, $ln(a, b)$ and $ln(c, d)$ are still different because they have been traced in a different way: drawing a line passing through a and b is an intensional act completely different from drawing the line passing through c and d.[30] Notice that a similar situation occurs in computability theory: it can be the case that two programs take the same arguments and give back the same values, but do it according to different algorithmic strategies.

More generally, the idea is that ln simulates a human geometer in the act of drawing a line through two distinct points in the same way as Turing's human computor can be simulated by a (closed) λ-term in the act of calculating the value of a given function; in particular, notice that λ-terms are just abstract representations of programs, and, like the latter, they are intensional objects par excellence (cf. Hindley and Seldin 2008, p. 76).

---

[30]More prosaically, the idea is that ln can be seen as a non-injective way to map points to lines, in the sense that there can be different ways to get the same line as a certain particular position on the plane.

Another essential feature of In is that its introduction into the language of the theory imposes a change of the logical order of the theory itself. More precisely, we pass from a theory expressed in a multisorted first-order language to a theory expressed in a typed language, where the types not only depend on the objects but also on the other types.[31] For instance, the type of In is $\Pi x : Pt.\Pi y : Pt.(x \# y \to Line)$; this type not only depends on the two chosen objects of type Pt, call them a and b, but also on the capacity of establishing that $a \# b$ is an inhabited type.

Now, thanks to In, we can reformulate **CNC** without using any disjunctions (Negri 2003, p. 399; von Plato 2007, p. 220):

$$\exists x \exists y \exists z (x \# y \land z \omega \, ln(x, y)) \quad \textbf{ET}$$

The form of **ET** leads us to devise its corresponding rule along the lines of the elimination rule for the existential quantifier, with assumptions containing eigenvariables. More precisely, we can follow Negri (2003) and Negri and von Plato (2011, §8) and transform axioms of the form

$$\forall \vec{x}(P_1 \land \ldots \land P_m \to \exists \vec{y}(Q_1 \land \ldots \land Q_s))$$

where none of the variables in the vector $\vec{y}$ are free in the P_i, into the rule-scheme

$$\cfrac{P_1 \ldots P_m \quad \begin{array}{c} [Q_1[\vec{z}/\vec{y}], \ldots, Q_s[\vec{z}/\vec{y}]]^{1.} \\ \vdots \\ C \end{array}}{C} \text{ GR 1.}$$

where the variables in the vector $\vec{y}$ have been replaced by the eigenvariables $\vec{z}$ not occurring free in C nor in any other assumptions. In this manner the axiom **ET** can be transformed into the rule

$$\cfrac{\begin{array}{c} [x \# y, z \omega \, ln(x, y)]^{1.} \\ \vdots \\ C \end{array}}{C} \text{ ET 1.}$$

where the eigenvariables x, y, and z are not free in C nor in any other assumptions.

In analogy with what we have just done with the rule GR, an axiom of the form

$$\forall \vec{x}(P_1 \land \ldots \land P_m \to Q_1 \lor \ldots \lor Q_n)$$

---

[31] In Barendregt's cube, which is a way of classifying theories in terms of their type dependencies, such a theory would be considered as belonging to the system $\lambda P\underline{\omega}$. Cf. Hindley and Seldin (2008, pp. 194, 200).

can itself be transformed into the rule-scheme (Negri and von Plato 2001, p. 209)

$$
\begin{array}{ccc}
& [Q_1]^{1.} & [Q_n]^{n.} \\
& \vdots & \vdots \\
P_1 \ldots P_m & \dot{C} & \cdots \quad \dot{C} \\
\hline
& C &
\end{array} \; GR^* \; 1.,\ldots,n.
$$

For instance, the constructive axiom of substitution for lines

$$\forall x \forall u \forall v (x \, \omega \, u \to u \, \# \, v \vee x \, \omega \, v) \quad \textbf{Csub}$$

is transformed into the rule

$$
\begin{array}{ccc}
& [u \# v]^{1.} & [x \, \omega \, v]^{2.} \\
& \vdots & \vdots \\
x \, \omega \, u & \dot{C} & \dot{C} \\
\hline
& C &
\end{array} \; Csub \; 1., 2.
$$

Note that a GR^* rule-scheme can be used in place of *Reg*, the latter being just a particular instance of the former, namely, when $n = 1$ and $C \equiv Q_n$.

As *Reg* rules, neither GR nor GR^* rules create new types of detours; nevertheless, they induce new permutative conversions. For instance, given a GR rule and an elimination rule for a certain logical connective ∘, we have that

$$
\cfrac{\cfrac{P_1 \ldots P_m \quad \cfrac{[Q]^{1.}}{\vdots} }{C} \; GR \, 1. \quad \cfrac{\vdots}{D}}{E} \circ_E \quad \underset{\leadsto \ldots \leadsto}{\text{converts to}} \quad \cfrac{P_1 \ldots P_m \quad \cfrac{\cfrac{[Q]^{1.}}{\vdots}{C} \quad \cfrac{\vdots}{D}}{E} \circ_E}{E} \; GR \, 1.
$$

Normalization for logical systems extended with GR and GR^* rules still holds but has to be adapted in order to take into consideration these new kinds of permutative conversions.

Exploiting the new rules, the proof of the existence of a point lying outside a given line l becomes

$$
\cfrac{z \, \omega \, ln(x,y) \quad [z \, \omega \, l]^{2.} \quad \cfrac{[ln(x,y) \# l]^{1.} \quad \cfrac{ln(x,y) = l}{\bot} \to_E}{\cfrac{z \, \omega \, l}{z \, \omega \, l} \; efq}}{z \, \omega \, l} \; Csub \; 1., 2.
$$

It is now possible to apply $\exists_I$, so to obtain $\exists w (w \, \omega \, u)$, and successively add the hypothesis $x \# y$ by weakening. But this does not yet allow applying ET, because the variable condition is not respected: x and y are free in the hypothesis

8 Constructibility and Geometry

$ln(x, y) = l$. Moreover, the term playing the role of the witness is a variable and thus we have no relevant information for identifying its corresponding object. Finally, the proof remains hypothetical: it depends on the hypothesis that l lies on the same place as the line passing through two distinct points x and y and not incident with a third point z.

Certainly, we could think of a more favorable situation in which $ln(x, y) = l$ depends on $x \# y$, namely,

$$\cfrac{z\omega\, ln(x,y) \quad \cfrac{[z\omega\, l]^{2\cdot} \quad \cfrac{[ln(x,y)\#l]^{1\cdot} \quad ln(x,y) = l}{\cfrac{\bot}{z\omega\, l}\ efq}\ \to_E}{Csub\ 1.,2.}}{z\omega\, l} \quad \begin{array}{c} x\#y \\ \vdots \\ \pi \\ \vdots \end{array}$$

As before, we can continue the proof by applying $\exists_I$. And, unlike the previous case, we are now entitled to apply the rule ET, so that the derivation can be closed. However, in this manner it would no longer be a direct proof, because it would not end with an introduction rule for the principal operator of the conclusion, i.e., $\exists_I$, but with the rule ET. Moreover, even if all the assumptions have been discharged, the derivation still remains hypothetical: it depends on the hypothesis of having a proof π of $ln(x, y) = l$ from $x \# y$, and since $ln(x, y) = l$ is not a theorem, this will depend in turn on the particular choice of x, y, and l.

In both cases, the possibility of closing the two derivations, and eventually finding the object corresponding to the witness of the existential quantifier, seems to depend on the possibility of having available certain objects possessing very specific properties. Evidently, the exhibition of a model could provide these objects and lead to the identification of the desired witness. In particular, once the model is given, there is no longer the need to appeal to ET, because all the assumptions of the proof would be satisfied; thus, the proof would be considered as closed and terminating with $\exists_I$. Moreover, all the free variables would be assigned to objects of the model, so that the witness could be exhibited: it would consist of the object corresponding to the interpretation of the variable z. However, this model cannot be a standard relational set-theoretical one. First of all, in the intuitionistic type theory that we used to present geometry, use is made of terms corresponding to proofs; therefore, the model has to contain objects that codify proofs. Secondly, the interpretation of ln cannot be a set of couples of objects; otherwise, ln would be interpreted as the graph of a function and its intensional character would be lost.[32] Nevertheless, even if such a model was at our disposal, those problems of circular explanation

---

[32]To be more precise, we should say that the model contemplated here is something like a realizability model.

that we already mentioned in the previous section would still remain open: the two given proofs operate as guides in the identification of the witness, but in fact, the witness was already there in the model and thus we cannot properly say that we have constructed it by the exclusive means of the proofs themselves. If we are not able to explain how the model has been given, or better, if we cannot provide an effective method for generating it, then the pure syntactic manipulations of (formal) proofs will not be sufficient to capture the notion of geometrical constructivity. But is this negative conclusion the only way to interpret the previous results?

We claim that this is not the case: in particular, by liberalizing the notion of proof considered hitherto, an alternative positive interpretation is possible. More precisely, if instead of working exclusively with closed proofs, we also accept *open* proofs, i.e., proofs having non-discharged assumptions, then the usual normalization procedures still apply.[33] However, if we do not want to work with a completely trivial notion of proof, we have to impose some limitations on the kind of open assumptions that can be allowed. Now, since we have reduced geometrical operations to syntactic rules acting on atomic sentences, it would be reasonable to limit open assumptions to atomic sentences and, possibly, to their duals, i.e., to the negations of atomic sentences. Under these conditions, it could be still possible to obtain open direct proofs of existential sentences. In particular, suppose we have a closed proof of an existential sentence terminating with a GR rule, e.g., ET. If we avoid the application of the GR rule, then some atomic premises would remain undischarged, but, according to what we have just said, we would still possess a proof. Moreover, when there are no other GR rules appearing in the proof, it is possible to transform this open proof into a proof terminating with $\exists_I$. The fundamental idea is to show that GR^* rules do not represent an obstacle to this enterprise. In particular, if all the occurrences of the premiss C in a GR^* rule coincide with an existential sentence obtained by the same instances of $\exists_I$,[34] then it is always possible to permute this introduction rule downward with respect to the GR^* rule, namely,

---

[33] Roughly speaking, the idea is that the reduction steps for detours involving logical connectives are defined on open proofs and not closed ones (cf. Schroeder-Heister 2006, §§3.1, 3.2). Moreover, as was said on p. 145, no new detours are created by mathematical rules. It remains to analyze the permutations between logical and mathematical rules, but we will see this on p. 150. It should be noticed that putting open proofs on the same level as closed ones, at least with respect to their semantic role, is far from trivial. Usually, in proof-theoretical semantics, the priority is given to closed proofs because the semantic key concept is that of validity, rather than that of computation, as it is in the case under analysis (cf. Schroeder-Heister 2006, §3.3; Schroeder-Heister 2014, §§2.2.2, 2.2.3).

[34] In order to obtain this kind of situation, some non-trivial work might be required. The main problem is that the term appearing as a premiss of the $\exists_I$ rules may not always be the same; in particular there could be a term t_i for each application of $\exists_I$. However, if a subterm property can be proved (cf. von Plato 2010a, §3), then each term t_i would appear in Q_i or in some other assumptions of the subtree concluding with $A[t_i/x]$. Thus, by means of appropriate substitutions, it would be possible to identify all the t_i. Note that this does not create captures of eigenvariables, because by hypothesis there are no more GR rules. In some special cases, it would be simply sufficient to add a sequence of equalities $t_i = t_j$ in the assumptions.

8 Constructibility and Geometry

$$\cfrac{P_1 \ldots P_m \quad \cfrac{[Q_1]^{1.}\!\!\!\vdots\\A[t/x]}{\exists x A(x)}\,\exists_I \quad \ldots \quad \cfrac{[Q_n]^{n.}\!\!\!\vdots\\A[t/x]}{\exists x A(x)}\,\exists_I}{\exists x A(x)}\,GR^*1,\ldots,n.$$

$$\overbrace{\rightsquigarrow \ldots \rightsquigarrow}^{\text{converts to}}$$

$$\cfrac{P_1, \ldots P_m \quad \cfrac{[Q_1]^{1.}\!\!\!\vdots\\A[t/x]}{} \quad \ldots \quad \cfrac{[Q_n]^{n.}\!\!\!\vdots\\A[t/x]}{}\,GR^*1,\ldots,n.}{\cfrac{A[t/x]}{\exists x A(x)}\,\exists_I}$$

However, working under open assumptions allows the witness to be an open term, namely, a term in which there appear variables already contained in these very same assumptions. Hence, if on the one hand we can say that proofs of this kind still provide means, via the normalization algorithm, for obtaining certain objects with well-determined characteristics, on the other hand, these means remain essentially hypothetical: they can be concretely exploited only when the open assumptions are satisfied. But, because we have *a priori* no certainty that these assumptions can really be satisfied, the methods assured by hypothetical proofs are only potential: they depend on the possibility of performing certain constructions in order to satisfy the hypothesis. To sum up, if we abandon any appeal to models, our formal reconstruction of geometrical practice is characterized by hypothetical and potential constructions: although the witnesses of existential quantifiers are not always concretely constructed, they remain nonetheless *constructible*. More precisely, when Euclidean geometry is studied from the point of view of the formal-logical analysis proposed in Sects. 8.1.2 and 8.1.3, then we are led to conceive geometrical objects not as constructed, but only as constructible.

8.5.3 Arithmetical vs Geometrical Objects

An immediate response to this kind of consideration may be to wonder what does determine the emergence of hypothetical and potential features inside geometrical activity. A possible answer can be given by drawing a comparison with arithmetic.

First, it has to be noticed that in arithmetic, the witness of the existential quantifier of a Π_2 sentence never remains undefined, in the sense that it can always be univocally identified without having to appeal to any notion of model external to the syntax. In particular, the method presented in Sect. 8.4.2 can always be used in order to extract a witness that corresponds to a closed term: it is sufficient to particularize the universal quantifier on a closed term and then apply the described procedure. And this is always possible, because in arithmetic each element of the intended model can already be constructed in the syntax itself by means of a finite set of signs and a finite set of algorithmic operations acting on them, namely, the primitive recursive functions of constant zero and successor, i.e., $0, s(0), \ldots, s(s(\ldots s(0)\ldots))$, etc. The syntactic objects thus obtained are objects for which we have cognitive evidence and for which we can decide whether the primitive relations holding among them are satisfied or not. In other words, these objects "can immediately be recognized as being equal or different" exactly because they are "concrete objects [...] existing in time and space" (Martin-Löf 1970, p. 9) and for which we have then an immediate sensible access. A direct consequence of this is that the normalization algorithm can always be concretely used for finding the witness, because the elements expected as inputs can themselves be effectively provided by means of algorithmic finitary methods. In other words, in the case of provable Π_2 sentences, their corresponding proofs can be used as programs because we can always have access to the data to let the algorithm run and terminate, giving other data as outputs.[35]

If we now turn to the analysis of the actions performable by a human geometer, as we noticed in Sect. 8.1.3, they are traditionally considered to take place in a continuous space, and not in a discrete one, as was the case for the human computer performing arithmetical calculations. The intended model of geometry seems then to be based on real numbers. But in general, independently of adopting either a classical or an intuitionistic definition, we have no guarantees of epistemic access to all real numbers. In particular, it is not the case that for all of them, a syntactic representation based on algorithmic finitary means is available, as is the case for arithmetical objects. A quite obvious solution would be to restrict the analysis exclusively to those real numbers that can be represented as sequences of rational numbers[36] generated by "effective rules for calculating each term" of these sequences (Dummett 1977, p. 37). The real numbers are then associated to a particular type of choice sequence usually called *lawlike* sequences, because the selection of each element is "wholly determined in advance by some effective rule" (Dummett 1977, p. 418). In this manner, even if composed of an infinite number of elements, a real number remains a syntactically presentable object, the properties of which do not go beyond the scope of our finite cognitive capacities. The reason is that according to the aforementioned definition, a real number comes along with an algorithmic method that allows exhibiting each of its elements in a finite number

---

[35] In fact, in this case, inputs and outputs are in the form of data structures, since they are given in a canonical normal form (see Martin-Löf 1984, p. 71), i.e., they are in the form

$$\underbrace{s(\ldots s(0)\ldots)}_{n \text{ times}}.$$

[36] Notice that rational numbers can be reduced to natural numbers by the usual techniques, such as Cantor's anti-diagonal argument. It follows that in order to speak of real numbers, we do not need to substantially expand the set of primitive signs we already used for arithmetic.

8 Constructibility and Geometry   155

of steps and that uses only the set of signs already adopted for natural numbers, so that potentially none of the elements remains out of reach. It becomes then natural to think of this kind of real number as nothing else but a program, having natural numbers as data inputs and rational numbers as data outputs. Consequently, proofs of Π_2 sentences applied to real numbers are programs acting on other programs. Hence, the ontology of the theory reduces to a particular kind of object which by definition does not transcend our epistemic capacities: programs can be seen as finite texts composed of sequences of instructions, and thus, they can be reduced to nothing else than finite linguistic objects. The latter are not only cognitively accessible almost by definition (as they are composed of a finite set of signs), but they are also epistemically transparent because, given a certain language (i.e., a set of signs plus a grammar[37]), someone in the presence of a finite text is always in a position to know that they are in such a situation.

However, this solution seems not to be compatible with the type of analysis we considered in Sect. 8.1.1. On the contrary, it seems to correspond to the adoption of a revisionist position. Instead of describing the activity performed by a human geometer, and fixing it at the syntactic level, this solution imposes a reformulation of the geometer's activity: new limitations and constraints are imposed, which can be justified only by shifting attention from the actions performed by the geometer to the modelization of the environment in which they are performed. More precisely, certain constructions are rejected not because they violate the set of the allowed actions, established before starting the production of figures, but because it is successively realized that some of these *a priori* permitted actions have been in fact performed on points not corresponding to algorithmically generated real numbers. Hence, the legitimacy of these constructions is not determined on the basis of a pure behaviouristic analysis, i.e., on the basis of the analysis of those human actions which are performable in a somehow intuitively conceived continuous space. On the contrary, their legitimacy is judged only *a posteriori*, after having made a well-determined choice concerning the form of the continuous space and after some actions have already been performed.[38]

Our initial proposal, on the other hand, was completely different. The idea was to give a linguistic-syntactic account of the actions performable by a human geometer by taking them basically at face value and not to (re)define this class of actions in a way depending on the particular representation of the continuum that has been chosen. In order to achieve such a syntactic representation of the geometer's actions, and at the same time leaving aside any commitment to the nature of the objects constituting the continuous space in which these actions take place, a valuable solution might be to work with hypothetical proofs. With this approach, it is indeed

---

[37]We should indeed assume this finite text to be correct from a grammatical point of view; otherwise, it could not convey any kind of contentful instruction.

[38]For instance, in the case under analysis, the principles leading to the choice of a particular form of the continuum consist both in its epistemic accessibility and in its syntactic representability, while structural properties, as, for example, cardinality or compactness, are left aside.

possible to deal with any kind of situation, even with those for which we lack an epistemic access to the objects involved. The idea is that if there are objects going beyond our cognitive capacities, we can nonetheless describe them by mentioning those specific characteristics in which we are interested in order to perform certain particular constructions. In this manner, even if we cannot have a direct access to such objects, nor a fortiori to the objects constructed starting from them, nonetheless we are still somehow directed towards them. In a nutshell, by making specific assumptions on certain variables (namely, the eigenvariables mentioned at the end of Sect. 8.5.2), via linguistic means we are pointing to some objects possessing well-determined characteristics, even if these objects are not, and never will be, epistemically accessible to us. Speaking in phenomenological terms, we could say that the open assumptions of hypothetical proofs can be seen as intentional acts that are not yet, or that can never be, fulfilled. By adopting such a perspective, we do not lose the constructive character of proofs, but only their general applicability: proofs of Π_2 sentences still codify algorithms for the extraction of a witness, but these are a sort of partial algorithms, because we are not always able to find the data to let them run and eventually terminate. In this sense, a geometrical construction can be seen in analogy to what in programming language is called a *procedure*, that is, a program (or a function) that does not necessarily return a value (Dowek 2009, p. 24). This means that the attention is focused on the instructions that must be carried out in order to perform a (geometrical) construction, rather than on the final result of this construction.[39] And this is nothing but another way of formulating the idea that what counts is the constructibility of geometrical objects rather than their actual constructions, as we already stated at the end of Sect. 8.5.2.

8.6 Conclusion: An Ontological Shift

By way of conclusion, we would like to attempt a radicalization of the positions discussed in Sects. 8.5.2 and 8.5.3, in order to sketch some possible philosophical consequences that can be drawn from the formal-logical reconstruction of geometrical activity we proposed.

We have argued that an axiomatic presentation of geometry represents an obstacle to the characterization of geometrical constructivity in terms of O1

---

[39] A similar comparison between geometry and arithmetic is drawn by Mumma (2012, §3). Roughly speaking, his idea is that while in arithmetic the arguments and values of constructions (functions) can always be put into canonical normal form (see note 35, *supra*), in geometry this is not the case, since this kind of data can vary continuously. Thus, while in arithmetic the study of the identity between constructions concerns also the identity between their respective values, in geometry such identities concern only the constructions, but not their objects (Mumma 2012, p. 117). Indeed, as noticed by Panza (2011, p. 48), in Euclidean geometry, questions about the identity of geometrical objects are ill-posed, since "there is no clear sense in which one could fix the reference of a singular term" for geometrical entities, like, for example, equilateral triangles, "in such a way that it be taken to refer to the same equilateral triangle in any one of its occurrences."

or O2. We want now to suggest that abandoning the axiomatic point of view goes with abandoning a referentialist position with respect to the understanding of mathematical theories. According to what we said in Sect. 8.3.1, if the role of axioms consists in the identification of certain specific mathematical structures, and if these structures are specified by set-theoretical means, then they can play the role of interpretational structures, i.e., of models. It follows that axioms themselves would be always interpreted as true in those structures that they identify; hence, an axiomatic proof can be seen as nothing but a manner to transmit the truth of the axioms to the conclusion. In contrast, when axioms are transformed into inference rules, a radical change takes place. No longer having axioms, there are no longer true sentences from which to start, neither are there implicit stipulations concerning the existence of some set of objects satisfying the axioms and playing the role of truthmakers. More precisely, transforming an axiom into an inference rule corresponds to transforming it into a postulate, that is, into the requirement that a certain action can be executed if certain conditions hold.[40] Following Proclus once again, we can say that the difference between axioms and postulates reflects the one between theorems and problems we analyzed in Sect. 8.4 (Proclus 1970, pp. 140–142): axioms convey a knowledge about the properties of certain objects considered as already existing, while postulates provide methods for producing these objects (Proclus 1970, p. 143). For example, the rule (8.3) gives a method for obtaining a line and thus of assuring that the domain of lines is not empty, once two distinct points are given. But nobody guarantees the nonemptiness of the domain of points. This shows how postulates do not rest on any kind of existential stipulations: when working with postulates, the initial domain of discourse does not need to be assumed as nonempty. Thus, if postulates are intended as the linguistic-inferential representations of the actions performable by a geometer, it follows that geometry itself must not be conceived of as an activity performed on an already given "receptacle" from which geometrical objects are made to emerge.

According to this conception, the assumption that a geometer's actions are performed on a continuous space can eventually be dispensed with.[41] In the end, geometry would consist in the stipulation of a set of algorithmic procedures that can be performed in principle, but for which the specific context in which they have to take place is not *a priori* established.[42] This entails a shift concerning

---

[40]The word "postulate" comes indeed from the Latin verb *postulare*, which means to demand, to require. Note that the Greek word used by Euclid, i.e., αἴτεμα, has basically the same meaning.

[41]The idea is that this assumption plays a similar role to Wittgenstein's ladder of §6.54 of the *Tractatus*. Assuming the geometrical space to be continuous is essential in order to ground our formal and logical reconstruction of Euclidean geometry on the analysis of the activity of an idealized human geometer, as described in Sect. 8.1.3. But once this reconstruction has been achieved, this assumption can finally be dropped, and the system obtained can be used as a non-interpreted formal system.

[42]A similar position is endorsed by von Plato (1995, p. 192), when he describes his system of constructive geometry as a system that "does not stipulate what the basic objects are, or how their basic relations are proved. In this sense it belongs to abstract mathematics, rather than to traditional

ontological questions. Unlike other mathematical theories, arithmetic in particular, geometry can be characterized by considering the actions that it allows and not the objects to which it commits. In this sense, if we accept the traditional Quinean idea that quantification is that through which one expresses ontological commitments, then we should conclude that in Euclidean geometry, questions about quantification become somehow meaningless, as is shown by the possibility of working under open assumptions containing eigenvariables (see Sect. 8.5.2). A similar idea is advocated by Panza (2011), when he says that the objects of Euclidean geometry "do not form a fixed domain of quantification and individual reference" (2011, p. 52) but "are merely objects that fall under some concepts and enter into particular arguments insofar they are represented, or supposed to be represented, by appropriate diagrams" (2011, p. 53). From this point of view, a property like the one expressed by Proposition 3 may lose sense, as it no longer represents a way to get access to the domain of interpretation and identify a particular object of this domain, since there is nothing such as this domain. We could thus sum up these ideas and say that Euclidean geometry can be seen as a theory the ontology of which is not composed of a set of extensional objects playing the role of references of language expressions but is composed instead of intensional objects represented by constructive actions corresponding to the algorithmic procedures mentioned in Sect. 8.5.3.

Acknowledgements I wish to thank Davide Crippa, Gilles Dowek, Pierluigi Graziani, Clément Houtmann, Jean-Baptiste Joinet, John Mumma, Marco Panza, Jan von Plato, Claudio Sacerdoti Coen, and Davide Rinaldi for useful discussions and insightful comments on earlier versions of the paper. I would also like to thank two anonymous referees for valuable suggestions and critiques which helped to improve my work.

References

Avigad, J., E. Dean, and J. Mumma. 2009. A formal system for Euclid's *Elements*. *Review of Symbolic Logic* 2: 700–768.
Awodey, S., and E.H. Reck. 2002. Completeness and categoricity. Part I: nineteenth-century axiomatics to twentieth-century metalogic. *History and Philosophy of Logic* 23: 1–30.
Beeson, M. 2010. Constructive geometry. In *Proceedings of the tenth Asian logic conference*, ed. T. Arai et al., 19–84. Singapore: World Scientific.
Bernays, P. 1930. Die Philosophie der Mathematik und die Hilbertsche Beweistheorie. [The philosophy of mathematics and Hilbert's proof theory] English trans. P. Mancosu. In Mancosu (1998), 234–265.
Bernays, P. 1967. David Hilbert. In *Encyclopedia of philosophy*, ed. P. Edwards, vol. 3, 496–505. New York: MacMillan.
Bishop, E. 1967. *Foundations of constructive analysis*. New York: McGraw-Hill.
Brouwer, L.E.J. 1924. Intuitionistische Zerlegung mathematischer Grundbgriffe. *Jahresbericht der Deutsche Mathematiker-Vereinigung* 33: 241–256.

---

constructive mathematics, where the aim has been to define once and for all the natural numbers and build all other mathematical structures upon them."

Dowek, G. 2009. *Principles of programming languages*. Berlin: Springer.
Dummett, M. 1963. Realism. In Id., *Truth and other enigmas*, 145–165. London: Duckworth.
Dummett, M. 1977. *Elements of intuitionism*. Oxford: Clarendon Press.
Engeler, E. 1968. Remarks on the theory of geometrical constructions. In *The syntax and semantics of infinitary languages*, Lecture notes in mathematics, vol. 72, ed. J. Barwise, 64–76. Berlin: Springer.
Engeler, E. 1993. *Foundations of mathematics: questions of analysis, geometry and algorithmics*, English trans. C.B. Thomas. Berlin: Springer.
Feferman, S. 1964. Systems of predicative analysis. *Journal of Symbolic Logic* 29: 1–30.
Feferman, S. 2005. Predicativity. In *The Oxford handbook of philosophy of mathematics and logic*, ed. S. Shapiro, 590–624. Oxford: Oxford University Press.
Friedman, H. 1999. A consistency proof for elementary algebra and geometry. Manuscript.
Gandy, R. 1988. The confluence of ideas in 1936. In *The universal turing machine: a half-century survey*, ed. R. Herken, 51–102. Berlin: Springer.
Givant, S., and A. Tarski. 1999. Tarski's system of geometry. *The Bulletin of Symbolic Logic* 5: 175–214.
Greenberg, M.J. 1993. *Euclidean and non-Euclidean geometries: development and history*, 3rd ed. New York: W.H. Freeman.
Hallett, M. 2008. Reflections on the purity of method in Hilbert's *Grundlagen der Geometrie*. In *The philosophy of mathematical practice*, ed. P. Mancosu, 198–255. Oxford: Oxford University Press.
Heyting, A. 1925. *Intuitionistische Axiomatik der Projectieve Meetkunde*. Groningen: P. Noordhoff.
Heyting, A. 1959. Axioms for intuitionistic plane affine geometry. In L. Henkin et al. (1959), 160–173.
Heyting, A. 1971. *Intuitionism: an introduction*. Amsterdam: North Holland.
Henkin, L., P. Suppes, and A. Tarski (eds.). 1959. *The axiomatic method, with special reference to geometry and physics*. Amsterdam: North Holland.
Hindley, J.R. and J.P. Seldin 2008. *Lambda-calculus and combinators: an introduction*. Cambridge: Cambridge University Press.
Kolmogorov, A. 1932. Zur Deutung der intuitionistischen Logik. [On the interpretation of intuitionistic logic], English trans. P. Mancosu. In Mancosu (1998), 328–333.
Kreisel, G. 1981. Constructivist approaches to logic. In *Modern logic, a survey: historical, philosophical, and mathematical aspects of modern logic and its applications*, ed. E. Agazzi, 67–91. Dordrecht: Reidel.
Lombard, M., and R. Vesley 1998. A common axiom set for classical and intuitionistic plane geometry. *Annals of Pure and Applied Logic* 95, 229–255.
Mancosu, P. (ed.). 1998. *From Brouwer to Hilbert: the debate on the foundations of mathematics in the 1920's*. Oxford: Oxford University Press.
Martin-Löf, P. 1970. *Notes on constructive mathematics*. Stockholm: Almqvist & Wiksell.
Martin-Löf, P. 1984. *Intuitionistic type theory*. Naples: Bibliopolis.
Mäenpää, P., and J. von Plato. 1993. The logic of Euclidean construction procedure. *Acta Philosophica Fennica* 49: 275–293.
Mints, G.E. 2000. Axiomatization of a Skolem function in intuitionistic logic. In *Formalizing the dynamics of information*, CSLI lecture notes, vol. 91, ed. M. Faller et al., 105–114. Stanford: CSLI Publications.
Moler, N., and P. Suppes. 1968. Quantifier-free axioms for constructive plane geometry. *Compositio Mathematica* 20: 143–152.
Mueller, I. 1981. *Philosophy of mathematics and deductive structure in Euclid's elements*. Cambridge, Mass.: MIT.
Mumma, J. 2012. Constructive geometrical reasoning and diagrams. *Synthese* 186: 103–119.
Negri, S. 2003. Contraction-free sequent calculi for geometric theories with an application to Barr's theorem. *Archive for Mathematical Logic* 42: 389–401.
Negri, S., and J. von Plato. 2001. *Structural proof theory*. Cambridge: Cambridge University Press.

Negri, S., and J. von Plato. 2005. The duality of classical and constructive notions and proofs. In *From sets and types to topology and analysis: towards practicable foundations for constructive mathematics*, ed. L. Crosilla and P. Schuster, 149–161. Oxford: Oxford University Press.
Negri, S., and von Plato, J. 2011. *Proof analysis: a contribution to Hilbert's last problem.* Cambridge: Cambridge University Press.
Quaife, A. 1989. Automated development of Tarski's geometry. *Journal of Automated Reasoning* 5: 97–118.
Pambuccian, V. 2004. Early examples of resource-consciousness. *Studia Logica* 77: 81–86.
Pambuccian, V. 2008. Axiomatizing geometric constructions. *Journal of Applied Logic* 6: 24–46.
Panza, M. 2011. Rethinking geometrical exactness. *Historia Mathematica* 38: 42–95.
von Plato, J. 1995. The axioms of constructive geometry. *Annals of Pure and Applied Logic* 76: 169–200.
von Plato, J. 2006. Normal form and existence property for derivations in Heyting arithmetic. *Acta Philosophica Fennica* 78: 159–163.
von Plato, J. 2007. In the shadows of the Löwenheim–Skolem theorem: early combinatorial analyses of mathematical proofs. *The Bulletin of Symbolic Logic* 13: 189–225.
von Plato, J. 2010a. Combinatorial analysis of proofs in projective and affine geometry. *Annals of Pure and Applied Logic* 162: 144–161.
von Plato, J. 2010b. Geometric proof theory of intuitionistic geometry. *Workshop on constructive aspects of logic and mathematics*, Kanazawa, 8–10 March 2010. <http://www.jaist.ac.jp/is/labs/ishihara-lab/wcalm2010/vonplato.pdf>
von Plato, J. 2013. *Elements of logical reasoning.* Cambridge: Cambridge University Press.
Poincaré, H. 1902. *La science et l'hypothèse*, 1968. Paris: Flammarion.
Prawitz, D. 1971. Ideas and results in proof theory. In *Proceedings of the second Scandinavian logic symposium*, J.E. Fenstad, 235–307. Amsterdam: North Holland.
Proclus 1970. *A commentary on the first book of Euclid's elements*, ed. G.R. Morrow. Princeton: Princeton University Press.
Rathjen, M. 2005. The constructive Hilbert program and the limit of Martin-Löf type theory. *Synthese* 147: 81–120.
Rybowicz, M. 2003. On the normalization of numbers and functions defined by radicals. *Journal of Symbolic Computation* 35: 651–672.
Schroeder-Heister, P. 2006. Validity concepts in proof-theoretic semantics. *Synthese* 148: 525–571.
Schroeder-Heister, P. 2014. Proof-theoretic semantics. In *The Stanford encyclopedia of philosophy* (Summer 2014 edition), ed. E.N. Zalta.
Seeland, H. 1978. *Algorithmische Theorien und konstruktive Geometrie.* Stuttgart: Hochschulverlag.
Sieg, W. 1994. Mechanical procedures and mathematical experiences. In *Mathematics and mind*, ed. A. George, 71–117. Oxford: Oxford University Press.
Szczerba, L.W. 1986. Tarski and geometry. *The Journal of Symbolic Logic* 51: 907–912.
Tarski, A. 1959. What is elementary geometry? In L. Henkin et al. (1959), 16–29.
Tarski, A., and S. Givant. 1999. Tarski's system of geometry. *The Bulletin of System Logic* 5: 175–214.
Troelstra, A.S. 1991. History of constructivism in the twentieth century. *ITLI Prepublication series*, ML-91-05, University of Amsterdam. <http://www.illc.uva.nl/Research/Reports/ML-1991-05.text.pdf>
Troelstra, A.S., and D. van Dalen. 1988. *Constructivism in mathematics: an introduction*, vol. 1. Amsterdam: North Holland.
Troelstra, A.S., and H. Schwichtenberg. 2000. *Basic proof theory*, 2nd ed. Cambridge: Cambridge University Press.
Turing, A.M. 1937. On computable numbers, with an application to the *Entscheidungsproblem*. *Proceedings of the London Mathematical Society* 42(s. 2): 230–265.
Van Bendegem, J.P. 2010. Finitism in geometry. In *The stanford encyclopedia of philosophy*, ed. E.N. Zalta, Spring 2010 ed. <http://plato.stanford.edu/archives/spr2010/entries/geometry-finitism>

Venturi, G. 2011. Hilbert, completeness and geometry. *Rivista Italiana di Filosofia Analitica Junior* 2: 82–104.

Vesley, R. 2000. Constructivity in geometry. *History and Philosophy of Logic* 20: 291–294.

Weyl, H. 1918. *Das Kontinuum: Kritischen Untersuchungen über die Grundlagen der Analysis.* Leipzig: Veit & Co.

Weyl, H. 1949. *Philosophy of mathematics and natural science*, English trans. O. Helmer. Princeton: Princeton University Press.

Chapter 9
A Cut-Like Inference in a Framework of Explicit Composition for Various Calculi of Natural Deduction

Michael Arndt and Laura Tesconi

9.1 Introduction

9.1.1 A Fundamental Principle of Abstract Reasoning

According to Paul Hertz, reasoning is concerned with relationships between (sets of) elements and a (single) element. More [conventionally], we shall here deal with symmetric relationships between formulae[1]: an *abstract derivation* from any number of assumptions Γ to any number of assertions Δ will be represented by the notation $\Gamma \rhd \Delta$ (which is suited to refer both to a derivation in natural deduction as well as to sequents[2]). An act of reasoning is, then, to obtain a previously unavailable (unknown) relationship of this kind by drawing on one (or several) of such that are already available (known).

One of the most fundamental principles of abstract reasoning is represented by the possibility of connecting different objects of reasoning by means of intermediary elements:

---

[1] Collections Γ, Δ are always considered to be multisets, that is, the multiplicity of formula occurrences is always accounted for, but not the order of those occurrences.

[2] About natural deduction, we refer to Prawitz (1965) explication of how to obtain deduction rules from inference rules by considering them as rules operating on pairs $\langle \Gamma, A \rangle$.

M. Arndt
WSI für Informatik, University of Tübingen, Germany
e-mail: arndt@informatik.uni-tuebingen.de

L. Tesconi (✉)
Department of Philosophy, University of Pisa, Pisa, Italy
e-mail: tesconi@fls.unipi.it

$$\frac{\Gamma \vartriangleright \Delta, A \qquad A, \Theta \vartriangleright \Lambda}{\Gamma, \Theta \vartriangleright \Delta, \Lambda}$$

This step amounts to the realization that A mediates between the relationships expressed by the abstract derivations $\Gamma \vartriangleright \Delta, A$ and $A, \Theta \vartriangleright \Lambda$, used as premises, and that by way of some generalized notion of transitivity, the relation expressed by the inferred abstract derivation $\Gamma, \Theta \vartriangleright \Delta, \Lambda$ may be obtained. Another way to put it is if some formula A occurs both as an assertion in an abstract derivation and also as an assumption in another abstract derivation, a new abstract derivation is obtained by joining the assumptions and the assertions of the premises and removing either references to A.

Whereas this principle is included in the sequent calculus as a proper rule—the cut rule—the calculus of natural deduction does not contain a formal inference rule that has this effect. Instead, the following (implicit) principle of *substitution*—indispensable for the meta-theory of natural deduction—is commonly employed as a notational convenience for the composition of derivations:

$$\begin{array}{c} \Gamma \\ \mathcal{D}_1 \\ A \\ \\ A \quad \Delta \\ \mathcal{D}_2 \\ C \end{array} \quad \text{is composed to} \quad \begin{array}{c} \Gamma \\ \mathcal{D}_1 \\ A \quad \Delta \\ \mathcal{D}_2 \\ C \end{array}$$

Actually, only a single occurrence of A is considered in the antecedent and succedent of the premises of cut, whereas this notation of substitution conventionally relates a single derivation of a formula A to multiple assumptions of that formula, so cut should rather be likened to a principle of *linear* substitution, that is, the composition of two abstract derivations on a single occurrence of a formula in the succedent of one and the antecedent of the other.

The cut rule and the substitution principle have been compared very often. They bear striking similarities, but also some significant differences. Firstly, substitution in natural deduction affects the composition of the two derivations by means of fusing the two occurrences of A into a single occurrence, which is retained in the resulting derivation, even though it no longer has the status of either assumption or assertion. Secondly, in no way can the principle of substitution can be stated as a proper rule of natural deduction, because it composes arbitrary derivations, whereas proper inference rules of the systems always add a single formula as new conclusion. However, when considering abstract derivations, implicit composition of derivations and cut become indistinguishable: thus, we shall consider cut as fundamental principle of reasoning corresponding to the joining of abstract derivations on complementary occurrences of the same formula.[3]

---

[3] It is for this reason that inference rules have to be taken to a different level of abstraction when talking about abstract derivations. Otherwise, the fundamental fact that cut is a rule of one calculus but not of the other could not be represented.

9.1.2 From Cut to Bidirectional Multicut

We shall now formulate a rule *multicut* that expresses the effect of subsequent applications of cut addressed to different elements of the antecedent in turn:

$$\left.\begin{array}{c} \Gamma_1 \triangleleft \Delta_1, A_1 \\ \vdots \\ \Gamma_k \triangleleft \Delta_k, A_k \end{array} \quad \begin{array}{c} A_1 \\ \vdots \\ A_k \\ \Theta \end{array}\right\} \triangleleft \Lambda$$
$$\overline{\Gamma_1, \ldots, \Gamma_k, \Theta \triangleleft \Delta_1, \ldots, \Delta_k, \Lambda} \text{ (mcut)}$$

The abstract derivations on the left are the *minor premises* of multicut, and the abstract derivation on the right is its *major premise*. The particular notation serves to emphasize how the minor premises are related to the cut formulae in the antecedent of the major premise.[4] The major premise acts as an accumulator for all of the minor premises by means of these cut formulae. Technically, it is a rule that has $k + 1$ premises, that is, it is schematic in that instances depend on the number of minor premises.

Moreover, in addition to the possibility to choose multiple cut formulae from the antecedent of the major premise, the possibility is open to choose multiple cut formulae also from the succedent of an abstract derivation. These generalizations are merged in a single rule called *bidirectional multicut*:

$$\left.\begin{array}{c} \Gamma_1 \triangleleft \Delta_1, A_1 \\ \vdots \\ \Gamma_k \triangleleft \Delta_k, A_k \end{array} \quad \begin{array}{c} A_1 \\ \vdots \\ A_k \\ \Theta \end{array}\right\} \triangleleft \left\{\begin{array}{c} C_1 \\ \vdots \\ C_l \\ \Lambda \end{array} \quad \begin{array}{c} C_1, \Xi_1 \triangleleft \Pi_1 \\ \vdots \\ C_l, \Xi_l \triangleleft \Pi_l \end{array}\right.$$
$$\overline{\Gamma_1, \cdots, \Gamma_k, \Theta, \Xi_1, \cdots, \Xi_l \triangleleft \Delta_1, \cdots, \Delta_k, \Lambda, \Pi_1, \cdots, \Pi_l} \text{ (bmcut)}$$

This rule singles out an intermediary derivation that acts as major premise in both directions, that is, several elements of *both* the antecedent and the succedent can be used as cut elements. Correspondingly, there are two groups of minor premises, left minor premises and right minor premises. As in the case of the unidirectional multicut, the conclusion of bidirectional multicut can be obtained by $k + l$ applications of the symmetric cut rule. The rule's unique feature lies in its ability to have a single abstract derivation mediate between two sets of abstract derivations by accumulating them from the left and from the right by means of cuts.

---

[4]Historically, the inference rule "*syllogism*", introduced by Paul Hertz in 1923, is the first formulation of multicut, see Hertz (1923). Apart from an additional side effect that contracts multiple occurrences of formulae in the antecedents of the major premise and the conclusion, it looks exactly as the rule that is presented here. Thus, multicut precedes the formulation of the cut rule. In fact, Gentzen introduced the cut rule in Gentzen (1932) and demonstrated how syllogism can be decomposed into a sequence of cuts. Moreover, see also Paul Bernays' definition of "*Syllogismus*" in Bernays (1965).

9.1.3 From Bidirectional Multicut to Explicit Composition

Bidirectional multicut expresses the mere idea of composing multiple derivations into a single one and is an absolutely generic rule; otherwise, its action depends exclusively on the choice of what is to be used as a main premise. For this reason, we shall use it as a foundation for a rule that will be called *explicit composition*.

Principles like that, governing abstract reasoning, are often employed in order to talk about consequence relations, and cut is usually just one of many means by which new pairs can be obtained. Instead, we shall take bidirectional multicut as the *only* reasoning principle that is to be employed on the level of abstract derivations so—since we wish to model reasoning steps by inference steps—explicit composition will be the only inference rule of the framework that we shall build.[5] As a consequence, whatever has to do with logical constants will have to enter in the manner of (instances of) very simple abstract derivations that address only logical constants and the constituents they are used with, insomuch as their configuration of formulae specifies the properties of the constants themselves. These simple abstract derivations are to play the role of major premises, which will accordingly exercise control over the manner in which the minor premises are to be composed.

The ultimate formulation of the explicit composition principle in the general case consists of a modification of bidirectional multicut, in which the major premise is required to be a clause, that we shall call the *control clause*, which expresses a relationships of formulae, whereas the minor premises are arbitrary abstract derivations. Specifically, it expresses a *local* relationship, in the sense that it is free of contextual formulae: only those formulae that are immediately relevant for the rule occur in the control clause, whereas all the contextual formulae reside in the abstract derivations that are composed by means of it. In its most general—though, as we shall see, very naive still—form, it is schematically given as

$$\dfrac{\begin{array}{ccc} \Gamma_1 \triangleright \Delta_1, A_1 & \left.\begin{array}{c} A_1 \\ \vdots \\ A_k \\ \Theta \end{array}\right\} & \blacktriangleright \left\{\begin{array}{c} C_1 \\ \vdots \\ C_l \\ \Lambda \end{array}\right. & \begin{array}{c} C_1, \Xi_1 \triangleright \Pi_1 \\ \vdots \\ C_l, \Xi_l \triangleright \Pi_l \end{array} \\ \vdots & & & \\ \Gamma_k \triangleright \Delta_k, A_k & & & \end{array}}{\Gamma_1, \ldots, \Gamma_k, \Theta, \Xi_1, \ldots, \Xi_l \triangleright \Delta_1, \ldots, \Delta_k, \Lambda, \Pi_1, \ldots, \Pi_l} \text{(EC)}$$

We shall divide the minor premises into those that are composed from the left and those that are composed from the right. A minor premise belonging to the former is called an *antecessor*, whereas one belonging to the latter is called a *successor*. In view of this terminology, the major premise can also be called the *intercessor*.

---

[5] An attempt of defining a most general rule of "composition" is already to be found in Gentzen (1936), where a *chain rule* is defined that, together with a specific set of basic logical sequents, allows one to reproduce almost all of the inference rules of the original system of natural deduction (in sequent style). See also the forthcoming Moriconi (2014).

It is allowed that $k, l, |\Theta|$ or $|\Lambda|$ (and combinations thereof) are 0, the former two resulting in an empty antecessor or, respectively, an empty successor. Since this inference rule is the only rule of this framework for composing abstract derivations, we will usually omit the label (EC).

9.2 The Rule of Explicit Composition

Explicit composition, as it is given above, is powerful enough to express some of the rules of both natural deduction and the sequent calculus by means of control clauses. But still, in order to accurately give account of all the effects that occur in the various rules of the calculi we are considering, additional features have to be available for the control clauses. We shall restrict ourselves to mentioning the limits of the formulation proposed so far and define the generalizations that will make up for them, together with some example for natural deduction. We refer to Arndt and Tesconi (2014) for details and examples for the sequent calculus.

9.2.1 Effective and Ineffective Formulae

Observe that natural deduction rules always compose abstract derivations from the left by means of their assertions: for this reason, for example, the control clause for natural deduction rule ($\supset$E)

$$\frac{\Gamma \triangleright A \qquad \Delta \triangleright A \supset B}{\Gamma, \Delta \triangleright B} \qquad \left.\begin{matrix} A \\ A \supset B \end{matrix}\right\} \blacktriangleright B$$

uses A and $A \supset B$ as substitution formulae, whereas a substitution on B should be explicitly forbidden.

The given example stresses the necessity to distinguish which formulae of a control clause are available for composition and which are not. For this purpose, we shall employ the following notation[6]:

$$\left.\begin{matrix} A_1 \\ \vdots \\ A_k \end{matrix}\right\} \Theta \blacktriangleright \Lambda \left\{\begin{matrix} C_1 \\ \vdots \\ C_l \end{matrix}\right.$$

---

[6]We are indebted to Peter Schroeder-Heister for suggesting this extremely succinct and elegant notation.

The position of Θ and Λ on the inside of the clause indicates that the formulae they contain are protected against composition, whereas formulae $A_1, \ldots, A_k$ are cut formulae for compositions of abstract derivations from the left, and formulae $C_1, \ldots, C_l$ are cut formulae for compositions of abstract derivations from the right. We call the formulae in Θ and Λ *ineffective* and the A_i and C_j *effective* in as far as their purpose to compose abstract derivations is concerned. The ineffective formulae are carried over into the conclusion of (EC), whereas the effective formulae will usually not reoccur in the conclusion, unless there are multiple occurrences thereof in the minor premises.

9.2.2 Near and Far Effects

It is now possible to express by means of control clauses all the inference rules that do not affect the assumptions of the derivations that are composed. However, there are inference rules that affect the antecedent of their premise in addition to its succedent. For example, in defining a control clause for rule $(\to\!I)$,

$$\frac{\begin{array}{c}[A]\\ B\end{array}}{A \to B}$$

one wants to express the fact that the assertion $A \to B$ is obtained and the assumption A is contextually discharged. In order to fulfill this requirement, a conceptual extension of (EC) is needed in as far as it must also allow for the cancellation of assumptions in abstract derivations. Specifically, it must allow for a control clause to cut a formula occurring in the succedent of an antecessor, the *near side* of some abstract derivation, and simultaneously remove another formula that occurs on the *far side* of the very same abstract derivation. A combination of such a *far effect* on a formula A in the antecedent of an antecessor and a *near effect* on a formula B in its succedent is expressed by the *effective pair* $\langle A | B \rangle$. Moreover, even though the standard calculi feature nothing like cancellation of assertions, we allow for the dual combination of near and far effects on the right-hand side of a control clause.[7]

The notation for control clauses is extended correspondingly:

$$\left.\begin{array}{c}\langle A_1 | B_1\rangle \\ \vdots \\ \langle A_k | B_k\rangle\end{array}\right\} \Theta \blacktriangleright \Lambda \left\{\begin{array}{c}\langle C_1 | D_1\rangle \\ \vdots \\ \langle C_l | D_l\rangle\end{array}\right.$$

---

[7]Such a notion is indeed present in the refutationistic calculus defined in Tranchini (2010).

Any of the A_i and any of the D_i can be vacuous, and the default cases that there should be no far effect are expressed by leaving the corresponding part of the combined effect empty, as in $\langle \cdot \,|\, B \rangle$ and $\langle C \,|\, \cdot \rangle$.

9.2.3 Iterated and Compound Effects

The additional features introduced above are not yet enough to define a correct control clause for the rule ($\rightarrow$I). Whereas control clause $\langle A \,|\, B \rangle\} \blacktriangleright A \rightarrow B$ appears to embody all the desired features, $\langle A \,|\, B \rangle$ expresses the removal of a single formula on each side of the abstract derivation it affects, whereas the cancellation expressed by the rule ($\rightarrow$I) is seen as merely allowing the removal of selected occurrences. This is accommodated by allowing effects to be marked by a Kleene star that indicates their *iteration*. The combined effect $\langle A^* \,|\, B \rangle$ on the left-hand side of a control clause can compose an abstract derivation $\Gamma \triangleright \Delta, B$ from the left by means of a near effect on B, that is, a cut with cut formula B, but at the same time *any number* of occurrences of A (even none at all) may be removed[8] from Γ. Using this generalized far effect, the natural deduction rule of ($\supset$I) may be finally expressed by the control clause $\{\langle A^* \,|\, B \rangle\} \blacktriangleright A \supset B$ which allows any instance (for $n \geq 0$) of the following inference:

$$\frac{\overbrace{A,\ldots,A}^{n}, \Gamma \triangleright \Delta, B \qquad \langle A^* \,|\, B \rangle\} \blacktriangleright A \supset B}{\Gamma \triangleright \Delta, A \supset B}$$

In a dual way, the combined effect $\langle C \,|\, D^* \rangle$ on the right-hand side of a control clause can compose an abstract derivation $C, \Gamma \triangleright \Delta$ from the right by means of a near effect on C, but at the same time, any number of occurrences of D (even none at all) may be removed from Δ. With a simple generalization, combinations of iterated near and far effects become available as well, as in $\langle A^* \,|\, B^* \rangle$, though they are not required for the present purpose.

A specialization of the general notion of iterated effects is that of *compound* effects. Instead of using the Kleene star, we use superscript letters to specify some exact multiplicity of formula occurrences. Thus, $\langle A^n \,|\, B^m \rangle$ expresses a compound effect resulting in the removal of n occurrences of the formula A from the antecedent of some abstract derivation and m occurrences of the formula B from its succedent. Analogously, $\langle C^n \,|\, D^m \rangle$ expresses a compound effect resulting in the removal of n occurrences of the formula C from the antecedent of some abstract derivation and

---

[8]Note that the Kleene star is not a logical symbol, but results from the omission of multiset parentheses for antecedent and succedent and also for effects. Indeed, if those were not omitted, $\langle A^* \,|\, B \rangle$ would have to be written $\langle \{A\}^* \,|\, \{B\} \rangle$.

m occurrences of the formula D from its succedent. Whereas the superscript n of far effects is ≥ 0—the case of $n = 0$ will yield the same result as an empty effect—the superscript m of near effects must be > 0. A compound effect can be considered as a mixing-style cut (in the case of a near effect) or as a multiple cancellation (in the case of a far effect). It corresponds to the simultaneous consideration in a single inference step of several occurrences of a formula, or even two or more different formulae: for example, the control clause

$$\left.\begin{array}{l}\langle \cdot | A \wedge B \rangle \\ \langle A^*, B^* | C \rangle\end{array}\right\} \blacktriangleright C$$

yields the general elimination rule ($\wedge$E), in which an arbitrary number of occurrences of assumptions A and B are simultaneously cancelled from one of the two minor premises. Thus, compound effects are not only limited to purely structural issues but also to a particular structural issue that is covered implicitly in certain logical rules.

9.2.4 Explicit Composition, Revisited Formulation

The various generalizations that were introduced so far can be summarized in the following formulation of explicit composition:

$$\frac{\left.\begin{array}{ll}\Sigma_1^\circ, \Gamma_1 \triangleright \Delta_1, \Phi_1^\circ & \langle \Sigma_1^\bullet | \Phi_1^\bullet \rangle \\ \vdots & \vdots \\ \Sigma_k^\circ, \Gamma_k \triangleright \Delta_k, \Phi_k^\circ & \langle \Sigma_k^\bullet | \Phi_k^\bullet \rangle\end{array}\right\} \Theta \blacktriangleright \Lambda \left\{\begin{array}{ll}\langle \Psi_1^\bullet | \Omega_1^\bullet \rangle & \Psi_1^\circ, \Xi_1 \triangleright \Pi_1, \Omega_1^\circ \\ \vdots & \vdots \\ \langle \Psi_l^\bullet | \Omega_l^\bullet \rangle & \Psi_l^\circ, \Xi_l \triangleright \Pi_l, \Omega_l^\circ\end{array}\right.}{\Gamma_1, \ldots, \Gamma_k, \Theta, \Xi_1, \ldots, \Xi_l \triangleright \Delta_1, \ldots, \Delta_k, \Lambda, \Pi_1, \ldots, \Pi_l} \text{(EC)}$$

All the multisets occurring in the abstract derivations—the Γ_i, Δ_i, Ξ_i and Π_i as well as the Σ_i°, Φ_i°, Ψ_i° and Ω_i°—are multisets of formulae, as are the ineffective formulae of Θ and Λ of the control clause, whereas multisets $\Sigma_i^\bullet$, $\Phi_i^\bullet$, $\Psi_i^\bullet$ and $\Omega_i^\bullet$ may also contain superscripted formulae. The relationship between $\circ$-ed and $\bullet$-ed multisets is given as follows:

1. Each occurrence of a formula A in $\Sigma_i^\bullet$ relates to exactly one occurrence of A in Σ_i°.
2. Each occurrence of a superscripted formula A^n in $\Sigma_i^\bullet$ relates to exactly n occurrences of A in Σ_i°.
3. Each occurrence of an asterisked formula A^* in $\Sigma_i^\bullet$ relates to an arbitrary number of occurrences of A in Σ_i° that is not otherwise accounted for.

That is to say, every $\circ$-ed multiset is a specific instance of a $\bullet$-ed multiset that may contain superscript specifications.

9 A Cut-Like Inference in a Framework of Explicit Composition

A control clause

$$\left.\begin{array}{c}\langle \Sigma_1^\bullet | \Phi_1^\bullet \rangle \\ \vdots \\ \langle \Sigma_k^\bullet | \Phi_k^\bullet \rangle \end{array}\right\} \Theta \blacktriangleright \Lambda \left\{\begin{array}{c}\langle \Psi_1^\bullet | \Omega_1^\bullet \rangle \\ \vdots \\ \langle \Psi_l^\bullet | \Omega_l^\bullet \rangle \end{array}\right.$$

will be called *bidirectional* when $k, l \geq 1$, that is, when it allows composition both from the left- and the right-hand side; analogously, it will be called *bioriented* when $|\Theta|, |\Lambda| > 1$, and if $|\Theta| = |\Lambda|$ and $k = l = 0$, then $\Theta \neq \Lambda$, that is, when it allows the addition of formulae both in the antecedent and the succedent of the conclusion of explicit composition, except for the trivial case of the same multiset of formulae. It will be called *left (right) directed* when $k \geq 1, l = 0$ ($l \geq 1, k = 0$), that is, when it allows composition only from the left (right)-hand side; analogously, it will be called *left (right) oriented* when $|\Theta| \neq 0, |\Lambda| = 0$ ($|\Theta| = 0, \Lambda \neq 0$), that is, when it allows addition of formulae only in the antecedent (succedent) of the conclusion of explicit composition. A set of control clauses is bidirectional (bioriented), when some of its control clauses are bidirectional (bioriented) or when some of its control clauses are left directed (left oriented) and some of its control clauses are right directed (right oriented).

Whereas the final formulation of (EC) no longer bears a close semblance to the original simple reasoning principle, we obtained it through successive generalizations, not only in the sense of simply allowing for multiple and bidirectional compositions, each focussing on a single formula, but instead allowing a much wider range of cut-like effects in each of the compositions that contribute to the conclusion of the rule. All of these effects were obtained from a close inspection of the rules of natural deduction and the sequent calculus.

With anything that specifically affects the premises or the conclusion being specified in the control clauses, the rule (EC) is much more general than, say, a logical or structural inference rule. Instead, it should be seen as a purely combinatorial operation that takes some number of abstract derivations and produces the composition thereof—with some minor but significant modifications that are determined by the control clause through its effective and ineffective formulae.

9.3 Control Bases

By allowing the wide spectrum of effects described in the previous section, control clauses can mimic, emulate, produce, generate, obtain? all the rules of the calculus of natural deduction (and all the structural and logical rules of the sequent calculus, using the multiplicative formulation of its rules, but this would lead us too far from the purposes of the present work) in the sense that (EC) has the premises of those

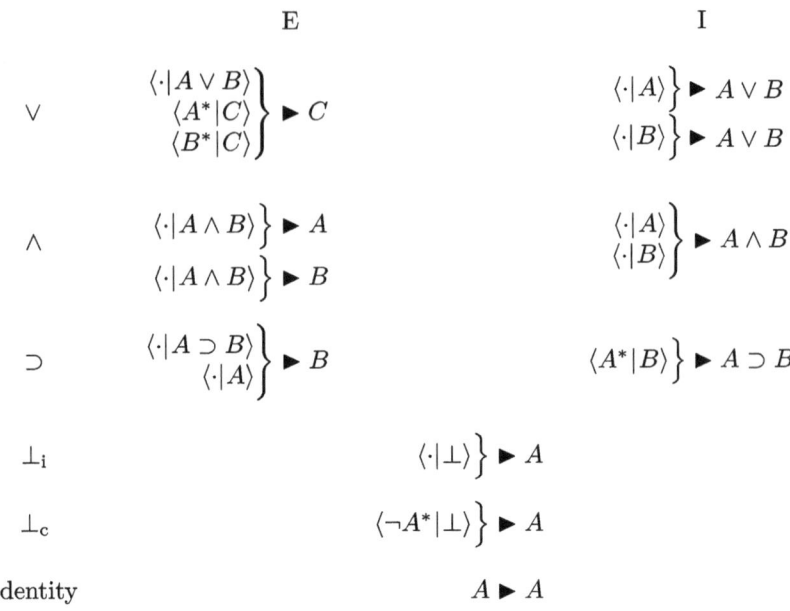

Fig. 9.1 Control base $\mathbb{B}_{ND}$ for standard natural deduction

rules as minor premises and their conclusions as its conclusion. The set of control clauses that correspond to the rules of a given system shall be called the *control base* for that system.

In this section, the control bases for standard natural deduction, natural deduction with general elimination rules and "bioriented" natural deduction[9] shall be defined in Figs. 9.1–9.3, respectively. The inherent properties of the calculi are reflected in the specific features of the corresponding control base, and this enables a direct comparison between the calculi that is based on the simple direct comparison of the control bases, rather than a comparison effected by some translation function. The advantages provided by this aspect of the framework will be fully appreciated once the control base for sequent calculus is defined as well (see Arndt and Tesconi 2014). In particular, the compared analysis of these control bases will emphasize that only the control base for the sequent calculus is bidirectional and bioriented, whereas an essential feature of control bases for the most conventional versions of natural deduction seems to be the left directionality—but modified versions will be defined already in the present work that show different properties.

---

[9]With "bioriented" natural deduction, we shall here refer to the calculus of bidirectional natural deduction proposed by Peter Schroeder-Heister in Schroeder-Heister (2009). The calculus of natural deduction with general elimination rules, instead, was first defined as such by Jan von Plato in von Plato (2001).

9 A Cut-Like Inference in a Framework of Explicit Composition

$$
\begin{array}{c|cc}
 & E & I \\
\hline
\vee & \left.\begin{array}{c}\langle\cdot|A\vee B\rangle\\\langle A^{*}|C\rangle\\\langle B^{*}|C\rangle\end{array}\right\}\blacktriangleright C & \left.\begin{array}{c}\langle\cdot|A\rangle\\\langle\cdot|B\rangle\end{array}\right\}\blacktriangleright A\vee B \\[2ex]
\wedge & \left.\begin{array}{c}\langle\cdot|A\wedge B\rangle\\\langle A^{*},B^{*}|C\rangle\end{array}\right\}\blacktriangleright C & \left.\begin{array}{c}\langle\cdot|A\rangle\\\langle\cdot|B\rangle\end{array}\right\}\blacktriangleright A\wedge B \\[2ex]
\supset & \left.\begin{array}{c}\langle\cdot|A\supset B\rangle\\\langle\cdot|A\rangle\\\langle B^{*}|C\rangle\end{array}\right\}\blacktriangleright C & \{\langle A^{*}|B\rangle\}\blacktriangleright A\supset B \\[2ex]
\bot_{i} & & \{\langle\cdot|\bot\rangle\}\blacktriangleright A \\
\bot_{c} & & \{\langle\neg A^{*}|\bot\rangle\}\blacktriangleright A \\
\text{identity} & & A\blacktriangleright A
\end{array}
$$

Fig. 9.2 Control base $\mathbb{B}_{\text{GND}}$ for natural deduction with general elimination rules

$$
\begin{array}{c|cc}
 & E & I \\
\hline
\vee & \left.\begin{array}{c}\langle A^{*}|C\rangle\\\langle B^{*}|C\rangle\end{array}\right\} A\vee B\blacktriangleright C & \left.\begin{array}{c}\langle\cdot|A\rangle\\\langle\cdot|B\rangle\end{array}\right\}\blacktriangleright A\vee B \\[2ex]
\wedge & \begin{array}{c}\{\langle A^{*}|C\rangle\}\,A\wedge B\blacktriangleright C\\\{\langle B^{*}|C\rangle\}\,A\wedge B\blacktriangleright C\end{array} & \left.\begin{array}{c}\langle\cdot|A\rangle\\\langle\cdot|B\rangle\end{array}\right\}\blacktriangleright A\wedge B \\[2ex]
\supset & \left.\begin{array}{c}\langle\cdot|A\rangle\\\langle B^{*}|C\rangle\end{array}\right\} A\supset B\blacktriangleright C & \{\langle A^{*}|B\rangle\}\blacktriangleright A\supset B \\[2ex]
\bot_{i} & & \{\langle\cdot|\bot\rangle\}\blacktriangleright A \\
\bot_{c} & & \{\langle\neg A^{*}|\bot\rangle\}\blacktriangleright A \\
\text{identity} & & A\blacktriangleright A
\end{array}
$$

Fig. 9.3 Control base $\mathbb{B}_{\text{BND}}$ for "bioriented" natural deduction

9.3.1 Some Properties of Control Clauses

We shall now give an overview of some general features of control clauses that concern the relationship between near and far effects and the bidirectionality that is intrinsically expressed in the explicit composition rule.

The first observation is that the combined effect $\langle A|B \rangle$ is directionally neutral in the sense that an abstract derivation $A, \Gamma \triangleright \Delta, B$ can either be composed from the left to a control clause via a combined effect $\langle A|B \rangle$ on its left-hand side or from the right via $\langle A|B \rangle$ on its right-hand side:

$$\frac{A, \Gamma \triangleright \Delta, B \qquad \langle A|B \rangle \} \Theta \blacktriangleright \Lambda}{\Gamma, \Theta \triangleright \Delta, \Lambda}$$

and

$$\frac{\Theta \blacktriangleright \Lambda \{ \langle A|B \rangle \qquad A, \Gamma \triangleright \Delta, B}{\Theta, \Gamma \triangleright \Lambda, \Delta}$$

It is to be noted that "directional neutrality" holds with respect to the relation of premises and conclusion, but, as regarding abstract derivations, the two control clauses refer to two different compositions.

The second observation requires the condition that succedents of abstract derivations never be empty—this is trivial when dealing with all of the control bases given so far. Then, when a combined effect $\langle A^*|B \rangle$ and an ineffective formula B both occur in the right side of a control clause, the same conclusion could be rendered by a control clause that does not employ combined effects.[10] In the derivation,

$$\frac{\Theta \blacktriangleright B, \Lambda \{ \langle A^*|B \rangle \qquad A, \Gamma \triangleright \Delta, B}{\Gamma, \Theta \triangleright B, \Delta, \Lambda}$$

the far effect of the control clause removes B from the antecedent of the abstract derivation, but at the same time, the ineffective formula B is retained in the conclusion. The same effect is obtained by a control clause that leaves the conclusion B intact:

$$\frac{\Theta \blacktriangleright \Lambda \{ \langle A^*|\cdot \rangle \qquad A, \Gamma \triangleright \Delta, B}{\Gamma, \Theta \triangleright B, \Delta, \Lambda}$$

These observations imply that a combined effect $\langle A^*|B \rangle$ on the left side of a control clause (in $\mathbb{B}_{ND}$, $\mathbb{B}_{GND}$ or $\mathbb{B}_{BND}$) that also has an ineffective formula B on its right side amounts to a near effect $\langle A^*|\cdot \rangle$ on the right side of the control clause:

---

[10] A dual phenomenon could be observed on the left-hand side, provided the fulfillment of the condition that antecedents of abstract derivations never be empty, but this obviously does not hold in any of the control bases defined in the present work.

9 A Cut-Like Inference in a Framework of Explicit Composition

$$\frac{A, \Gamma \triangleright \Delta, B \qquad \{\langle A^*|B\rangle\} \blacktriangleright B}{\Gamma \triangleright \Delta, B}$$

and

$$\frac{\blacktriangleright \{\langle A^*|\cdot\rangle\} \qquad A, \Gamma \triangleright \Delta, B}{\Gamma \triangleright \Delta, B}$$

However, these control clauses are not at all equivalent to each other, as they exercise different controls over the premises that are to be composed by (EC). In fact, regardless of the fact that they obtain the same abstract derivation as a conclusion, they express different rules with different directionality.

9.3.2 Modified Control Bases

We shall now exploit these properties in order to "invert" the directionality of control clauses for elimination rules of $\mathbb{B}_{GND}$ and $\mathbb{B}_{BGND}$ —except for the case of disjunction that we shall not consider here because it would require a limitation of (EC), in order to prevent that more than one formula appear in the succedent of a conclusion (further details will be included in an already mentioned following work). The "modified" control bases $\mathbb{B}_{MGND}$ and $\mathbb{B}_{MBND}$ are shown in Figs. 9.4 and 9.5, respectively, and, opposite to the original control bases, are both bidirectional.

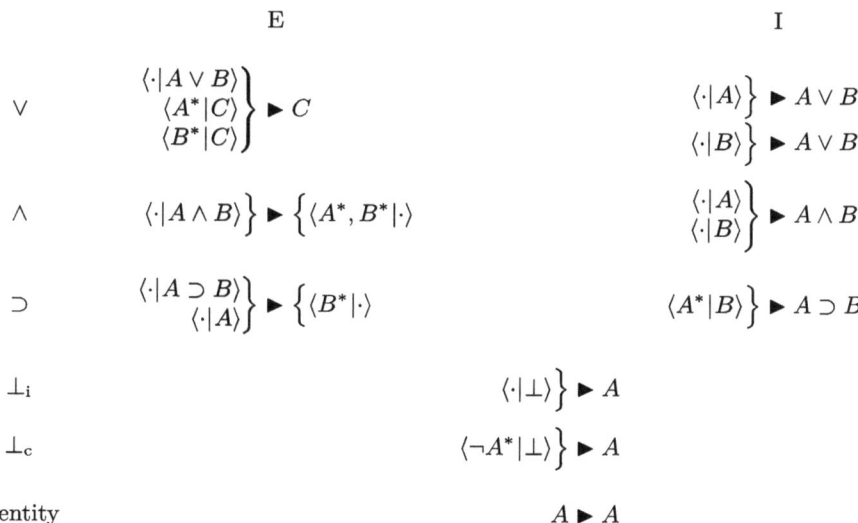

Fig. 9.4 Control base $\mathbb{B}_{MGND}$ for modified natural deduction with general elimination rules

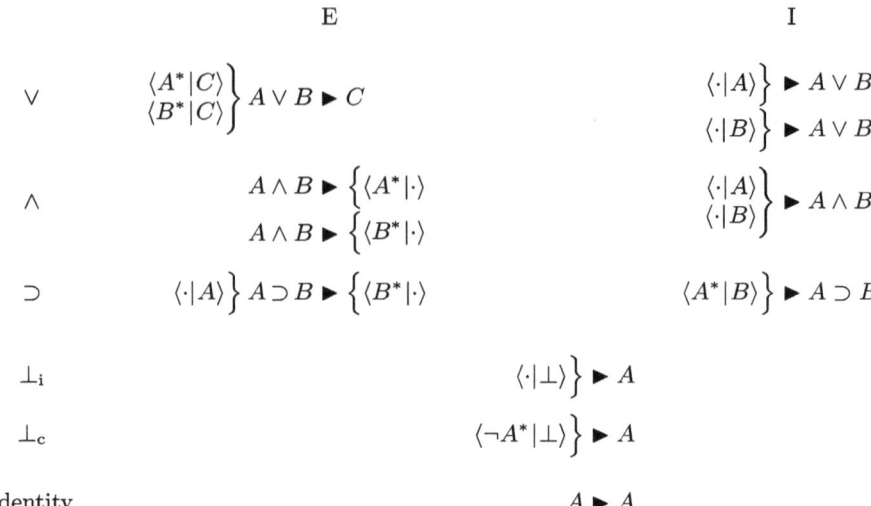

Fig. 9.5 Control base $\mathbb{B}_{\text{MBND}}$ for modified "bioriented" natural deduction

9.4 Co-Identity and Its Effects on Control Bases for Natural Deduction Systems

9.4.1 The Control Clause for Co-Identity

All of the control bases defined so far share the control clause for *identity*

$$A \blacktriangleright A,$$

which trivially introduces the formula A both in the antecedent and the succedent of the conclusion. This corresponds to the initial sequent $A \Rightarrow A$, in the sequent calculus, or the derivation of A by its assumption in natural deduction.

Its "dual", the control clause for *co-identity*

$$\langle \cdot | A \rangle \} \blacktriangleright \{\langle A | \cdot \rangle$$

instead has the effect of restoring the restriction performed by ineffective formulae on the character of bidirectionality.

This clause forces (EC) to simply connect the antecessor to the successor, which in this case are just two abstract derivations:

$$\frac{\Gamma \triangleright \Delta, A \qquad \langle \cdot | A \rangle \} \blacktriangleright \{\langle A | \cdot \rangle \qquad A, \Xi \triangleright \Pi}{\Gamma, \Xi \blacktriangleright \Delta, \Pi}$$

thus expressing in the setting of abstract derivations the principle of linear substitution: in fact, in the context of the sequent calculus this control clause renders the cut rule. However, it is not included in the control bases for any of the versions of natural deduction considered.

Co-identity recovers the possibility of composition on formulae that were introduced by means of ineffective formulae in previous inference steps. Compare how the following two derivations yield the same conclusion from the premises $\Gamma \triangleright \Delta, A$ and $C, \Xi \triangleright \Pi$:

$$\frac{\Gamma \triangleright \Delta, A \qquad \langle \cdot | A \rangle \} \blacktriangleright \{ \langle A | \cdot \rangle \quad \dfrac{A \blacktriangleright \{ \langle C | \cdot \rangle \quad C, \Xi \triangleright \Pi}{A, \Xi \triangleright \Pi}}{\Gamma, \Xi \triangleright \Delta, \Pi}$$

and

$$\frac{\dfrac{\Gamma \triangleright \Delta, A \quad \langle \cdot | A \rangle \} \blacktriangleright C}{\Gamma \triangleright \Delta, C} \qquad \langle \cdot | C \rangle \} \blacktriangleright \{ \langle C | \cdot \rangle \quad C, \Xi \triangleright \Pi}{\Gamma, \Xi \triangleright \Delta, \Pi}$$

In the first derivation, the upper application of (EC) employs the schematic control clause $A \blacktriangleright \{\langle C | \cdot \rangle\}$ to compose $C, \Xi \triangleright \Pi$ on formula C, which—by means of the ineffective formula A—introduces A into the antecedent of its conclusion. The subsequent composition that uses co-identity has the effect of a simple cut with cut formula A. In contrast to this, the second derivation first composes $\Gamma \triangleright \Delta, A$ on A through a schematic control clause $\{\langle \cdot | A \rangle\} \blacktriangleright C$, thereby introducing C, and the second composition, employing co-identity, then acts as cut on the formula C.

9.4.2 The Addition of Co-identity to Given Control Bases

Although the observations up to this point might seem to suggest a substantial correlation between co-identity and bidirectionality, there are other aspects to consider. It is indeed true that co-identity confers an explicit feature of bidirectionality to a given control base it is added to, but—as a first thing, and trivially enough—there are bidirectional control bases that do not contain co-identity among their control clauses: consider, for instance, the control base $\mathbb{B}_{MGND}$ which contains the bidirectional control clauses for conjunction elimination and implication elimination. Then, on the one hand, we shall see that the addition of co-identity to a unidirectional control base does not necessarily imply that the resulting system has more compositional possibilities than the original one—this is the case, for example, of control bases $\mathbb{B}_{ND}$, $\mathbb{B}_{GND}$ and $\mathbb{B}_{MGND}$, as we shall demonstrate shortly. On the other hand, the addition of co-identity to a control base that already enjoys bidirectionality may imply that the resulting system has more compositional possibilities than the original one. This is the case for control bases $\mathbb{B}_{BND}$ and $\mathbb{B}_{MBND}$.

Control bases $\mathbb{B}_{ND}$ and $\mathbb{B}_{GND}$ are strictly unidirectional and right oriented. That is, control clauses in $\mathbb{B}_{ND}$ and $\mathbb{B}_{GND}$ always compose abstract derivations from the left and all of the ineffective formulae occur in their right-hand sides. In these systems, all formulae might be considered "joining knots" of different derivation fragments, so the addition of co-identity to the corresponding control bases will not actually add the possibility of the join itself but rather the possibility of giving information about which formula we want to be considered as "joining knot". In the framework of explicit composition for the mentioned control bases, co-identity may perform a cut on the same formula, in different points of the derivation or even on different formulae, without altering neither the conclusion nor the order of the other inferences. We shall now show that, in the framework of explicit composition for $\mathbb{B}_{ND}$, adding the possibility of using co-identity does not add the possibility of new combinations of derivation fragments. In other words, any two derivation fragments on which co-identity could be applied could be combined, *with the same result*, without resorting to co-identity, via the control clauses of $\mathbb{B}_{ND}$ alone. In fact, consider a derivation that ends with

$$\text{(co-id)} \frac{\begin{array}{c} \mathcal{D}_1 \\ \Gamma \triangleright B \end{array} \quad \langle \cdot|B\rangle\} \blacktriangleright \{\langle B|\cdot\rangle \quad \begin{array}{c} \mathcal{D}_2 \\ B, \Delta \triangleright C \end{array}}{\Gamma, \Delta \triangleright C}$$

Inferences leading to $B, \Delta \triangleright C$ may as well be performed, one by one, starting from $\Gamma \triangleright B$ instead of $B \triangleright B$—which results from identity—without any application of co-identity. This is proved by a trivial induction on the length of derivation $\mathcal{D}_2$, and we will only provide some representative examples of the cases. In particular, only for the most trivial case described below, we will give the example of a derivation that does not end with co-identity, in order to show how the procedure affects—or better: *does not* affect—the subsequent applications of explicit compositions.

If the co-identity is performed on a derivation of minimal length, as in

$$\frac{\frac{\frac{\Gamma \triangleright A \to B \quad \langle \cdot|A \to B\rangle}{\Delta \triangleright A \quad \langle \cdot|A\rangle}\} \blacktriangleright B}{\Gamma, \Delta \triangleright B} \quad \langle \cdot|B\rangle\} \blacktriangleright \{\langle B|\cdot\rangle \quad \frac{B \blacktriangleright B}{B \triangleright B}}{\frac{\Gamma, \Delta \triangleright B}{\Sigma \triangleright B \to D}} \quad \langle \cdot|B \to D\rangle \}\blacktriangleright D}{\Gamma, \Delta, \Sigma \triangleright D} \quad \langle \cdot|D\rangle\} \blacktriangleright D \vee C}{\Gamma, \Delta, \Sigma \triangleright D \vee C}$$

then it can be simply eliminated:

$$\frac{\frac{\frac{\Gamma \triangleright A \to B \quad \langle \cdot|A \to B\rangle}{\Delta \triangleright A \quad \langle \cdot|A\rangle}\} \blacktriangleright B}{\frac{\Gamma, \Delta \triangleright B}{\Sigma \triangleright B \to D}} \quad \langle \cdot|B \to D\rangle \}\blacktriangleright D}{\Gamma, \Delta, \Sigma \triangleright D} \quad \langle \cdot|D\rangle\} \blacktriangleright D \vee C}{\Gamma, \Delta, \Sigma \triangleleft D \vee C}$$

9 A Cut-Like Inference in a Framework of Explicit Composition

This case corresponds to composing on an assumption:

$$
\begin{array}{cc}
\Delta & \Gamma \\
\vdots & \vdots \\
A \to B & A \\
\hline
B & \Sigma \\
\text{comp} & \vdots \\
\hline
B & B \to D \\
\hline
D \\
\hline
D \vee C
\end{array}
\quad
\begin{array}{c}
\text{may be} \\
\text{transformed in\_to\_?}
\end{array}
\quad
\begin{array}{ccc}
\Delta & \Gamma & \\
\vdots & \vdots & \Sigma \\
A \to B & A & \vdots \\
\hline
B & & B \to D \\
\hline
D \\
\hline
D \vee C
\end{array}
$$

Here the notation $*B*$ emphasizes that the composition takes place on the derivation ending with B, that is, B itself.

If the co-identity is performed on a derivation of non-minimal length, as in

$$
\cfrac{\cfrac{\Gamma \rhd A \quad \Delta \rhd A \to (B \vee C) \quad \langle \cdot | A \to (B \vee C) \rangle \quad \langle \cdot | A \rangle}{\Gamma, \Delta \rhd B \vee C} \} \blacktriangleright B \vee C \quad \cfrac{\langle \cdot | B \vee C \rangle \} \blacktriangleright \{(B \vee C | \cdot) \quad \cfrac{B \vee C \rhd E \vee F \quad E \rhd D \quad F \rhd D}{B \vee C \rhd D} \quad \cfrac{\langle \cdot | E \vee F \rangle \langle E^* | D \rangle \langle F^* | D \rangle}{} \} \blacktriangleright D}{\Gamma, \Delta \lhd D}
$$

then it can be permuted over the last inference of its successor in order to be able to apply the induction hypothesis:

$$
\cfrac{\cfrac{\Gamma \rhd A \quad \Delta \rhd A \to (B \vee C) \quad \langle \cdot | A \to (B \vee C) \rangle \quad \langle \cdot | A \rangle}{\Gamma, \Delta \rhd B \vee C} \} \blacktriangleright B \vee C \quad \cfrac{\langle \cdot | B \vee C \rangle \} \blacktriangleright \{(B \vee C | \cdot) \quad B \vee C \rhd E \vee F}{\cfrac{\Gamma, \Delta \rhd E \vee F \quad E \rhd D \quad F \rhd D}{\Gamma, \Delta \rhd D}} \quad \cfrac{\langle \cdot | E \vee F \rangle \langle E^* | D \rangle \langle F^* | D \rangle}{} \} \blacktriangleright D
$$

This case corresponds to composing on an assumption of a derivation:

$$
\begin{array}{cccc}
\Gamma & \Delta & & \\
\vdots & \vdots & & \\
A & A \to (B \vee C) & & \\
\hline
B \vee C & & & \\
\text{comp} & & & \\
B \vee C & E & F & \\
\vdots & \vdots & \vdots & \\
E \vee F & D & D & \\
\hline
D & & &
\end{array}
\quad
\begin{array}{c}
\text{may be} \\
\text{be transformed in}
\end{array}
\quad
\begin{array}{cccc}
\Gamma & \Delta & & \\
\vdots & \vdots & & \\
A & A \to (B \vee C) & & \\
\hline
B \vee C & & & \\
\text{comp} & & & \\
B \vee C & E & F & \\
\vdots & \vdots & \vdots & \\
E \vee F & D & D & \\
\hline
D & & &
\end{array}
$$

Here the notation $*D*$ and $*E \vee F*$ emphasizes that the compositions take place on the fragments of derivation ending with D and $E \vee F$, respectively. We have thus proved that any derivation of $\mathbb{B}_{\text{ND}}$ plus co-identity can be transformed into a derivation of $\mathbb{B}_{\text{ND}}$ of the same conclusion where the order of the other inferences is not altered.[11]

---

[11] Without going into details of a formal definition, we chose to give the inferences a leftmost, topmost order.

The case of $\mathbb{B}_{\text{GND}}$ is completely analogous and we shall only demonstrate a single example:

$$\dfrac{\begin{array}{ll}\Gamma \triangleright A \to B & \langle \cdot | A \to B\rangle \\ \Delta \triangleright A & \langle \cdot | A\rangle \\ B \triangleright E \to F & \langle B^* | E \to F\rangle\end{array} \bigg\} \blacktriangleright E \to F}{\Gamma, \Delta \triangleright E \to F} \qquad \langle \cdot | E \to F\rangle \} \blacktriangleright \{\langle E \to F | \cdot\rangle \qquad \dfrac{\begin{array}{ll}E \to F \triangleright E \to F & \langle \cdot | E \to F\rangle \\ \Sigma \triangleright E & \langle \cdot | E\rangle \\ F \triangleright C & \langle F^* | C\rangle\end{array}\bigg\} \blacktriangleright C}{E \to F, \Sigma \triangleright C}$$

$$\overline{\Gamma, \Delta, \Sigma \triangleright C}$$

first becomes

$$\dfrac{\begin{array}{ll}\Gamma \triangleright A \to B & \langle \cdot | A \to B\rangle \\ \Delta \triangleright A & \langle \cdot | A\rangle \\ B \triangleright E \to F & \langle B^* | E \to F\rangle\end{array}\bigg\} \blacktriangleright E \to F}{\Gamma, \Delta \triangleright E \to F} \qquad \langle \cdot | E \to F\rangle \} \blacktriangleright \{\langle E \to F | \cdot\rangle \qquad \begin{array}{l}E \to F \blacktriangleright E \to F \\ E \to F \triangleright E \to F\end{array}$$

$$\dfrac{\begin{array}{l}\Gamma, \Delta \triangleright E \to F \\ \Sigma \triangleright E \\ F \triangleright C\end{array} \qquad \begin{array}{l}\langle \cdot | E \to F\rangle \\ \langle \cdot | E\rangle \\ \langle F^* | C\rangle\end{array} \bigg\} \blacktriangleright C}{\Gamma, \Delta, \Sigma \triangleright C}$$

and eventually becomes

$$\dfrac{\begin{array}{ll}\Gamma \triangleright A \to B & \langle \cdot | A \to B\rangle \\ \Delta \triangleright A & \langle \cdot | A\rangle \\ B \triangleright E \to F & \langle B^* | E \to F\rangle\end{array}\bigg\} \blacktriangleright E \to F}{\begin{array}{l}\Gamma, \Delta \triangleright E \to F \\ \Sigma \triangleright E \\ F \triangleright C\end{array}} \qquad \begin{array}{l}\langle \cdot | E \to F\rangle \\ \langle \cdot | E\rangle \\ \langle F^* | C\rangle\end{array} \bigg\} \blacktriangleright C$$

$$\overline{\Gamma, \Delta, \Sigma \triangleright C}$$

This case corresponds to

$$\begin{array}{ccc} \Gamma & \Delta & B \\ \vdots & \vdots & \vdots \\ A \to B & A & E \to F \end{array}$$
$$\dfrac{E \to F}{\text{comp}} \qquad \begin{array}{cc} \Sigma & F \\ \vdots & \vdots \\ E & C \end{array}$$
$$\dfrac{E \to F}{*C*}$$

first being considered as

$$\begin{array}{ccc} \Gamma & \Delta & B \\ \vdots & \vdots & \vdots \\ A \to B & A & E \to F \end{array}$$
$$\dfrac{E \to F}{\text{comp}} \qquad \begin{array}{cc} \Sigma & F \\ \vdots & \vdots \\ E & C \end{array}$$
$$\dfrac{*E \to F*}{C}$$

which may be transformed in to (???)

$$\begin{array}{ccc} \Gamma & \Delta & B \\ \vdots & \vdots & \vdots \\ A \to B & A & E \to F \end{array} \qquad \begin{array}{cc} \Sigma & F \\ & \vdots \\ E & C \end{array}$$
$$\dfrac{E \to F}{C}$$

9 A Cut-Like Inference in a Framework of Explicit Composition 181

Here the notation $*C*$ and $*E \to F*$ emphasizes that the compositions take place on the fragments of derivation ending with C and $E \to F$, respectively.

The modified control base $\mathbb{B}_{MGND}$ is still right oriented, but it is bidirectional already in the absence of co-identity. As for its conventional version, the addition of this control clause has no other real effect than providing information about the exact point in a derivation where a composition takes place. Consider, for example, the derivation

$$\dfrac{\dfrac{\Gamma \triangleright A \to B \quad \langle \cdot | A \to B \rangle}{\Delta \triangleright A \quad \langle \cdot | A \rangle} \} \blacktriangleright \{\langle B | \cdot \rangle\} \quad B \triangleright E \to F}{\Gamma, \Delta \triangleright E \to F} \quad \dfrac{\langle \cdot | E \to F \rangle\} \blacktriangleright \{\langle E \to F | \cdot \rangle\} \quad \dfrac{E \to F \triangleright M \lor N \quad \langle \cdot | M \lor N \rangle}{\substack{M \triangleright D \\ N \triangleright D}} \dfrac{\langle M | D \rangle}{\langle N | D \rangle}\} \blacktriangleright D}{E \to F \triangleright D}}{\Gamma, \Delta \triangleright D}$$

Because of bidirectionality, co-identity can always be permuted over the last inference of its successor, as in the procedure exploited before, or its antecessor, provided that the control clause of its last inference admits a successor, as in the following derivation:

$$\dfrac{\dfrac{\Gamma \triangleright A \to B \quad \langle \cdot | A \to B \rangle}{\Delta \triangleright A \quad \langle \cdot | A \rangle}\} \blacktriangleright \{\langle B | \cdot \rangle\} \quad \dfrac{B \triangleright E \to F \quad \langle \cdot | E \to F \rangle\} \blacktriangleright \{\langle E \to F | \cdot \rangle\} \quad E \to F \triangleright M \lor N}{B \triangleright M \lor N}}{\dfrac{\Gamma, \Delta \triangleright M \lor N \\ M \triangleright D \\ N \triangleright D}{\Gamma, \Delta \triangleright D} \quad \dfrac{\langle \cdot | M \lor N \rangle}{\langle M | D \rangle} \dfrac{}{\langle N | D \rangle}\} \blacktriangleright D}$$

Both possibilities decrease the length of premises of co-identity; the latter is obviously never available for $\mathbb{B}_{ND}$, $\mathbb{B}_{GND}$ and $\mathbb{B}_{BND}$ because their control clauses do not admit any successors.

Control base $\mathbb{B}_{BND}$, instead, is unidirectional, and it is bioriented as well. That is, control clauses in $\mathbb{B}_{BND}$ always compose abstract derivations from the left, but ineffective formulae occur in both sides. In this case, adding the possibility of using co-identity truly adds the possibility of new combinations of fragments of derivation. In other words, it is not the case that any two fragments of derivation on which co-identity could be applied could be combined together, *with the same result*, already without resorting to co-identity, via the control clauses of $\mathbb{B}_{BND}$. Specifically, the addition of co-identity allows composition on ineffective formulae on the left-hand side of control clauses for elimination rules, something that would not be possible otherwise.

In fact, observe that a derivation $A, \Delta \triangleright \Gamma$ cannot be a successor of any other control clause than co-identity, since they all have ineffective formulae on their right side; this is true for control bases $\mathbb{B}_{ND}$ and $\mathbb{B}_{GND}$ as well, but whereas in the this latter case co-identity can be permuted without altering the order of other inferences until it reaches the point where it is performed on an abstract derivation obtained from the identity, it is not always the case for control base $\mathbb{B}_{BND}$, where some control clauses have ineffective formulae on the left-hand side that introduce elements in to the antecedent of the conclusion. The following derivation will serve as a counterexample:

$$\frac{A, \Gamma \rhd B \quad \langle A^*|B\rangle\} \blacktriangleright A \to B}{\Gamma \blacktriangleright A \to B} \quad \langle \cdot|A \to B\rangle\} \blacktriangleright \{\langle A \to B|\cdot\rangle \quad \frac{\begin{array}{c}\Delta \rhd A \\ B \rhd C\end{array} \quad \langle \cdot|A\rangle \atop \langle B^*|C\rangle\} A \to B \blacktriangleright C}{\Delta, A \to B \rhd C}$$
$$\Gamma, \Delta \rhd C$$

The same obstacle arises for the modified control base $\mathbb{B}_{\text{MBND}}$ which is both bidirectional and bioriented. That is, when the last inference of co-identity's successor is left oriented, it is not possible to permute co-identity over it. As it happens for the modified control bases $\mathbb{B}_{\text{MGND}}$, because of the gained feature of bidirectionality, in some cases co-identity can be permuted over the last inference of its antecessor (namely, when the last inference of its control clause admits a successor) instead, but not always, as the following example shows:

$$\frac{A, \Gamma \rhd B \quad \langle A^*|B\rangle\} \blacktriangleright A \to B}{\Gamma \blacktriangleright A \to B} \quad \langle \cdot|A \to B\rangle\} \blacktriangleright \{\langle A \to B|\cdot\rangle \quad \frac{\Delta \rhd A \quad \langle \cdot|A\rangle\} A \to B \blacktriangleright \{\langle B|\cdot\rangle \quad B \rhd C}{A \to B, \Delta \rhd C}$$
$$\Gamma, \Delta \rhd C$$

9.4.3 Conclusions

The examples proposed suggest that the addition of co-identity to a single-oriented control base, regardless of its directionality, does not open more possibilities of combination of abstract derivations, whereas its addition to a bioriented control base does. In this latter case, in fact, when a formula is introduced among the assumptions by means of a certain rule, it appears as a leaf in the derivation tree, but it is not available before the application of the rule itself. Thus, it is not always possible that it is (???), at the same time and in that very same point of the derivation, introduced as a conclusion by some other inference. Another way to put it is that in bioriented systems not all formulae of a derivation can be considered "joining knots" of different fragments of derivations.

As a conclusion, explicit composition turns out to provide a framework that proves to be very suited to compare the effects of a cut-like inference, embodied by co-identity, on various calculi of natural deduction. In particular, it becomes evident that the absence of such inferences in single-oriented systems, like natural deduction or natural deduction with general elimination rules, does not entail an absence of "detours", whereas it does for bioriented systems—indeed, absence of co-identity in bioriented systems entails the very impossibility of even expressing "detours" (in the same way as the impossibility of expressing "detours" is entailed by the absence of cut in the sequent calculus). This observation provides evidence against a quite hasty account of the cut rule, according to which it can be matched either to a redundancy or to composition of derivations, indiscriminately (thus, sometimes, erroneously establishing a link between the two notions). Instead, the cut rule should be rather read as a principle of reasoning, whose potential depends very much on the formal context within which it is employed and whose correspondence to the idea of redundancy is only a—very interesting indeed—particular case of a much broader range of effects.

References

Arndt, Michael and Laura Tesconi. 2014. The role of cut as a principle of explicit composition. In *Second Pisa Colloquium in Logic, Language and Epistemology*, ed. E. Moriconi and L. Tesconi. ETS.

Bernays, Paul. 1965. Betrachtungen zum Sequenzen-Kalkul. In *Contributions to logic and methodology in honor of J.M. Bochenski*, ed. A.-T. Tymieniecka. Amsterdam: North-Holland.

Gentzen, Gerhard. 1932. Über die Existenz unhabängiger Axiomensysteme zu unendlichen Satzsysteme. *Mathematische Annalen* 107:329–350.

Gentzen, Gerhard. 1936. Die Widerspruchsfreiheit der reine Zahlentheorie. *Mathematische Annalen* 112:493–565.

Hertz, Paul. 1923. Über Axiomensysteme für beliebige Satzsysteme. II. Sätze höheren Grades. *Mathematische Annalen* 89:76–102.

Moriconi, Enrico. 2014, In preparation. *Early structural reasoning. Gentzen 1932*.

Prawitz, Dag. 1965. *Natural deduction. A proof-theoretic study*. Stockholm: Almqvist & Wiksell.

Schroeder-Heister, Peter. 2009. Sequent calculi and bidirectional natural deduction: On the proper basis of proof-theoretic semantics. In *The Logica Yearbook 2008*, ed. M. Peliš. London: College Publications.

Tranchini, Luca. 2010. Refutation: a proof-theoretic account. In *First Pisa Colloquium in Logic, Language and Epistemology*, ed. C. Marletti. Pisa: ETS.

von Plato, Jan. 2001. Natural deduction with general elimination rules. *Archive for Mathematical Logic* 40:541–567.

Chapter 10
On the Distinction Between Sets and Classes: A Categorical Perspective

Samuele Maschio

10.1 Introduction

10.1.1 The Thorny Relation Between Categories and Sets

The relation between sets and categories is a thorny and intriguing topic (see, e.g., Blass 1984). The set-theoretical foundations of category theory were discussed by some mathematicians, e.g., by Feferman (1977), Engler (1969), and Lolli (1977). On the contrary, the categorical foundation of set theory is not yet a completely clear matter. Categorical logic allows us to talk about the categorical models of mathematical theories; in particular, it makes possible to talk about the categorical models of set theories, for example, *IZF*, *CZF*, and *ZF*. In 1995, Joyal and Moerdijk in their *algebraic set theory* (Joyal and Moerdijk 1995) proposed a technique to ensure the existence of internal models of ZF and IZF in a category. Before this, mathematicians had proposed categorical models of set theory within categories. However, these categories were still built using a model of set theory. Joyal and Moerdijk's approach has thus two main advantages:

1. It allows to prove the existence of a model of *IZF* inside a category by simply checking some purely categorical properties of a family of arrows.
2. It works for a relatively large family of categories.

Their book led to the large spread of several works (see, e.g., Awodey et al. (2007) or Simpson (1999)). Algebraic set theory was expected to be not only a useful tool for mathematics but also something interesting for foundational studies.

S. Maschio (✉)
Research Assistant, Dipartimento di Matematica, Università di,
Via Trieste, 63, 35121 Padova, Italy
e-mail: maschio@math.unipd.it

Category theory provides a foundation for mathematics: this is maybe one of the most striking and imprecise sentences (see, e.g., Marquis 1995; Landry and Marquis 2005). In fact this is not necessarily wrong, but it is indispensable to clearly explain the meaning of the words *foundation for mathematics*. Kreisel observed (see e.g., Kreisel 1971) that category theory provides a powerful tool to organize mathematics; this is also what many mathematicians interested in category theory think, considering it as a foundation.

However, there are also some proposals to use category theory to build a theory of sets based on functions. On such a theory, it is possible to found the whole mathematical building: the typical examples are the axiomatic systems proposed by Lawvere (2005) and philosophically analyzed by McLarty, for example, in McLarty (2004).

This paper is in accordance with the first vision of *category theory as a foundation*: we think that category theory *describes* mathematics in a very effective way. In the following pages, this descriptive power will be shown by using category theory to explain the role of classes in the practice of ZF set theory. Category theory—both in its usual and in its internal version—will be used to represent the relation between mathematics and metamathematics.

10.1.2 A Nontrivial Question

The original title of this paper was *What is the real category of sets?*. Since it was a little pretentious and too vague, we changed it into the present one. However, the question *What is the real category of sets?* is strongly connected with the content of the paper and has been the motivation for the work. In fact, in the practice of mathematics, we often refer to *the* category of sets, even if we use *a* category of sets. Actually we typically use categories of sets which are built starting from a given model of ZF set theory. So the question should be rephrased as follows: *Is there anything that is uniquely determined and that can be called the real category of sets?* If such a category exists, then it must be the common core of all the different categories of sets. This is why we chose to propose an answer considering the syntactic category of definable classes of ZF and its internal category of sets (which are in fact mere syntactical constructions). This is motivated by a fundamental correspondence between functors defined on the syntactic category of a theory with values in a category equipped with appropriate structures and models of the same theory in it. This particularly applies to ZF.

10.1.3 Classes and ZF

Set theory is mainly ZF (or—in other words—most set theorists study ZF). ZF is a classical first-order theory with equality, and its language has (countably many) individual variables, no functional symbols, and a binary relational symbol $\in$; its specific axioms are the following ones:

1. $\forall x \forall y (\forall t (t \in x \leftrightarrow t \in y) \rightarrow x = y)$;
2. For every formula P with free variables $p_1, \ldots, p_k, t$,

$$\forall p_1 \ldots \forall p_k \forall x \exists y \forall t (t \in y \leftrightarrow (t \in x \wedge P));$$

3. $\forall x \forall y \exists z \forall t (t \in z \leftrightarrow (t = x \vee t = y))$;
4. $\forall x \exists z \forall t (t \in z \leftrightarrow \exists y (t \in y \wedge y \in x))$;
5. $\forall y \exists z \forall x (x \in z \leftrightarrow x \subseteq y)$;
6. For every formula F with free variables $p_1, \ldots, p_k, x, y$,

$$\forall p_1 \ldots \forall p_k \forall z (\forall x (x \in z \rightarrow \exists! y F) \rightarrow \exists z' \forall x (x \in z \rightarrow \exists y (y \in z' \wedge F)));$$

7. $\exists x (0 \in x \wedge \forall t (t \in x \rightarrow \sigma(t) \in x))$;
8. $\forall x (x \neq 0 \rightarrow \exists z (z \in x \wedge \forall t (t \in x \rightarrow t \notin z)))$;

where, as usual, $x \subseteq y$ is a shorthand for $\forall t (t \in x \rightarrow t \in y)$, $\sigma(t)$ stays for $t \cup \{t\}$, and 0 is a shorthand for the empty set. In detail, for every formula $P = P(x)$,

$$P(\sigma(t)) \text{ means } \forall x (\forall s (s \in x \leftrightarrow ((s = t) \vee (s \in t))) \rightarrow P(x)) \text{ and}$$

$$P(0) \text{ means } \forall x (\forall t (t \notin x) \rightarrow P(x)).$$

Set theorists work with sets, but they also talk about classes. However, classes do not exist in ZF. In practice, set theorists consider classes as shorthands or formal writings. If P is a formula with a distinguished variable x, then a class is a formal writing $\{x | P(x)\}$, and the expression

$$t \in \{x | P(x)\}$$

is simply a shorthand for $P[t/x]$. For example, set theorists write $t \in \mathcal{V}$, as a shorthand for $t \in \{x | x = x\}$, which is a shorthand for $t = t$. They also write $t \in \mathcal{ON}$ as a shorthand for $t \in \{x | ON(x)\}$, which is a shorthand for $ON(t)$, where $ON(x)$ is the formula expressing that x is an ordinal:

$$\forall s \forall s' \forall s'' ((s \in t \wedge s' \in t \wedge s'' \in t \wedge s \in s' \wedge s' \in s'') \rightarrow s \in s'') \wedge$$

$$\wedge \forall s \forall s' ((s \in t \wedge s' \in t \wedge s \neq s') \rightarrow (s \in s' \vee s' \in s)) \wedge \forall s (s \in t \rightarrow s \subseteq t).$$

It is clear that classes are not mathematical, but metamathematical objects. However, these shorthands share some properties with sets. For instance, the notion of intersection of classes or the notion of subclass can be given a meaning. This accidental fact leads of course to some confusion! As a matter of fact, the metamathematical level is often mixed up with the mathematical one. As a consequence, impressive (but at the same time highly incorrect) assertions are not uncommon. This is a typical example:

Sets are exactly those classes for which the comprehension axiom is true.(∗)

Sentences like this can be used as a sort of *convincing arguments* to give students a hint of the notion of set. Nonetheless, in the context of ZF, these statements sound quite dangerous rather than useful.

Set theorists easily use classes, which are formal objects, as quasi-sets, for two reasons. Firstly, we have already seen that their syntactical structure (and in particular the use of connectives and quantifiers) makes this legitimate for many operations. Secondly, set theorists know that the theory of classes *NBG* is a conservative extension of *ZFC*. Nevertheless, in this context, classes are *nothing more than syntactical objects*.

In the following sections, we will see how the use of category theory can help to describe the relation between metamathematics and mathematics. Furthermore, category theory makes easier the distinction between real sentences about sets and external (naïve) ones.

10.2 The Syntactic Category of ZF

In this section, we will study the category of ZF definable classes and its relevant subcategories. First of all, we want to define the syntactic category of ZF (see Johnstone (2002) for the general construction). We will indicate it with $\mathbb{ZF}$. Its objects are formulas in context, i.e., the following formal writings:

$$\{x_1, \ldots, x_n | P\},$$

where $x_1, \ldots, x_n$ is a (possibly empty) list of distinct variables and P is a formula which has free variables among $x_1, \ldots, x_n$. An arrow from a formula in context $\{\mathbf{x}|P\}$ to another formula in context $\{\mathbf{y}|Q\}$ is an equivalence class of formulas in context

$$[\{\mathbf{x}', \mathbf{y}'|F\}]_\equiv,$$

where $\mathbf{x}'$ is a list of variables having the same length as $\mathbf{x}$ and $\mathbf{y}'$ is a list of variables having the same length as $\mathbf{y}$, which satisfies the following requirements:

1. $F \vdash_{ZF} P[\mathbf{x}'/\mathbf{x}] \wedge Q[\mathbf{y}'/\mathbf{y}]$.
2. $F \wedge F[\mathbf{y}''/\mathbf{y}'] \vdash_{ZF} \mathbf{y}' = \mathbf{y}''$.
3. $P[\mathbf{x}'/\mathbf{x}] \vdash_{ZF} \exists \mathbf{y}' F$.

The equivalence relation is given by

$$\{x', y'|F\} \equiv \{x'', y''|F'\} \text{ iff } \vdash_{ZF} F \leftrightarrow F'[x'/x'', y'/y''].$$

For the composition, it is enough to consider the composable arrows $[\{x', y'|F\}]_\equiv$ and $[\{y', z'|F'\}]_\equiv$ with all distinct variables (this doesn't determine a loss of generality). In this case, the composition is given by

$$[\{x', z'|\exists y'(F \wedge F')\}]_\equiv.$$

For very general reasons, the category we obtain is regular and Boolean (see Johnstone (2002) for a proof). In particular we recall the fact that the product of two objects $\{x|P\}$ and $\{y|Q\}$ is given by

$$\{x, y|P \wedge Q\},$$

where we have supposed (without loss of generality) that all variables involved are distinct. Moreover, terminal objects are given for example by

$$\{\ |\forall x(x = x)\}, \text{ either } \{x|\forall t(t \notin x)\}.$$

We can also easily prove that $\mathbb{ZF}$ is extensive, thanks to the fact that

$$\vdash_{ZF} 0 \neq 1,$$

where 1 stays for $\sigma(0)$.

Initial objects are given, for example, by

$$\{\ |\exists x(x \neq x)\}, \text{ either } \{x|x \neq x\}.$$

$\mathbb{ZF}$ is also an exact category; the proof of it is based on *Scott's trick* (see Rosolini 2011; Maschio 2012). Finally, $\mathbb{ZF}$ has also a subobject classifier that is given by

$$[\{x, x'|x = 0 \wedge x' = 1\}]_\equiv : \{x|x = 0\} \to \{x|x = 0 \vee x = 1\}.$$

10.2.1 Definable Classes

The category of definable classes of ZF, which we denote with $\mathbb{DCL}[ZF]$, is the full subcategory of $\mathbb{ZF}$ which is determined by all those objects having a list of variables with length 1. This is clearly conceived of as the category of classes, in the context of the practice of ZF.

Although $\mathbb{DCL}[ZF]$ is a subcategory of $\mathbb{ZF}$, it is provable to be equivalent to it. This results from the possibility to represent ordered pairs in ZF:

$$x = <x_1, x_2> \equiv^{def} x = \{\{x_1\}, \{x_1, x_2\}\}.$$

An arbitrary object of $\mathbb{ZF}$, $\{x_1, .., x_n | P\}$, is isomorphic to

$$\{x | \exists x_1 \ldots \exists x_n (x = << \ldots < x_1, x_2 >, \ldots >, x_n > \wedge P)\},$$

where x is a variable not included in $x_1, \ldots, x_n$.

10.2.2 The Categorical Side of Class Operations

At this point, we would like to give an example of the categorical interpretation of the practical operations between classes. Our aim is to prove how adequate $\mathbb{DCL}[ZF]$ is for the description of the category of formal classes. On this purpose, we will focus on intersection. We consider two definable classes $\{x|P\}$ and $\{y|Q\}$ (we can assume that x and y are distinct without loss of generality); we usually consider their intersection as $\{x | P \wedge Q[x/y]\}$. However, this operation can be expressed in mere categorical terms. In fact $\{x | P \wedge Q[x/y]\}$ is isomorphic in $\mathbb{DCL}[ZF]$ to the object we obtain by considering the following pullback.

$$\begin{array}{ccc} \{x|P\} \cap \{y|Q\} & \longrightarrow & \{x | x = x\} \\ \downarrow & & \downarrow {\scriptstyle [\{x,x',y' | x=x' \wedge x=y'\}]_\equiv} \\ \{x,y | P \wedge Q\} & \xrightarrow[{[\{x,y,x',y' | P \wedge Q \wedge x=x' \wedge y=y'\}]_\equiv}]{} & \{x, y | x = x \wedge y = y\} \end{array}$$

10.2.3 Definable Sets

Now we will consider the full subcategory $\mathbb{DST}[ZF]$ of $\mathbb{DCL}[ZF]$ determined by the objects $\{x|P\}$ for which

$$\vdash_{ZF} \exists z \forall x (x \in z \leftrightarrow P).$$

This is the category of sets we have in mind, when we think sentence (∗) is true.

We obtain the following result:

Theorem 10.2.1. $\mathbb{DST}[ZF]$ *is a Boolean topos.*

Proof. Finite limits exactly correspond to finite limits in $\mathbb{ZF}$: a terminal object is given by $\{x|x=0\}$, equalizers are exactly equalizers in $\mathbb{ZF}$, while the product of two definable sets $\{x|P\}, \{y|Q\}$ is given by

$$\{z|\exists x \exists y (z =<x,y> \wedge P \wedge Q)\}$$

with the obvious projections (we are assuming x and y to be distinct without loss of generality). The subobject classifier is exactly the subobject classifier of $\mathbb{ZF}$, while exponentials $\{y|Q\}^{\{x|P\}}$ are given by

$$\{f|\text{Fun}(f) \wedge \forall s(s \in \text{dom}(f) \leftrightarrow P(s)) \wedge \forall s'(s' \in \text{ran}(f) \rightarrow Q(s'))\},$$

with the evaluation arrow given by

$$[\{F, y| \exists f \exists x (F =<f,x> \wedge \text{Fun}(f) \wedge \forall s(s \in \text{dom}(f) \leftrightarrow P(s)) \wedge \\ \wedge \forall s'(s' \in \text{ran}(f) \rightarrow Q(s')) \wedge <x,y> \in f)\}]_\equiv.$$

□

10.3 Algebraic Set Theory in the Syntactic Category of ZF

In order to better understand the relation between formal classes, definable sets, and sets, we are now going to talk a little about algebraic set theory and the syntactic category of ZF following Rosolini's example in Rosolini (2011); in fact the notion of definable sets is strictly connected with a specific notion of small maps in $\mathbb{ZF}$.

10.3.1 Simpson's Axioms for Algebraic Set Theory

In this section, we will introduce a list of axioms for algebraic set theory proposed by Simpson (see Simpson 1999). Every axiomatization of algebraic set theory is based on a pair $(\mathbb{C}, \mathcal{S})$ in which $\mathbb{C}$ is a category and $\mathcal{S}$ is a family of arrows of $\mathbb{C}$ (called family of *small maps*). The axioms are the following ones:

1. $\mathbb{C}$ is a regular category.
2. The composition of two arrows in $\mathcal{S}$ is in $\mathcal{S}$.
3. Every mono is in $\mathcal{S}$.
4. (Stability) If $f \in \mathcal{S}$ and f' is a pullback of f, then $f' \in \mathcal{S}$.

5. (Representability) For every X, there exists an object $\mathcal{P}_{\mathcal{S}}(X)$ and an arrow $e_X : \in_X \to X \times \mathcal{P}_{\mathcal{S}}(X)$ with $\pi_2 \circ e_X \in \mathcal{S}$ so that, for every $\psi : R \to X \times Z$ with $\pi_2 \circ \psi \in \mathcal{S}$, there exists a unique arrow $\rho : Z \to \mathcal{P}_{\mathcal{S}}(X)$ that fits in a pullback as follows:

$$\begin{array}{ccc} R & \longrightarrow & \in_X \\ \psi \downarrow & \lrcorner & \downarrow e_X \\ X \times Z & \xrightarrow{id_X \times \rho} & X \times \mathcal{P}_{\mathcal{S}}(X) \end{array}$$

6. (Power Set) $\sqsubseteq_X : \sqsubseteq_X \to \mathcal{P}_{\mathcal{S}}(X) \times \mathcal{P}_{\mathcal{S}}(X)$ satisfies $\pi_2 \circ \sqsubseteq_X \in \mathcal{S}$,
where $\sqsubseteq_X$ is determined by the following property:
An arrow $f = \langle f_1, f_2 \rangle : Z \to \mathcal{P}_{\mathcal{S}}(X) \times \mathcal{P}_{\mathcal{S}}(X)$ *factorizes through* $\sqsubseteq_X$ *if and only if considering the following pullbacks*

$$\begin{array}{ccccc} P & \longrightarrow & \in_X & \longleftarrow & Q \\ \pi \downarrow & \lrcorner & \downarrow e_X & \llcorner & \downarrow \pi' \\ X \times Z & \xrightarrow{id \times f_1} & X \times \mathcal{P}_{\mathcal{S}}(X) & \xleftarrow{id \times f_2} & X \times Z \end{array}$$

the arrow π factorizes through π'.

We also want to recall some definitions:

Definition 10.3.1. An object U in a category $\mathbb{C}$ is *universal*, if for every object X in $\mathbb{C}$, there exists a mono $j : X \to U$.

Definition 10.3.2. An object U in a regular category $\mathbb{C}$ with a class of small maps $\mathcal{S}$ is a *universe* if there exists a mono $j : \mathcal{P}_{\mathcal{S}}(U) \to U$.

Definition 10.3.3. A ZF-*algebra* for a regular category $\mathbb{C}$ with a class of small maps $\mathcal{S}$ is an internal sup-semilattice $(U, \subseteq)$ together with an arrow $\sigma : U \to U$, so that for every $\lambda : B \to U$ and for every $j : B \to A \in \mathcal{S}$, there exists $\sup_j(\lambda) : A \to U$ so that for any $j' : B' \to A$ and $\lambda' : B' \to U$, once we consider the following pullback

$$\begin{array}{ccc} P & \xrightarrow{\pi_2} & B \\ \pi_1 \downarrow & \lrcorner & \downarrow j \\ B' & \xrightarrow{j'} & A, \end{array}$$

we have that

$$\sup_j(\lambda) \circ j' \subseteq \lambda' \text{ if and only if } \lambda \circ \pi_2 \subseteq \lambda' \circ \pi_1.$$

The *morphisms* of ZF-algebras are the morphisms between internal sup-semilattices that preserve $\sup_j(\lambda)$ along $j \in \mathcal{S}$ and commute with the arrows σ. An *initial* ZF-algebra is an initial object in the category of ZF-algebras and morphisms between them.

10.3.2 Small Maps in $\mathbb{ZF}$

We now want to define a class $\mathcal{S}_{ZF}$ of small maps in $\mathbb{DCL}[ZF]$. An arrow

$$\{x|P\} \xrightarrow{[\{x,y|F\}]} \{y|Q\}$$

is in $\mathcal{S}_{ZF}$ if and only if

$$\vdash_{ZF} \forall y \exists z \forall x (F(x,y) \leftrightarrow x \in z).$$

The class $\mathcal{S}_{ZF}$ is a class of small maps, as Simpson means (Simpson 1999):

Lemma 10.3.4. *Every mono is in $\mathcal{S}_{ZF}$.*

Proof. This is obtained by using axiom 2 to obtain the existence of an empty set and axiom 3 to prove the existence of singletons, once we notice that an arrow

$$[\{x,y|F\}]_\equiv$$

is mono if and only if

$$F \wedge F[x'/x] \vdash_{ZF} x = x'$$

□

Lemma 10.3.5. *Every composition of arrows in $\mathcal{S}_{ZF}$ is in $\mathcal{S}_{ZF}$.*

Proof. This is obtained by using axioms 6 and 4. □

Lemma 10.3.6. *The stability axiom is satisfied by $\mathcal{S}_{ZF}$.*

Proof. This is obtained by using axiom 6. □

Lemma 10.3.7. *The representability axiom is satisfied by $\mathcal{S}_{ZF}$.*

Proof. A definable class $X = \{x|P\}$ is fixed. Then, the definable class $\mathcal{P}_{\mathcal{S}_{ZF}}(X)$ is given by

$$\{y | \forall x (x \in y \to P)\},$$

while the definable class $\in_X$ is given by

$$\{z | \exists x \exists y (z = <x,y> \wedge \forall t (t \in y \to P(t)) \wedge x \in y)\},$$

and the arrow $e_X \colon \in_X \to X \times \mathcal{P}_{\mathcal{S}_{ZF}}(X)$ is given by

$$[\{z, z' | \exists x \exists y (z = <x,y> \wedge \forall t (t \in y \to P(t)) \wedge x \in y) \wedge z = z'\}]_\equiv.$$

Now if the following arrow

$$\{z|\exists x\exists y(z = \langle x,y\rangle \wedge R(x,y))\} \xrightarrow{[\{z,z'|\exists x\exists y(z=\langle x,y\rangle \wedge R(x,y))\wedge z=z'\}]_{\equiv}} \{x|P\} \times \{y|Q\}$$

satisfies $R(x,y) \vdash_{ZF} P(x) \wedge Q(y)$ and $\vdash_{ZF} \forall y \exists y' \forall x(R(x,y) \leftrightarrow x \in y')$, which means that it represents (without loss of generality) a relation with the second component in $\mathcal{S}_{ZF}$, then its representing arrow from $\{y|Q\}$ to $\mathcal{P}_{\mathcal{S}_{ZF}}(X)$ is given by

$$[\{y,y'|Q \wedge \forall x(R(x,y) \leftrightarrow x \in y')\}]_{\equiv}.$$

□

Lemma 10.3.8. *The power set axiom is satisfied by $\mathcal{S}_{ZF}$.*

Proof. The subset relation for $\{x|P\}$ is given by the following arrow:

$$[\{z,z'|\exists y \exists y'(z =< y,y' > \wedge y \subseteq y' \wedge \forall x(x \in y' \to P(x))) \wedge z = z'\}]_{\equiv}$$

from $\{z|\exists y \exists y'(z =< y,y' > \wedge y \subseteq y' \wedge \forall x(x \in y' \to P(x)))\}$ to $\mathcal{P}_{\mathcal{S}_{ZF}}(X) \times \mathcal{P}_{\mathcal{S}_{ZF}}(X)$.

This relation is in $\mathcal{S}_{ZF}$ by virtue of axiom 5. □

10.3.3 An Algebraic Set-Theoretical Light on Definable Sets

First of all, the small definable classes are exactly those classes $\{x|P\}$ for which the unique arrow to 1, which is $[\{x,y|P \wedge y = 0\}]$, is small. This means that

$$\vdash_{ZF} \forall y \exists z \forall x(x \in z \leftrightarrow (P(x) \wedge y = 0)).$$

Now we know that

$$\vdash_{ZF} \exists y(y = 0)$$

and so the previous condition is equivalent to

$$\vdash_{ZF} \exists z \forall x(x \in z \leftrightarrow P(x)).$$

This means that the small definable classes are exactly the definable sets. Furthermore, $\{x|N(x)\}$ is a small definable class, where $N(x)$ is the formula saying that x is a finite ordinal: this follows from axioms 7 and 2.

Finally, $\{x|x = x\}$ is a universal definable class (and so also a universe), because, for every definable class $\{x|P\}$, the arrow $[\{x,x'|P \wedge x = x'\}]_{\equiv}$ is a mono from $\{x|P\}$ to $\{x|x = x\}$.

Last but not least, there exists an explicit (and obvious) representation for an initial ZF-algebra: this is given by

$$(\{x|x=x\}, \sqsubseteq_{\{x|x=x\}}, [\{x,z|z=\{x\}\}]_\equiv).$$

If

$$[\{z,z'|F(z,z')\}]_\equiv : \{z|P\} \to \{z'|Q\}$$

is small in $\mathbb{DCL}[ZF]$, and $[\{z,x|\lambda(z,x)\}]_\equiv$ is an arrow from $\{z|P\}$ to $\{x|x=x\}$, then

$$\sup\nolimits_{[\{z,z'|F(z,z')\}]}([\{z,x|\lambda(z,x)\}]_\equiv)$$

is given by the arrow

$$[\{z',x|Q \wedge \forall t(t \in x \leftrightarrow \exists z(F(z,z') \wedge \lambda(z,t)))\}]_\equiv.$$

10.4 The Internal Category of Sets

In this section, we will introduce internal category theory to build an internal category of sets in $\mathbb{ZF}$. This will be the internal topos induced by the initial ZF-algebra for $\mathcal{S}_{ZF}$.

10.4.1 Internal Category Theory

First of all, we need to recall the notion of internal category, which is the generalization of the notion of small category. Although we can define an internal category in an arbitrary category (requiring the existence of certain pullbacks), we prefer considering the case of a category $\mathbb{C}$ with all finite limits.

Definition 10.4.1. An *internal category* of $\mathbb{C}$ is a 6-ple

$$(C_0, C_1, \delta_0, \delta_1, ID, \square)$$

in which C_0, C_1 are objects of $\mathbb{C}$ and $\delta_0, \delta_1 : C_1 \to C_0$, $ID : C_0 \to C_1$,

$$\square : C_1 \times_\square C_1 \to C_1$$

are arrows of $\mathbb{C}$, where $C_1 \times_\square C_1$ fits in the following pullback.

$$\begin{array}{ccc} C_1 \times_\square C_1 & \xrightarrow{p_1} & C_1 \\ {\scriptstyle p_0}\downarrow & & \downarrow{\scriptstyle \delta_0} \\ C_1 & \xrightarrow{\delta_1} & C_0 \end{array}$$

The 6-ple $\mathbb{C}$ must satisfy the following requirements:

1. $\delta_1 \circ ID = \delta_0 \circ ID = id_{C_0}$.
2. $\delta_0 \circ \square = \delta_0 \circ p_0$ and $\delta_1 \circ \square = \delta_1 \circ p_1$.
3. $\square \circ \lceil ID \circ \delta_0, id_{C_1} \rceil = \square \circ \lceil id_{C_1}, ID \circ \delta_1 \rceil = id_{C_1}$.
4. $\square \circ \lceil \square \circ p_0, p_1 \rceil = \square \circ \lceil p_0, \square \circ p_1 \rceil \circ \lceil p_0 \circ p_0, \lceil p_1 \circ p_0, p_1 \rceil \rceil : (C_1 \times_\square C_1) \times_\square C_1 \to C_0$.

Here we denote with $\lceil f, f' \rceil$ the unique arrow that exists for the definition of pullback, and $(C_1 \times_\square C_1) \times_\square C_1$ is the pullback of $\delta_1 \circ \square$ and δ_0.

Before going to the next section, we show a way to externalize internal categories. This is done in a very natural way by means of global elements.

Proposition 10.4.2. *If* $\mathcal{C} = (C_0, C_1, \delta_0, \delta_1, ID, \square)$ *is an internal category of* $\mathbb{C}$ *and* I *is an object of* $\mathbb{C}$, *then*

$$\text{Hom}(I, \mathcal{C}) := (\text{Hom}(I, C_0), \text{Hom}(I, C_1), \delta_0 \circ (-), \delta_1 \circ (-), ID \circ (-), \square \circ \lceil (-)_1, (-)_2 \rceil)$$

is a category.

Proof. Every point of the definition of category follows immediately because of the relative point in the definition of internal category. This is possible as the functor $\text{Hom}(I, \bullet)$ preserves all finite limits. □

We will denote by $\Gamma(\mathcal{C})$ the category $\text{Hom}(1, \mathcal{C})$, where 1 is a (selected) terminal object of $\mathbb{C}$.

Remark 10.4.3. We should notice that an internal category in a category $\mathbb{C}$ with all finite limits is nothing more than a model of the first-order theory of categories (see Johnstone 2002) in $\mathbb{C}$.

10.4.2 The Real Category of Sets: $\mathcal{SET}$

We will now define an internal category of $\mathbb{ZF}$ (or equivalently of $\mathbb{DCL}(ZF)$), called $\mathcal{SET}$, which is built on the initial ZF-algebra for $\mathcal{S}_{ZF}$. This category is given by the following assignments:

1. $\mathcal{SET}_0 := \{x | x = x\}$.
2. $\mathcal{SET}_1 := \{F | \exists f \exists z (F = <f, z> \wedge \text{Fun}(f) \wedge \text{ran}(f) \subseteq z)\}$.

3. $\delta_0 := [\{F, x | \exists f \exists z (F =< f, z > \wedge \text{Fun}(f) \wedge \text{ran}(f) \subseteq z \wedge \text{dom}(f) = x)\}]_\equiv$.
4. $\delta_1 := [\{F, z | \exists f (F =< f, z > \wedge \text{Fun}(f) \wedge \text{ran}(f) \subseteq z)\}]_\equiv$.
5. $ID := [\{x, F | \exists f (F =< f, x > \wedge \forall t (t \in f \leftrightarrow \exists s (s \in x \wedge t =< s, s >)))\}]_\equiv$.
6.

$$\square := [\{J, G | \exists f \exists f' \exists z \exists f''(J =< f, < f', z >> \wedge \text{Fun}(f) \wedge \text{Fun}(f')$$
$$\wedge \text{ran}(f) \subseteq \text{dom}(f') \wedge \text{ran}(f') \subseteq z \wedge G =< f'', z > \wedge$$
$$\wedge \forall t (t \in f'' \leftrightarrow (\exists s \exists s' \exists s''(< s, s' > \in f \wedge < s', s'' > \in f' \wedge t =< s, s'' >))))\}]_\equiv,$$

once we easily realized that the object of composable arrows is given by

$$\{J | \exists f \exists f' \exists z (J =< f, < f', z >> \wedge \text{Fun}(f) \wedge$$
$$\wedge \text{Fun}(f') \wedge \text{ran}(f) \subseteq \text{dom}(f') \wedge \text{ran}(f') \subseteq z)\}.$$

We obtain that

Theorem 10.4.4. $\mathcal{SET}$ *is an internal category.*

Proof. See Johnstone (1977). □

We also derive that this is an internal topos, as every construction for a topos can be done in $\mathbb{ZF}$. This follows from a well-written proof of the fact that sets and functions form a topos formalized in ZF.

Moreover, we should notice that this internal category exactly corresponds to the canonical interpretation (according to model theory) of the first-order theory of categories in ZF set theory.

10.5 Definable Sets and Global Elements

We have just well explained what we mean by the internal category $\mathcal{SET}$; we now would like to study the category $\Gamma(\mathcal{SET})$. Following in detail the definition, the objects of $\Gamma(\mathcal{SET})$ are the equivalence classes $[\{x | P(x)\}]_\equiv$ of definable classes with the property that $\vdash_{ZF} \exists! x P(x)$. The arrows of $\Gamma(\mathcal{SET})$ are the classes of equivalence $[\{f | P(f)\}]_\equiv$ which satisfy $\vdash_{ZF} \exists! f P(f)$, and

$$P(f) \vdash_{ZF} \exists f' \exists z (f =< f', z > \wedge \text{Fun}(f') \wedge \text{ran}(f') \subseteq z).$$

As a consequence, we have the following theorem:

Theorem 10.5.1. $\mathbb{DST}(ZF)$ *and* $\Gamma(\mathcal{SET})$ *are equivalent.*

Proof. We need to consider the following functors:

1. $\mathbf{P} : \mathbb{DST}(ZF) \to \Gamma(\mathcal{SET})$ is given by
   $\mathbf{P}(\{x|P(x)\}) := [\{z|\forall x(x \in z \leftrightarrow P(x))\}]_\equiv$
   $\mathbf{P}([\{x,y|F(x,y)\}]_\equiv) :=$
   $:= [\{f'|\exists f \exists z(f' =< f,z> \wedge \forall t(t \in f \leftrightarrow \exists x \exists y(t =< x,y> \wedge F(x,y))) \wedge$
   $\wedge \forall y(y \in z \leftrightarrow Q(y)))\}]_\equiv$
2. $\mathbf{P}' : \Gamma(\mathcal{SET}) \to \mathbb{DST}[ZF]$ is given by
   $\mathbf{P}'([\{z|P(z)\}]_\equiv) := \{x_0|\exists z(P(z) \wedge x_0 \in z)\}$
   $\mathbf{P}'([\{f'|Q(f')\}]_\equiv) := [\{x,x'|\exists f'(Q(f') \wedge \exists z \exists f(f' =< f,z> \wedge < x,x'>$
   $\in f))\}]_\equiv$
   where x_0 is a fixed variable. (We can think of it as the first variable. The condition for this is that the variables of the language of ZF are presented in a countable list.)

We can then immediately see that $\mathbf{P} \circ \mathbf{P}'$ is the identity functor for $\Gamma(\mathcal{SET})$, while there is a natural isomorphism from the identity functor of $\mathbb{DCL}[ZF]$ to $\mathbf{P}' \circ \mathbf{P}$, which is given by the arrows

$$[\{x,x'|P(x) \wedge x = x'\}]_\equiv : \{x|P(x)\} \to \{x_0|\exists z(x_0 \in z \wedge \forall x(x \in z \leftrightarrow P(x)))\}$$

$\square$

We can think about objects of $\Gamma(\mathcal{SET})$ as *sets with a name*, because they are exactly determined by a class of formulas $[P]$ with one free variable; the name of the object of $\Gamma(\mathcal{SET})$ determined by $[P]$ could be, for example, $[[P]]$.

10.6 Final Remarks

In our attempt to clarify the relation between (formal) classes and sets and between metamathematics and mathematics, by means of a unique mathematical structure, we started by introducing the syntactic category $\mathbb{ZF}$ (that we proved to be equivalent to the category of definable classes of ZF). This category shapes the metamathematical level: its objects are classes as are usually (see, e.g., Jech 2003) introduced in the set theorists' practice. Moreover, this category has an important full subcategory: the category of definable sets, whose objects are those definable classes $\{x|P\}$ for which

$$\vdash_{ZF} \exists z \forall x(x \in z \leftrightarrow P).$$

This corresponds to the naïve category of sets. Obviously, this is *not* the real category of sets. In this context, the *real* category of sets is $\mathcal{SET}$. However, this is not a category: it is an *internal category* in $\mathbb{ZF}$. The relation between metamathematics and mathematics in this context exactly corresponds to the relation between categories and internal categories. Mathematical concepts are represented through internal categories, and external (or metamathematical) concepts are expressed on the categorical level. In our opinion, the most interesting result described in the previous sections is the equivalence of the two most (at least in our opinion) natural ways to give an external account of the notion of set. We proved that the category of definable sets is equivalent to the category obtained by global sections on $\mathcal{SET}$: *the classes that satisfy the comprehension axiom are exactly those classes that can be named.*

Acknowledgements The author would like to acknowledge G. Rosolini and T. Streicher for very useful and fruitful discussions. The *diagrams* package by P. Taylor was used in this article.

References

Awodey, S., C. Butz, A. Simpson, and T. Streicher. 2007. Relating topos theory and set theory via categories of classes. *Bulletin of Symbolic Logic* 13(3): 340–358.
Blass, A. 1984. The interaction between category theory and set theory. *Contemporary Mathematics* 30: 5–29.
Engler, E., and H. Röhrl. 1969. On the problem of foundations of category theory. *Dialectica* 23(1): 58–66
Feferman, S. 1977. Categorical foundations and foundations of category theory. In *Logic, foundations of mathematics, and computability theory*, The University of Western Ontario series in philosophy of science, vol. 9, ed. R.E. Butts and J. Hintikka, 149–169. Dordrecht: Springer.
Jech, T. 2003. *The third millennium edition, revised and expanded.* Springer monographs in mathematics. Berlin/New York: Springer.
Johnstone, P. 1977. *Topos theory.* London/New York: Academic Press.
Johnstone, P.T. 2002. *Sketches of an elephant: A topos theory compendium, vol. 2*, Volume 44 of Oxford logic guides. Oxford: The Clarendon Press/Oxford University Press.
Joyal, A., and I. Moerdijk. 1995. *Algebraic set theory*, Volume 220 of London mathematical society lecture note series. Cambridge: Cambridge University Press.
Kreisel, G. 1971. Observations of popular discussions on foundations. In *Axiomatic set theory*, Proceedings of symposia in pure mathematics, 1, ed. D.S. Scott and T.J. Jech, 183–190. Providence: American Mathematical Society.
Landry, E., and J.P. Marquis. 2005. Categories in context: historical, foundational and philosophical. *Philosophia Mathematica (3)* 13(1): 1–43.
Lawvere, F.W. 2005. An elementary theory of the category of sets (long version) with commentary. *Reprints in Theory and Applications of Categories* 11: 1–35.
Lolli, G. 1977. Categorie, universi e principi di riflessione. *Bollati Boringhieri.*
Marquis, J.P. 1995. Category theory and the foundations of mathematics: philosophical excavations. *Synthese* 103(3): 421–447.
Maschio, S. 2012. Aspects of internal set theory. PhD thesis.
McLarty, C. 2004. Exploring categorical structuralism. *Philosophia Mathematica (3)* 12: 37–53.
Rosolini, G. 2011. La categoria delle classi definibili in IZF è un modello di AST. *XXIV incontro di logica AILA.*
Simpson, A. 1999. Elementary axioms for category of classes. In *Logic in computer science*, Trento, 77–85

Part III
Philosophy and Mathematics

Chapter 11
Structure and Applicability

Michele Ginammi

11.1 Introduction

In this article I am going to discuss the possibility of solving the applicability problem of mathematics by means of the notion of structure. First of all, what is the applicability problem? As Steiner (1998) points out, there is not *one* single problem that may be called 'the applicability problem', but *many* problems that must be faced in different ways.[1] I am going to focus on what could be called the problem of the *representative effectiveness* of mathematics: how is it possible that we can employ mathematics to *represent* physical systems? Or, better, how does mathematics play a part in making *descriptive* claims about the world?[2]

Based on the talk for the conference 'Filosofia della matematica: dalla logica alla pratica. Giovani studiosi a confronto', held in Pisa at the Scuola Normale Superiore, 24–26 September 2012. I would like to thank all the conference participants for their valuable and stimulating contributions and particularly Marina Imocrante for her precious bibliographical suggestion. I would also like to thank Gabriele Lolli, Christopher Pincock, Sorin Bangu, Gabriele Galluzzo, Giulia Felappi, Roberto Gronda, Gian Maria Dall'Ara and two anonymous referees, who read the previous drafts of this paper and helped me with insightful comments.

[1] Steiner distinguishes four applicability problems: (1) a *semantic* problem, (2) a *metaphysical* problem, (3) a *descriptive* problem and (4) a *heuristic* problem. The specific problem I am going to discuss in this paper would fall into (3), but I will consider a wider problem than Steiner's.

[2] A bit of terminological clarification is needed. Let us say that A mathematically represents B when A is a mathematical model such that by observing it we can draw conclusions that are reasonably true in B. Following Pincock (2012, pp. 25–26), I will mean by 'model' an entity (either concrete or abstract) by means of which we aim to represent a physical domain and by 'representation', a model with a content, where the content is what provides us with the conditions under which the representation is accurate.

M. Ginammi (✉)
Ph.D. student, Scuola Normale Superiore, Pisa, Italy
e-mail: michele.ginammi@sns.it; mginammi@gmail.com

One of the most interesting accounts of this problem appeals to the notion of structure. According to it, the representative effectiveness of mathematics can be satisfactorily accounted for if we say that, when a mathematical structure satisfies certain conditions, it can be considered as a good description of a physical system. A very naive formulation of this structural account is the following: mathematics is so useful and effective as a representative tool in science because (A) mathematics studies structures and (B) these structures can be found in nature and their description is hence part of the scientific picture of reality.[3] We can find an example of this structural account in Shapiro (1997), when he says that:

> the contents of the nonmathematical universe exhibit underlying mathematical structures in their interrelations and interactions. According to classical mechanics, for example, a mathematical structure much like the inverse-square variation of real numbers is exemplified in the mutual attraction of physical objects. In general, physical laws expressed in mathematical terms can be construed as proposals that a certain mathematically defined structure is exemplified in a particular area of physical reality. (see also Shapiro 1983, p. 248)

Further and more sophisticated examples can be found in French (2000), Pincock (2004), van Fraassen (2008) and others.[4] The key idea of this account is that the conditions that a mathematical representation has to satisfy in order to properly describe a physical system are *structural*; namely, a mathematical scientific representation presupposes the existence of a structural relation between a mathematical structure and the arrangement of some properties, quantities and entities in a given scientific domain. The representation is appropriate if there is a 'structural similarity' between the target (the physical domain) and the representing tool (the mathematical structure).

Such an account can be variously stated. For example, one may focus, as Pincock (2004) does, on the conditions that a statement of applied mathematics (a 'mixed' statement) must satisfy in order to be considered as true: 'According to the mapping account of applications, the truth of a statement of applied mathematics (or 'applied statement') depends on the existence of a mapping of a certain kind

---

[3]This formulation suggests that the structural account of applicability implies in some sense some version of mathematical structuralism. This is nevertheless false, since it simply implies that we can consider mathematics as the study of abstract structures, but it does not imply that this is the only way to understand mathematics, or that it is the best. Analogously, this formulation also suggests that the structural account implies in some sense some version of structural scientific realism. To see whether this other implication holds is a more complicated problem, which depends on what one means by 'structural scientific realism', but in general this implication does not hold, either, since the structural account implies only that structures are *part* of the scientific picture of reality.

[4]Also Steiner (1998) mentions a 'distressingly common "explanation" for the effectiveness of mathematics in physics' according to which 'mathematics studies 'structures', and these structures are displayed in nature where they can be studied by physics' (p. 6), but he rejects this account since he thinks the currency of this explanation just stems from a confusion among different senses of the term 'applicability'. I think the way in which Steiner gets rid of the structural solutions to the applicability problems is offhand. In this paper I will not take care of replying to Steiner's claim, but I hope to show that the structural account *can* offer a valuable contribution to the solution of (if not all the problems, at least) the problem of the *representative* effectiveness of mathematics.

from a physical situation to a mathematical domain' (p. 69). Thus, if such a mapping exists, we are entitled to say that our 'applied statements' are true. However, my focus in this article will be on the *epistemic* conditions that enable us to consider a mathematical structure as a good representation of a given physical system (the target). In other words, how do we come to know that a certain mathematical structure can be used to represent a physical system in such a way that we can employ that mathematical structure to gain (possibly new) knowledge about the physical domain at issue?

However, before facing these epistemic problems, it is important to understand what this 'structural similarity' consists in. The mathematical representation has obviously to preserve the structure of the physical domain, but this is not something that can be easily done when we do not completely know the target or when our knowledge of it is still tentative. Actually, several problems arise when we observe that these mathematical scientific representations are frequently effective not only in describing a physical domain but also in broadening our knowledge of that domain. Thus, the structural account we have just considered should be able to account also for a possible 'heuristic' role of mathematics.

In this article I will take into consideration the actual effectiveness of this structural account to make these questions clearer. As I will try to show, the structural account needs to be improved in several senses in order to be considered a satisfactory account.

11.2 The Structural Account

11.2.1 Setting Out the Problem

The first problem to solve is to understand how to define this 'structural relation' in a rigorous manner. As a first attempt, let us say that M is a mathematical structure and S a structured physical domain or a 'physical structure'.[5] We will say that S is homomorphic to M *iff* there exists a $\phi: S \to M$ such that:

1. For any function f_A in S, there exists a correspondent f_B in M such that, for any $x_1, \ldots, x_n \in dom(S)$,

$$\phi(f_A(x_1, \ldots, x_n)) = f_B(\phi(x_1), \ldots, \phi(x_n)).$$

---

[5]I take a mathematical structure to be defined, as usually, as an ordered quadruple

$$M = \langle dom(M), \{R_i\}_{i \in I}, \{f_j\}_{j \in J}, \{c_k\}_{k \in K} \rangle,$$

where $dom(M)$ is a non-empty set of objects (this is our 'universe'; it must be noted that the nature of these objects is absolutely irrelevant); $\{R_i\}_{i \in I}$ is a non-empty set of n_i-ary relations defined on $dom(M)$; $\{f_j\}_{j \in J}$ is a set of functions defined on $dom(M)$; and $\{c_k\}_{k \in K}$ is a set of special elements belonging to $dom(M)$ (including, the unity element). I, J, K are three disjoint sets of indices. I will leave open, for the moment, the question concerning what a 'physical' structure is.

2. For any relation R_a in S, there exists a correspondent R_b in M such that, for any $(x_1, \ldots, x_m) \in dom(S)$, $(x_1, \ldots, x_m) \in R_a$ iff $(\phi(x_1), \ldots, \phi(x_m)) \in R_b$.
3. For any constant c_α in S, there exists a c_β in M such that $\phi(c_\alpha) = c_\beta$.

With this definition in mind, we can now say that when ϕ is injective we have a monomorphism, when ϕ is surjective we have an epimorphism and when ϕ is both injective and surjective we have an isomorphism.

11.2.2 Minimal Condition

All these structural relations guarantee some kind of structure preservation. The problem now is to understand whether there is a 'minimal' structural relation that an applied mathematical structure has to satisfy in order to guarantee its descriptive effectiveness. At a first glance it seems that isomorphism is the best situation to work with, since it grants a perfect correspondence between the two domains. Of course it is, but it is also a very tight condition to be imposed as a minimal request. It is not difficult to find a case of mathematical representation which does not meet this condition and nevertheless is effective in representing a physical domain.[6] On the other side, the loosest condition we can impose is the homomorphic one. However, in this way we run the risk that the condition we impose does not suffice to secure the representation with a content. Namely, given a fixed physical domain, there will always exist a homomorphism from it to a mathematical structure that has no representative content. Let us take, as an example, the trivial homomorphism that collapses all the elements in $dom(S)$ to one element in $dom(M)$ (the unity element, if it exists) and all the relations in R_S to the identity relation. This is a homomorphism, but we all agree that this would not be a good mathematical representation. It seems that a minimal condition should be found between these two extreme cases, isomorphism and (trivial) homomorphism, and that the way to articulate such a minimal condition is by taking into consideration the notion of 'content' of a representation.

Let us take the example of a city map. What we desire is that the map reproduces the relevant aspects of the geographical area at issue, where by 'relevant aspects' I mean those peculiarities (streets, distances, corners, eventually altitudes and so on) that we are interested in knowing and that make the map *effective* for the aims it has been created for. Also in this case we have to avoid the case of the trivial homomorphism; namely, we want that every relevant aspect of the represented area

---

[6] A typical example is offered by the Navier-Stokes equations. Here we use a set of equations to successfully model the behaviour of a fluid substance. These equations implicitly assume that the substance at issue is *continuous*, although our best theories say that the ultimate composition of matter is *not* continuous. So, assuming that our best theories are true, it does not seem to be possible to set up an isomorphism between the mathematical structure and the physical one, since the former seems to be 'richer' than the latter. Nonetheless, these equations are effectively employed in representing the behaviour of the phenomenon at issue.

be *distinctly* represented by a mark (a line, a dot, or whatever else) on the map. One might think that this amounts to demanding that the map be such that there is a *monomorphic* relation from the land to the map.[7] But the monomorphism is such only when we have fixed what we mean by 'relevant aspect'. If we are interested only in distances, we can reasonably avoid any indication about the altitude; but if we want to know also how much exertion we will have to bear in going from point A to point B, we will need in addition an indication of the altitudes and of the differences in elevation. Similarly, the same considerations can be made for the case of a mathematical representation. Also in this situation we want that the mathematical representing structure be able to grasp univocally and distinctively every single *relevant* element of the physical domain we are going to represent. We want that the mathematical structure be able to grasp all the relevant elements and all the relevant facts and relations in the physical system, *without any loss of information*—and it seems that the only way to grant this is to impose that the homomorphism is injective, i.e. that it is a monomorphism. Thus, it seems that the minimal condition we are searching for is the monomorphism condition—but such a monomorphic relation can be defined only when we have clearly in mind what the relevant aspects of the physical domain are.

Bueno and Colyvan (2011) consider the same problem, although under a different perspective. As they say, 'It would seem that the mapping employed will depend on the richness of the two structures in question' (p. 348), S and M. If S is richer than M (i.e. there are more objects or more structural relations between them in the physical domain than can be represented in the mathematical structure), then ϕ can be a simple homomorphism (neither injective nor surjective) or an epimorphism. But if M is richer than S, then we must consider a third possibility, monomorphism. But we also want, in order for mathematics to be useful, that ϕ be invertible so that we can 'move freely into and out of the mathematics, just as we can move freely between our street directory and the city' (pp. 348–349)—and this implies that ϕ must be a monomorphism, with the further consequence that it is apparently not possible for the physical structure to be richer than the mathematical one. Although Bueno and Colyvan (2011) bring forward examples of mathematical representations in which this is exactly what seems to be the case, it seems to me that this is the consequence of their overlooking the fact that, when we represent a physical domain, we are interested only in some *relevant* aspects of that domain. If we keep this point into consideration, cases in which S is structurally richer than M are simply ruled out. For if we admit that M is a good representation of S and nevertheless S is richer than M, this means that there are objects or relations in S that simply are irrelevant for the aims for which we made the representation: if the mathematical representation is not able to grasp them and remains effective, then these elements cannot be considered as relevant. But the question is much more complicated than this.

---

[7]Injective homomorphisms are also usually called 'embeddings', but in this paper I will continue talking of monomorphisms instead, since in the literature on the argument 'embedding' is often used as a generic term (see, e.g. Pincock 2004).

11.2.3 Some Remarks on the Structural Account

First of all, we said that the minimal condition should be articulated around the notion of 'content' of a representation, but this notion immediately leads us to the highly pragmatic (and highly problematic) notion of 'relevant aspects' of a physical domain. The representation is made with a certain aim in view, and it is this aim that determines which is the relevant content of the representation. This is a very delicate point. For example, if I want to give a mathematical representation of the kinematics of a system of bodies, I will ignore the colours of the bodies, since they are irrelevant to my representation. So, I will consider a physical structure S that has no colours among its objects (or among its relations).[8] The problem is: how can I select the relevant aspects of the piece of nature that I am trying to represent? It seems that this task should *precede* the application of mathematics, but it often happens that it is just mathematics that helps us in selecting the relevant aspects of a physical system.[9] This is a very complicated issue which is not possible to discuss extensively here, but I will shortly try to say something more. However, this remark seems to suggest that there is no hope to articulate an account of representative effectiveness of mathematics in *purely* structural terms, since the pragmatic component seems to be irreducible.[10]

That an account of the representative effectiveness of mathematics cannot be achieved in purely structural terms is also shown by the fact that (whatever be the minimal condition to be prescribed) the mathematical structure *can* be richer than the targeted physical domain (and that is what usually happens). So, as Bueno and

---

[8] An anonymous referee made me notice that in this case the physical domain (the bodies of the system and the relations among them) is (non-trivially) homomorphic to the mathematical structure (the additive structure of real numbers), but is not injective, since two bodies can correspond to the same mass-number—and hence the monomorphism rule seems to be infringed. However, in this case the monomorphism should be not from the physical system to the additive structure of real numbers, but rather from the physical system to a vector space, whose number of dimensions depends on the number of bodies composing the system (in the Euclidean space, for each body we will have three dimensions for its position, three for its velocity and one for its mass), on which a relation of correspondence between these vectors is fixed. In this way the monomorphism is preserved. I recall I am dealing only with the *representative* effectiveness of mathematics and that the monomorphism condition I am trying to defend as the proper *minimal* request refers just to this particular effectiveness of mathematics. Measuring is surely an important way in which we 'apply' mathematics, but it is different from representation—even if representation, of course, subsumes measuring practices. Even if measuring can be accounted for in structural terms as much as mathematical representations can, it will satisfy different conditions.

[9] Batterman (2002), for example, stresses the importance of 'asymptotic reasoning' in removing explanatory non-relevant aspects. The problem is here interwoven with mathematical explanatory power in science.

[10] Bueno and Colyvan (2011) come to the same conclusion. They also say that Pincock (2004) aim was precisely to account for the applicability of mathematics in *purely* structural terms, but I must disagree with them on this point, since in no place does Pincock make such a claim. The same remark is made by Batterman (2010, p. 8n).

Colyvan (2011) rightly point out, we may be confronted with situations in which our mathematical structure predicts more than one possible solution and not all these solutions have an empirical counterpart. Let us take the case of a quadratic equation used to predict where a projectile will land. Such an equation will have two solutions, and it may happen that these two solutions will not coincide. If, for example, we want to predict the landing of a projectile launched from the cliff of a mountain, one of the two solutions will be negative and will have no physical interpretation. In other cases, however, mathematical solutions previously considered as physically meaningless are suddenly and unexpectedly reassessed in virtue of a new interpretation of them. A typical example of this is the postulation of the existence of *antielectrons* (or *positrons*), made by the physicist Paul A.M. Dirac on the basis of an original interpretation of his relativistic version of the Schrödinger wave equation for electrons. Now, the problem is: how can we distinguish, within a mathematical structure, its representing from its non-representing parts? The problem is quite simple for the case of a projectile, but what about more complex and intricate cases? According to Bueno and Colyvan (2011):

> such crucial information required to solve this physical problem *is not part of the mapping between mathematical structure and physical structure*. In short, the mapping account of mathematical applications is incomplete.
>
> [...]. Moreover, the incompleteness of the mapping account is seen clearly as a result of problems relating to the specifications of the mappings in question. (pp. 349–350)

The structural account is not able, *by itself*, to tell us which of the possible choices is the right one (or are the right ones) and which is not. What is more, it is even unable to *justify* why, in the projectile example, the positive solution is the right one, while the negative one is not. From a *purely* structural perspective, both of them are indistinguishable as far as their rightness is concerned. The lesson to be learned is our account cannot avoid taking *non*structural components into consideration.

Secondarily, this way to set out the issue presupposes that we already have a structured physical domain. That is just what permits us to say that the mathematical description of that domain is effective *because* the mathematical structure employed is a 'good copy', or a 'good representation', of the physical structure underlying the physical domain. But what is this 'physical' structure, and in which sense is the physical domain 'already structured'? Moreover, that the physical domain has to be in some sense pre-structured seems to suggest that we already know at least what elements compose this domain. But in many contexts this is not the case, and nevertheless mathematics helps us to deepen our knowledge of the elements of the domain. For example, in particle physics mathematical models play a very valuable role in discovering new particles.[11] How is that possible? I will try to say something more about this aspect in Sect. 11.3.

---

[11]An interesting case in this regard is the discovery of the omega minus particle, made by Gell-Mann and Ne'eman in 1963. For a historical and philosophical reconstruction of the episode, see Bangu (2008, 2012, cap. 5).

Thirdly, the considerations presented until now justify us in saying that, whatever the minimal structural condition be, this condition is just *necessary*, but not *sufficient* in order to account for the representative effectiveness of mathematics. In other words, *if* M is effective in representing S, *then* we can say that there is a monomorphism from S to M—but the converse is not true. In order to prove the converse, we should prove that any monomorphism from S to M makes of M an *effective* representation of S. But we have already noticed that the monomorphism must go from the *relevant* elements and relations in S to the elements and relations in M. So, in order to prove the converse, we should already know what the relevant aspects of a physical domain are.

Finally, the structural account, as we have seen, says that a mathematical representation is effective only if there exists a preserving structure relation (whatever it be: monomorphism, homomorphism or anything else) from the physical domain to the mathematical structure. But how do we come to say that such a structure-preserving relation really subsists? To say that we have to know in advance how the physical domain is structured—we have at least to know what elements compose it and how they interact. However, as I have already noted, mathematical representations are usually employed to *discover* new entities or new relations in the physical domain.[12] In this case, there is a difficulty about how we can come to know that a particular monomorphism really subsists.[13]

To sum up, we saw that a mathematical structure M seems to be successfully applicable only if there exists a monomorphism from the (relevant) physical structure S to M. This seems to be a necessary condition, since, if such a monomorphism does not exist, some elements of S (objects or relations) will be identified by the representing mathematical structure M and this would be a loss of information. Yet, this cannot be taken to be a sufficient condition, for the reason I have just pointed out. Moreover, the structural account suffers from different problems that must be solved in some way. In the next sections I will try to give my modest contribution to make the structural account more satisfying, but it seems clear since now that it will be impossible to exclude from the account some nonstructural, pragmatic component.

---

[12] I have already mentioned the discovery of the omega minus particle. Other cases that could exemplify this kind of situation are the discovery of the planet Neptune and the discovery of the neutrino.

[13] It must be noted that it is not sufficient that *there is* a monomorphism: we must be able to spell it out in order to make the representation effective—i.e. in order to make the representation a useful tool to make verifiable predictions and hypotheses.

11.3 Physical and Mathematical Structures

11.3.1 The Problem of the Coordination Revised

In the previous section, I have pointed out that the physical domain has to be already, in some sense, 'structured' in order to define a structure-preserving relation from it to a mathematical structure. Namely, on the physical side we should already have something like a 'physical' structure. But what does this expression mean? What is a physical structure? One might say that a physical structure is just a mathematical structure embedded in nature so that the only difference between the two is that the first is abstract and the second is a model for it. Well, but how do we know that the latter is just a model for the former? What we have not considered, up to now, is the fact that the physical structure is something that we do not know, something hidden in the phenomena of nature. The mathematical structure is often just a 'tool' by means of which we manage to grasp this hidden physical structure. But, given that the monomorphic relation is the only way for us to be sure that the mathematical structure can be effectively and successfully used to know the physical structure, how can we set up such a monomorphic relation if one of the two terms of the relation is unknown? And how can we know that such a relation *really subsists*?

Actually, there are two problems here that we should keep distinct: the former concerns *how* we can set up a relation between physical and mathematical structures; the latter concerns how we can understand that such a relation *is actually* a monomorphism. Let us start with the first problem. The point is: how can we fix a relation (whatever it is: isomorphism, homomorphism or monomorphism) between a mathematical structure and a physical structure *that we do not know*? Or, to use a different terminology, how can we compare a structure to a (alleged) model of it if we do not know the proper interpretation that links the former to the latter—and we do not even know the structure?[14]

The problem just sketched can be seen as a variation of the well-known problem of the coordination raised by Reichenbach (1965).[15] 'The mathematical object of knowledge — he says — is uniquely determined by the axioms and definitions of mathematics' (p. 34). On the contrary:

> The physical object cannot be determined by axioms and definitions. It is a thing of the real world, not an object of the logical world of mathematics. Offhand it looks as if the method of representing physical events by mathematical equations is the same as that

---

[14]This point is strictly linked to the fact that a mathematical representation can be (and often *is*) useful also in fostering new discoveries. Bueno and Colyvan (2011) do not pay any attention to this point, and it seems to me that an account of the representative effectiveness of mathematics which fails in accounting for it should be considered unsatisfying.

[15]On this parallelism with Reichenbach's problem of coordination, see also van Fraassen (2006, 2008).

of mathematics. Physics has developed the method of defining one magnitude in terms of others by relating them to more and more general magnitudes and by ultimately arriving at "axioms", that is, the fundamental equations of physics. Yet what is obtained in this fashion is just a system of mathematical relations. What is lacking in such system is a statement regarding the significance of physics, the assertion that the system of equations is true for reality. (p. 36)

If, in a certain sense, mathematical truths are granted by the internal *coherence* of a mathematical structure, the same cannot be said for the physical relations. In this case we need something like a 'coordination': 'physical things are coördinated to equations. Not only the totality of real things is coördinated to the total system of equations, but *individual* things are coördinated to *individual* equations' (p. 37). Of course, one might say that what we are looking for is just a function (a mapping, an interpretation) between mathematical objects (and operations, and relations) and physical objects (and operations and relations)—so, what is the problem? Well, in the specific case of the physical coordination, the matter is much more complex than this:

[...], if two sets of points are given, we establish a correspondence between them by coordinating to every point of one set a point of the other set. For this purpose, the elements of each set must be defined; that is, for each element there must exist another definition in addition to that which determines the coordination to the other set. Such definitions are lacking on one side of the coordination dealing with the cognition of reality. Although the equations, that is, the conceptual side of the coordination, are uniquely defined, the "real" is not. (p. 37)

In short, baldly stated, the problem is that if the target of the representation is not a mathematical object, then we do not have a well-defined range for the function.

The situation is restated and analysed by van Fraassen (2006) as follows:

a function that relates A [*the phenomenon*] and B [*the mathematical structure*] must have a set as its domain. If A is, for example, a thunderstorm or a cloud chamber — a physical process, event, or object — then A is not a set. *Fine*, the realist answers, but A has parts and the function's domain is the set of these parts. Moreover, there are specific relations between these parts, and these relations have as their extensions sets of sequences in that domain. The function provides a proper matching provided that the images of these relations are relevant relations in the model. (p. 540)

So, we have A (the phenomenon at issue), B (the mathematical representation of it) and something that we could call $S(A) = \langle SA_1, SA_2 \rangle$, where SA_1 is the set of parts of A and SA_2 is the family of sets that are extensions of relations on these parts. In this account, $S(A)$ is actually a mathematical structure, and the Reichenbach problem in comparing $S(A)$ with B is ruled out, for both are abstract (mathematical) structures. But in this way we are just pushing the problem one step back; now the problem is: how can we compare A with $S(A)$? But that is not all. We have several ways to divide up A. Which of these is the right one? Realist might answer that the right one is that 'carves nature at joints', but this sounds more like beating about the bush, since we do not know what the proper way is to 'carve nature at joints'.

11.3.2 Phenomena, Data Models and Theory Models

So, let us take one step back, and try to find a way out. There are some points that should be noted and that may permit us to deflate (at least part of) the problem. The problem at hand is clearly a problem of representation: how can we represent a concrete physical system by means of an abstract structure such as a mathematical one? Now, as van Fraassen properly points out:

> The question of how a specific mathematical object can be used to represent some specific phenomena makes sense only in a context in which some description of the latter is at hand. Reichenbach, it seems to me, mistakenly pursues the 'profound' 'foundational' question of how such use is possible *outside any such context* — as if theories are received by babies or primitives before the acquisition of language.
>
> [...].
>
> He is explicitly addressing a situation in which there is no description at hand for what is to be represented. (van Fraassen 2006, pp. 541–542)

But we actually *have* a description at hand for physical entities, and this is just that offered by our own language. All this, evidently, involves a large amount of pragmatic elements, but it is quite obvious that mathematics is not an instrument that we apply in an aseptic context. Therefore, all things considered, the question of how we can represent a physical phenomenon by means of abstract objects such as mathematical ones is not really problematic. At least, it is no more problematic than the problem of denotation in philosophy of language. In other words, there is nothing problematic in using mathematical objects in order to *denote* parts of a physical phenomenon.

However, there are still two problems to solve: (1) since there are several ways to divide up a phenomenon into parts, which of these ways is the right one, and how can we recognize it? (2) How can we understand that a certain mathematical structure has a monomorphic relation to the target phenomenon (so that we can say that such a mathematical structure is a suited and effective representation of it)?

Questions (1) and (2) are particularly tangled up. As van Fraassen (2008) interestingly points out:

> [...] the assertion or denial of isomorphism depends on a certain selection on our part. In the case of two mathematical objects we can make the selection in a straightforward way, since they are already 'given' in a format which lends itself to us. Given a particular Hilbert space and a family of operators on it singled out by some equations, the relevant questions can obviously be formulated: for example, does this family contain an element I such that for all its members X, $IX = XI = X$? But how do I formulate questions of this sort for a part of nature, without using a selective description of it that already rests on a 'mathematization'? (p. 366, note 7)

Now, van Fraassen analyses the issue as follows. In concrete settings, the structure-preserving relation intervenes not between the phenomenon and the theory model but between the *data model* and the theory model. When we collect the data, what we have is already a mathematical structure, and the mathematical structure of the theory model tries to embed the mathematical structure of the data model.

However, according to van Fraassen, this does not push the problem one step back. Why? Because the 'construction of a data model is precisely the selective relevant description (by the user of the theory) required for the possibility of representation of the phenomenon' van Fraassen (2006, p. 544). The point that van Fraassen emphasizes is that:

> *There is nothing in an abstract structure itself that can determine that it is the **relevant data model**, to be matched by the theory.* That is why our talk of data models 'between' the theoretical model and the phenomena does not simply push Reichenbach's question one step back, to be faced all over again in the same way. [...].
>
> That is, the phenomenon, what it is like, taken by itself, does not determine which structures are data models for it. That depends on our selective attention to the phenomenon, and our decisions in attending to certain aspects, to represent them in certain ways and to a certain extent. (pp. 544–545)[16]

In representing a phenomenon, there is an ineliminable indexical component: the representation is a representation of something (as thus and so) *by somebody*. The data model is the phenomenon as represented by someone, and when we want to check a claim of adequacy, we will compare the theory model with the data model. But to say that the theory is adequate to the phenomena *as represented by us* (viz. our data model) is the same, *for us*, as to say that the theory is adequate to the phenomenon *tout court*. This last point is, on van Fraassen's point of view, a *pragmatical tautology*, and 'Appreciating that this equivalence for us is a pragmatical tautology removes the basis for the challenge' (p. 545)—that is, removes the necessity to push Reichenbach's question one step back.

11.4 In Search of a Way Out

11.4.1 A Proposal of Integration for the Structural Account

Van Fraassen's solution, by appealing to the notion of 'pragmatical tautology', sounds quite tricky, and not many would follow him on this road. Nevertheless, there are two points in van Fraassen's analysis that I want to retain and emphasize: (A) there is nothing in the phenomenon that determines which structures are data models for it, namely, which mathematical structures are able to capture its relevant aspects, and (B) when we want to check a claim of adequacy for a theory, we compare the theory model *with the data model* (and not with the phenomenon itself). So, the effectiveness of mathematics is not simply a matter of *conditions* that we have pre-emptively to satisfy for its application. What we need is a system that permits us to check the adequacy of a mathematical representation in the double sense of: (I) checking the adequacy of the model theory as properly embedding

---

[16]Italics and bolds are in the text.

(viz. monomorphically, as we saw) the data model and (II) checking the adequacy of the data model in representing the relevant aspects of the phenomenon at issue. Moreover, all these considerations show that the structural account—as presented so far—is not enough in order to account for the applicability of mathematics. It can constitute, at best, a good starting point, but it needs an integration, and such an integration has to consider the pragmatical elements that intervene in the realization of a mathematical model for a physical phenomenon.

In this section I am going to propose an integration of the structural account that, I think, can solve part of these problems and can also answer questions (2), which I left open ('how can we understand that a certain mathematical structure has a monomorphic relation to the target phenomenon so that we can say that such a mathematical structure is a suited and effective representation of it?'). My idea aims to fuse together the structural account with the DDI model account proposed by Hughes (1997), in a fashion that resembles the 'Inferential Account' proposed by Bueno and Colyvan (2011), yet different from it on some important points. The DDI model is a general account of modelling in physics. According to it, modelling is conceived as consisting of three main components: denotation, demonstration and interpretation (from which the name "DDI"). As Hughes clarifies, he is:

> not arguing that denotation, demonstration and interpretation constitute a set of speech acts individually necessary and jointly sufficient for an act of theoretical representation to take place. [He is] making the more modest suggestion that, if we examine a theoretical model with these three activities in mind, we shall achieve some insight into the kind of representation that it provides. Furthermore, we shall rarely be led to assert things that are false. (p. 329)

In short, I can sum up the DDI account as follows. A model *denotes* a physical phenomenon; it is a symbol for it and stands for it. In this manner, denotation plays the fundamental representative role, in accordance with Goodman's (1968) dictum that 'denotation is the core of representation and is independent of resemblance' (p. 5). Such a representation has also an internal dynamics, which permits us to make *demonstration* within it and make novel predictions. But the predictions that we draw in the model remain predictions *about* the model if we do not intervene with a process of *interpretation* that permits us to move back from the model to the phenomenon at issue.

Now, my idea consists in integrating this account by means of our previous considerations about the structural account and the data models, *plus* some extra considerations about the way in which we can recognize the existence of a monomorphism between a mathematical structure and the structure of a data model. The following figure gives a visual representation of my proposal (Fig. 11.1).

The DDI account is preserved, as evident by the presence of the denotation-demonstration-interpretation triad. Yet, whereas in the DDI scheme, denotation and interpretation occurred between the phenomenon and the model; here the 'model' is composed by two parts: the data model and the theory model. According to van Fraassen's considerations, the phenomenon is represented by the data model (and only derivatively, or secondarily, by the theory model), and this representative role is centred on the notion of denotation. Still, we have to add an important

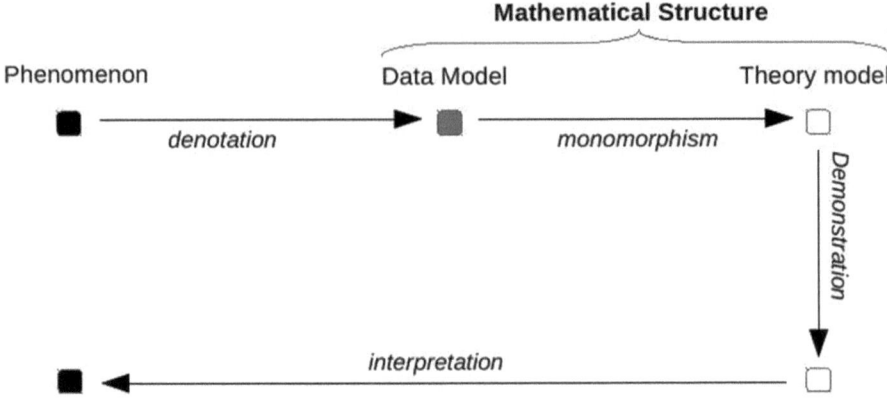

Fig. 11.1 Integration of the structural account

remark: such a denotation is only 'partial'; namely, the denotation does not give a complete coordination between elements of the phenomenon and terms of the data model. The only elements for which we give a denotation are those that appear in the data model. So, for example, if we are trying to describe the interrelations among pressure, volume and temperature in a thermic system, we will register the value for each of these values in different situations, and we will associate a number for any measurement. So, in our data model, we will denote pressure, volume and temperature by means of positive numbers. But this denotation is only 'partial': nothing is said about, for example, the denotation of negative numbers. We do not know if the structure that we are delineating in this way will be defined only on positive numbers or also on negative numbers. It will be a matter of interpretation to understand whether and how we will have to interpret negative numbers appearing in the mathematical structure as values for temperature, or pressure, or volume.

The data model, obviously, is not a complete mathematical structure. It is rather a *clue*, or a trace, of mathematical structure. Our task consists in finding the proper mathematical structure according to which the consequents in the physical phenomenon—as represented by the data model—are always a consequence in our mathematical structure, which corresponds to the monomorphic condition.[17] Now, the problem is that, given this condition, we cannot take for granted any consequence that we can demonstrate in the mathematical structure. For the monomorphic condition guarantees that if something is part of the hidden structure of the data model, then such a part will be surely deducible within our mathematical structure;

---

[17] On this regard, I think that the famous quotation from Hertz (1956)—'the necessary consequents of the images in thought are always the images of the necessary consequents in nature of the things pictured' (p. 1)—should be revised in the following way: *the necessary consequents of the things pictured should always be the necessary consequents of the images*. The original claim corresponds to the isomorphic condition, but we saw that such a condition is too restrictive.

but it does not guarantee that *every* consequence deducible from our mathematical structure is a consequence in the data model (and, hence, also in the phenomenon represented by it). So, what can we do? The answer is that every consequence that we can demonstrate in the mathematical structure must be empirically checked by means of an interpretation that permits us to go from the mathematical side to the empirical one. But this leads to two possibilities. Let us suppose that γ is a mathematical term deduced from the mathematical structure M, and suppose that we *do not know* how to interpret such a term. The two possibilities are either to hazard an interpretative guess at γ and then check its correctness or, if we have no idea concerning how such a term could be interpreted, to say that γ is just a mathematical sign having no interpretation, depending on the fact that the structure-preserving relation is not an isomorphism but a monomorphism.

Let me make a toy example to better explain this point. Let us take the following puzzle: five men find themselves shipwrecked on an island, with nothing edible in sight but coconuts, plenty of these, and a monkey. They agree to split the coconuts into five equal integer lots, any remainder going to the monkey. Man 1 suddenly feels hungry in the middle of the night and decides to take his share of coconuts at that very moment. He finds the remainder to be one after division by five, so he gives this remaining coconut to the monkey and takes his fifth of the rest, lumping the coconuts that remain back together. A while later, man 2 also wakes up hungry and does exactly the same thing: takes a fifth of the coconuts, gives the monkey the remainder, which is again one, and leaves the rest behind. So do men 3, 4, and 5, too. In the morning they all get up, and no one mentions anything about his coconut affair on the previous night. So they share out the remaining lot in five equal parts finding, once again, a remainder of one left for the monkey. Find the initial number of coconuts.[18] There are in fact an infinite number of solutions to this problem, but obviously we are asked to find the smallest number of coconuts that satisfies the condition. The answer is 15,621 (the reader may check by himself the correctness of this solution). However, a story says that it was Paul Dirac who notes that such a puzzle may have also another solution: -4 coconuts! This answer is right: each time a man arrives at the heap of coconuts, he finds -4 coconuts; since $-4 \div 5 = -1$ with a reminder of $+1$,[19] he takes away the remainder from the heap and gives it to the monkey (i.e. he gives $+1$ to the monkey); what remains in the heap is -5 coconuts; his one-fifth share is -1, which he takes, leaving -4 coconuts behind for the next man and so on, till the final division.[20] The point is that when we set up the equation for the solution of the problem, we have a linear diophantine equation.

---

[18]Reported in Barrow (1988, p. 254).

[19]Typically, quotient and remainder functions are defined only for natural numbers; hence, such an expression makes no sense until we define these functions for negative numbers too. Alternatively, one can check the validity of this negative solution by substituting -4 in the diophantine equation, assuming that the equation is defined also on negative numbers.

[20]The story also says that, by thinking about this problem, Paul Dirac came to the idea of the anti-matter. Such a story is quoted also by Barrows, but I was not able to check its reliability.

Now, in a problem like this we are obviously led to search positive solutions, since we are interpreting numbers as set of coconuts, and intuitively we do not handle sets of coconuts having negative cardinality. We *know* that there could be negative solutions, but we consider this only as a mathematical fact *without interpretation against reality*. However, if we can find a possible interpretation for such negative numbers, there is nothing to stop us to check this interpretation in reality and to examine whether it could be fruitful. In the case of the coconut puzzle, our experience of the macroscopic world suggests that such a negative interpretation will not be very fruitful; but in more abstract physical contexts (e.g. quantic world), this fact could be less obvious, and by giving credit to this interpretation, we could be led to new interesting discoveries.

I think there is nothing 'miraculous' here, as Wigner (1960) alleged. It is just a matter of interpretation: some interpretations could be entirely unproductive, some other interpretations could be very fruitful; but whether it is the first or the second case that occurs, it is not (only) a matter of which mathematical structure we adopt but also of whether the interpretation that we stick to the mathematical structure passes the empirical test or not.[21]

The way in which we set up our interpretation is quite complex and involves different aspects. The basis is obviously given by the initial denotation on which we have built up our data model. As far as the terms, for which we have given a denotation, are concerned, the interpretation is simply the inverse of the denotation. But then we have to extend this initial, partial denotation. In general, empirical verification is the main judge for interpretation; so, there are no tight prescriptions to be rigorously followed.[22] However, we usually extend the interpretation in conformity with a general principle of 'coherence'. It is hard to precisely clarify what this principle is; perhaps an example can be more helpful. Let us suppose we denote the velocity of a body along a certain direction by means of a vector v_1. If, very trivially, this vector is then transformed into a different vector v_2 in virtue of a certain relation describing the kinematics of that body in the system at issue, we will coherently interpret this new vector as the velocity of that body at a later instant

---

[21] As it has been often noted, Wigner's analysis seems to have been led astray by an overestimation of successful over unsuccessful cases of mathematical application in physics (see, e.g. Azzouni 2000; Pincock 2012). A closer attention to unsuccessful cases would have probably pointed out that the apparent 'miracle' of applied mathematics is often just the result of a long chain of failure in finding the proper interpretation for the suitable mathematical structure.

[22] In some cases, the theory model can refer to the past. In these cases it would be impossible to *empirically* test the 'predictions' of such a theory, because the initial conditions cannot be recreated at this time. The evaluation of this question is quite complex and cannot be examined in this paper. I would say that in these cases, all we can do is to rely on an inference to the best explanation or hope that some other experiment could offer an *indirect* confirmation of the theory. Thanks to an anonymous referee for bringing this point to my attention.

of time. We will not interpret it as, for example, the mass, or the acceleration of the body, since this interpretation would be incoherent with the original one.[23]

Moreover, the extension of the original interpretation will have to be made so as to maintain a certain coherence also with our experience. In the previous coconut case, for example, experience suggests that it is meaningless to extend the interpretation for negative numbers, because in our empirical experience we never meet with a set of -4 objects. However, in some cases, what we need for advancement is just the breaking of such an apparent coherence—and the more we go far from our immediate experience, the less problem we will have to 'force' this coherence in order to explore new possibilities.

It is important to note that the nonexistence of coherent interpretations for some statements about a mathematical structure does not invalidate the whole structure *as a suitable representation*. We can abandon a mathematical structure for several reasons. We can do it because the mathematical structure is too complicated and we are not able to handle it; or, we can abandon it because it is not complicated enough and we are not able to draw interesting predictions from it; or, we can abandon it because, at a certain point, we realize that a relevant fact or a relevant property of the phenomenon at issue is not (properly) represented by it. However, the fact that a mathematical structure, in a sense, represents more than what it is asked to represent is not a real problem; potentially, it is rather a source of richness and novelty which we should keep into consideration. If our mathematical structure makes a prediction which is not verified, we can try to shuffle off the guilt of such a failure upon the interpretation. If we can do it in a coherent way, we can continue working with that mathematical structure; but if we cannot do it, then we have to revise our mathematical structure and find a more suitable one.

11.4.2 How Do We Detect the Monomorphism?

There is still a question I left open: the question concerning how we understand that a mathematical structure is monomorphic to a data model. Indeed, if we *do not* know the structure that the data model partially represents, how can we say that such a structure is actually monomorphic to the mathematical structure under

---

[23] A less trivial and more interesting example of the 'principle of coherence' is offered, once again, by the previous mentioned case of the discovery of the omega minus particle. Here we have that each particle belonging to a particular class (the spin-$\frac{3}{2}$ baryons), along with its properties, is represented by means of a position in the $S - I3$ (strangeness-isospin) plane (according to the $SU(3)$ formalism). Given that the nine already known (in 1962) spin-$\frac{3}{2}$ baryons determined that this class of particles would have formed a decuplet scheme in such a plane, a tenth position was still vacant. So, since all the other 'positions' in the scheme represented a certain particle of the class, a general principle of coherence led Gell-Mann and Ne'emann to think that this position represented a particle too—a new particle still to be discovered. The new particle was then effectively detected 1 year later.

scrutiny? We know the mathematical structure, but we do not know the structure partially revealed by the data model (which, in our intentions, should represent the phenomenon). In some cases we know in advance that the mathematical structure we are going to adopt is richer than the phenomenon we aim to represent, and we test the mathematical structure assuming that the data model is at least monomorphic to it. This is the case, for example, of the Navier-Stokes equations, where we know in advance that the continuous mathematics we are going to apply is richer than the ultimate discrete dynamic of the phenomenon at issue. In some cases, yet, we have no idea about it. So, what can we do?

In these cases, it seems to me, there is no alternative but to *suppose* that the mathematical structure we want to adopt is actually *isomorphic* to the data model structure. If this hypothesis were true, then it would be true that—to repeat Hertz's words—'the necessary consequents of the images in thought are always the images of the necessary consequents of the things pictured'. In other words, if this hypothesis were true, we could be sure that any consequence in the mathematical structure has some correspondence in the data model structure and hence that it represents a fact in the phenomenon. So, if we run into a difficulty (e.g. when my mathematical structure tells me that a possible solution for the coconut puzzle is −4), we can shuffle off the guilt of such a failure upon the isomorphism hypothesis: the solution suggested by the mathematical structure does not take place in the real world because our mathematical structure is too rich, and the data model is only monomorphic (not isomorphic) to it. We do not have to abandon the mathematical structure, since we can simply abandon the hypothesis of isomorphism. We have thus a proof that the mathematical structure is not isomorphic to the data model, but in spite of this we can still rely on the hypothesis that the data model is monomorphic to our mathematical structure. Of course, not all the difficulties can be settled in this way. In the coconut example, we can proceed in this way because the mathematical structure gives us also a *valid* solution (i.e. 15,621). But if the only solution given by our mathematical structure were not valid, then we could not shuffle off the guilt upon the hypothesis of isomorphism, but directly upon the mathematical structure itself. In that case the mathematical structure would not be suitable, and we would have to find a more suitable one.

The following diagram, showing the possible working flow of a scientist engaged in representing a physical domain by means of a mathematical structure, should help in clarifying the previous considerations. At the beginning of the process, the scientist has a mathematical structure M that she thinks is a good candidate for representation, but she does not know whether it *perfectly* (viz. isomorphically) represents the data model or not. So, she initially supposes that the data model is *iso*morphic to the mathematical structure.[24] This could turn out to be a good hypothesis, and then she can go on working with such an assumption. But such a hypothesis may bring to a failure. In that case, depending on the kind of failure

---

[24]I am not saying that this hypothesis is *always* necessary. In some cases she already knows that M does not perfectly represent the phenomenon, and then the isomorphism hypothesis need not to be assumed, of course.

Fig. 11.2 Working flow of a scientist engaged in representing a physical domain by means of a mathematical structure

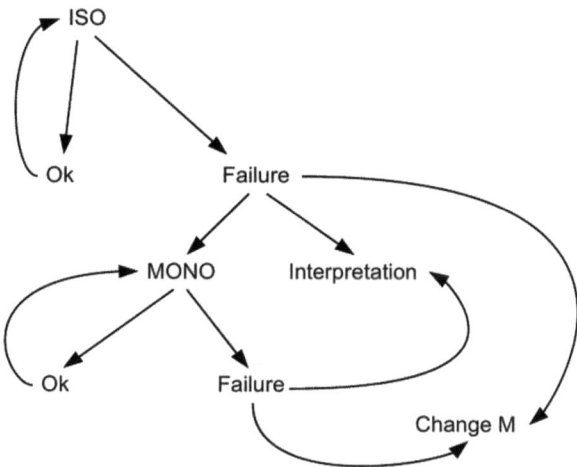

at issue, she can shift to a different assumption, i.e. that the data model is rather *mono*morphic to the mathematical structure; otherwise, she can shift to a different interpretation; or she can shift to a completely different mathematical structure to work with. Also the monomorphism hypothesis can turn out either satisfying or not. Again, just like before, depending on the kind of failure, she can attempt to shift to a different interpretation or to a different mathematical structure. In different words, she makes a strong hypothesis (isomorphism), and if this hypothesis turns out to be too strong, she tries to come to a compromise, by weakening the hypothesis, by modifying the interpretation or by shifting to a different mathematical structure (Fig. 11.2).

11.5 Conclusions

In this paper, I presented the structural account for the representative effectiveness of mathematics, and I took into consideration some difficulties related to it. Part of these difficulties can be settled, as we saw; but some criticalities remain. This shows that the structural account needs to be integrated in order to better explain the representative applicability, especially where mathematical representations seem to play an active role in making new discoveries and to foster new advancements in scientific research. I tried to provide the required integration, by considering the problems and the deficiencies emerged by means of the DDI model proposed by Hughes's (1997) and van Fraassen's considerations. It seems to me that the resulting account is in a better condition to make the applicability of mathematics less mysterious and miraculous than it is often presented. In this integrated account, a major role is played by a certain number of pragmatical and contextual elements. Such a role is not always schematizable in a rigorous manner, but I think this is not an imperfection. Rather, I think that all these bestow upon the account an amount of dynamism and plasticity, which is good for understanding, for example, the evolution of different theories and theory changes.

References

Azzouni, Jody. 2000. Applying mathematics: an attempt to design a philosophical problem. *Monist* 83(2): 209–227.
Bangu, Sorin. 2008. Reifying mathematics? Prediction and symmetry classification. *Studies in History and Philosophy of Modern Physics* 39: 239–258.
Bangu, S. 2012. *The applicability of mathematics in science: indispensability and ontology*. Basingstoke: Palgrave Macmillan.
Barrow, John D. 1988. *The World within the World*. London: Oxford University Press. Trad. it. *Il mondo dentro il mondo*. Adelphi, Milano, 1991.
Batterman, Robert W. 2002. *The devil in the details: asymptotic reasoning in explanation, reduction and emergence*. New York: Oxford University Press.
Batterman, R.W. 2010. On the explanatory role of mathematics in empirical science. *The British Journal for the Philosophy of Science* 61(1): 1–25.
Bueno, Octavio and Colyvan, Mark. 2011. An inferential conception of the application of mathematics. *Noûs* 45(2): 345–374.
French, Steven. 2000. The reasonable effectiveness of mathematics: partial structures and the application of group theory to physics. *Synthese* 125: 103–120.
Goodman, Nelson. 1968. *Languages of art*. Indianapolis: Bobbs Merrill.
Hertz, Heinrich. 1956. *The principles of mechanics presented in a new form*. New York: Dover. English translation of the original edition: Hertz, Heinrich. 1894. *Die Prinzipien der Mechanik in neuem Zusammenhange dargestellt*. Leipzig: J. A. Barth.
Hughes, Richard I.G. 1997. Models and representation. *Philosophy of Science* 64: 325–335.
Pincock, Christopher. 2004. A revealing flaw in Colyvan's indispensability argument. *Philosophy of Science* 71: 61–79.
Pincock, Christopher. 2012. *Mathematics and scientific representation*. Oxford: Oxford University Press.
Reichenbach, Hans. 1965. *The theory of relativity and a-priori knowledge*. Berkeley/Los Angeles: University of California Press. English translation of the original edition: Reichenbach, Hans. 1920. *Relativitätstheorie und Erkenntnis a priori*. Berlin: Springer.
Shapiro, Stewart. 1983. Mathematics and reality. *Philosophy of Science* 50: 523–548.
Shapiro, Stewart. 1997. *Philosophy of mathematics. Structure and ontology*. Oxford: Oxford University Press.
Steiner, Mark. 1998. *The applicability of mathematics as a philosophical problem*. Cambridge, MA: Harvard University Press.
van Fraassen, Bas C. 2006. Representation: the problem for structuralism. *Philosophy of Science* 73(5): 536–547.
van Fraassen, Bas C. 2008. *Scientific representation: paradoxes and perspectives*. Oxford: Clarendon.
Wigner, Eugene. 1960. The unreasonable effectiveness of mathematics in the natural sciences. *Communications in Pure and Applied Mathematics* 13(1): 1–14. Reprinted in *Symmetries and reflections*. Bloomington: Indiana University Press, 1967.

Chapter 12
Defending Maddy's Mathematical Naturalism from Roland's Criticisms: The Role of Mathematical Depth

Marina Imocrante

12.1 Maddy's Mathematical Naturalism

A naturalistic approach generally rejects the possibility of *a priori* philosophical inquiries. In Quine's words, naturalism is "the recognition that it is within science itself, and not in some prior philosophy, that reality is to be identified and described" (Quine 1981, p. 21).

Following Quine, Maddy's naturalism[1] does not extend to posing philosophical questions "from some special vantage point outside of science, but as an active participant, entirely from within" (Maddy 2011, p. 39). Maddy's naturalism "takes the correctness of successful scientific practice as a datum for philosophical theorizing rather than something susceptible to philosophical challenge" (Linnebo 2012, p. 134).

From this perspective, philosophical positions defined as naturalistic must state their theories not only as a matter of a simple deference to authoritative scientific statements but also for internal scientific reasons, which means grounding them on experiment and well-confirmed scientific theories (Maddy 2011, p. 39).

According to Maddy's naturalism, the appropriate method of investigating a particular domain of reality is by means of the science which specifically addresses

---

[1] Penelope Maddy's approach to the philosophy of mathematics has evolved from early cognitive realism (Maddy 1997) to her present mathematical naturalism. In this chapter, I focus on her present naturalistic account as presented in (Maddy 2007a, 2007b, 2011).

M. Imocrante (✉)
Vita-Salute San Raffaele University, Milan, Italy

Institut d'Histoire et de Philosophie des Sciences et des Techniques, Université Paris 1 Panthéon-Sorbonne, Paris, France
e-mail: marina.imocrante@gmail.com

this domain. With regard to mathematics, first and foremost it is worth highlighting that Maddy understands mathematics "as a human practice" (Maddy 2007b, p. 361). From this practical perspective, Maddy affirms the autonomy of mathematics from both philosophical and scientific considerations:

> My naturalist [...] begins, as Quine's does, within empirical science, and eventually turns, as Quine's does, to the scientific study of that science. She is struck by two phenomena: first, most of her best theories involve at least some mathematics, and many of her most prized and effective theories can only be stated in highly mathematical language; second, mathematics, as practice, uses methods different from those she's turned up in her study of empirical science. She could, like the Quinean, ignore those distinctive methods and hold mathematics to the same standards as natural science, but this seems to her misguided. The methods responsible for the existence of the mathematics she now sees before her are distinctively mathematical methods; she feels her responsibility is to examine, understand, and evaluate those methods on their own terms; to investigate how the resulting mathematics does (and doesn't) work in its empirical applications; and to understand how and why it is that a body of statements generated in this way can (and can't) be applied as they are. (Maddy 2007a, p. 448)

Maddy's suggestion is that mathematical objects and practice should be investigated with methods derived from mathematical practice itself.

In the same manner, the ontological questions concerning mathematical objects and statements must be answered within mathematics itself. Indeed, mathematical objects should not receive the same epistemological treatment reserved for physical objects. While for physical objects we need higher introduction and confirmation standards (e.g., identification through empirical instruments), standards for introducing mathematical objects are different: their role in successful mathematical theories is the only element we ought to use in confirming their existence.

Due to this epistemic disanalogy between mathematical and scientific objects, and due to the constitutive autonomy of mathematics with respect to philosophical concerns, Maddy does not think that a naturalistic investigation into the foundations of mathematics necessarily leads to a realist ontological position:

> [...] for my naturalist natural science is the final arbiter of what there is, and it doesn't seem to support its mathematical ontology [...]. Mathematics itself offers no ontological guidance beyond the minimal "mathematical things exist" [...]. In fact, I suspect that a decision on these matters will have more to do with the theory of truth than with the methodological or naturalized philosophical facts about mathematics or natural science. (Maddy 2007a, pp. 456–457)

It is for this reason that the second philosophy proposed by Maddy (2007b) ceases to give prominence to the defense of a particular position in the ontology of mathematics:

> Does mathematics have a subject matter like physics, chemistry, or astronomy? Are mathematical claims true or false in the same sense? If so, by what means do we come to know these things? What makes our methods reliable indicators of truth? The answers to these questions will not come from mathematics itself - which presents a wonderfully rich picture of mathematical things and their relations, but tells us nothing about the nature of their existence [...]. (Maddy 2007b, p. 361)

In considering mathematical objects and theories by looking at their role in mathematics as practiced by working mathematicians, her methodological naturalist focuses more on epistemological issues (how do we build mathematical theories? How do we account for them? "How can we properly determine if a new sort of entity is acceptable or a new method of proof reliable?" (Maddy 2011, p. 31)), than on ontological and semantical ones (do mathematical objects exist? Are mathematical theories true?). Maddy's only proviso is that, no matter which ontological position we endorse, it should not contradict second-philosophical methods of inquiry.

This is why, regarding the ontology of mathematics, Maddy "does not address alternative theories of ontology she does not find in that practice" (McLarty 2013, p. 390), that is, the day-to-day practice of mathematics, and adopts quite indifferently what she calls *Thin Realism* (the thesis that mathematical objects exist, but they only have the properties ascribed to them by mathematical theories, any other question about their nature being irrelevant) or *Arealism* (the thesis that mathematical objects do not exist).[2] These two ontological positions are taken to be "equally accurate, second-philosophical descriptions of the nature of pure mathematics" (Maddy 2011, p. 112).

12.2 Roland's Objections

Before Maddy (2011) appeared, Jeffrey W. Roland (2007) charged Maddy's account of mathematics with failing to be naturalistic: in his opinion, Maddy would be unable to explain the reliability of mathematical beliefs without breaking one of the main principles of naturalism.

As I have explained above, Maddy's latest position with respect to mathematics is compatible with "there being no fact of the matter regarding the truth and falsity of mathematical claims" (Roland 2007, p. 425). But, in Roland's opinion, "if there is no fact of the matter with respect to truth and falsity in mathematics, that undermines the project of giving an epistemology of mathematics" (Roland 2007, p. 425).

Roland writes:

> Epistemology is centrally concerned with systematic connections between justification and truth. If there is no fact of the matter as to whether claims concerning Fs are true or false, then there simply is no question of systematic connections between what justifies our F-beliefs and the truth about Fs. (Roland 2007, p. 425)

In Roland (2009), arguing for the impossibility of naturalizing any epistemology of mathematics, Roland explains in the following manner what, in his opinion, is essential for the possibility of an epistemology of any discipline, including mathematics, that is, the "truth-conduciveness" of beliefs:

> Suppose we have an epistemology E that ratifies our acceptance of pure mathematics as justified. [...] The notion of justification endorsed by E must be truth directed; i.e. it must be such that beliefs justified according to that notion tend to be true. [...] What makes a

---

[2]Arealism is taken as different from nominalism: Maddy states the difference in Maddy (2011, pp. 96–98), claiming that "[...] if Arealism is to be considered a version of nominalism, it certainly isn't the 'stereotypical' variety" (Maddy 2011, pp. 97–98).

conception C of justification a conception of *epistemic* justification is at least in large part that beliefs which are justified according to C tend to be true, i.e. that there is some sort of systematic connection between beliefs justified according to C and what is actually the case. Moreover, endorsing the truth-directedness of epistemic justification [...] is to recognize a widely accepted conviction that an epistemic notion of justification must be systematically connected to truth, i.e. truth-conducive. (Roland 2009, p. 71)

Roland claims that a naturalistic account is able to answer the epistemological question of "how we are justified in believing what we (justifiably) do about the world?" (Roland 2007, p. 430) because of the two positions it entails: *disciplinary holism* and ontological realism.

What Roland calls *disciplinary holism* is the "cross-discipline criticism and support allowed for" by naturalism (Roland 2007, p. 431):

[...] the family of disciplines that fall under the heading 'science' is large enough and varied enough that meaningful criticism of one discipline can be mustered in another while remaining within science (broadly construed to include natural and social sciences plus the mathematics and logic applied in the practice of these sciences). [...] While science as a whole is insulated from outside criticism on the naturalist's view, individual branches of science [...] are not insulated from each other. (Roland 2007, p. 430)

As to the naturalist's ontological realism, Roland explains why he thinks that the epistemology of naturalism is linked to realism in the following terms:

Naturalism has it that our inductive practices are underwritten by our appreciation, conscious or not, of natural kinds. Successful inductions are those done on projectible[3] properties of (predicates applied to singular terms denoting) objects, and the naturalist, following Quine, holds that 'a projectible predicate is one that is true of all and only the things of a natural kind' (Quine 1969, p. 116). Thus, our ability to successfully engage in induction is linked to our ability to tell projectible predicates from nonprojectible ones, which is in turn linked to our ability to track general features of the world [...]. So since naturalists are generally realists about natural kinds, naturalism, in its account of our inductive practices, takes a realist stance toward the general *prima facie* subject matter of the sciences. (Roland 2007, p. 431)

In particular, Roland claims that a causal form of realism is essential to a naturalistic epistemology:

[...] An account of the reliability of perception [in the case of natural science] must bridge theory and the world. This bridge is provided by a causal theory of detection [...]. (Roland 2007, p. 433)

A causal theory of perception is "the ground level of detection" (Roland 2007, p. 433), and, due to disciplinary holism, all sciences are rooted in this ground level:

This is the sense in which it is reasonable to say that physics and physiology, in addition to biology, chemistry, psychology, neuroscience - even sociology and economics - ultimately depend on perception. The experience on which empirical science depends is perceptual

---

[3] According to Roland, projectibility is a property of predicates that measures the degree to which past instances can be taken as guides to future ones.

experience, broadly construed to include detection (indirect perception) by instruments, but empirical science only fulfills its primary mission, i.e. to tell us about the world, if that experience is causally connected to the world. (Roland 2007, p. 433)

So, the naturalist's account of the reliability of scientific beliefs requires the commitment to a realist conception of causation.

> Reliability has to do with truth [...]. An adequate explanation of the reliability of certain types of interactions in terms of causation (i.e. causal powers, processes, or structure) must give us reason to think that beliefs formed as a result of the right types of interactions are true in a robust sense. A realist conception of causation can do this [...]. (Roland 2007, p. 435)

These two features of naturalism, that is, disciplinary holism and ontological realism, enable us to explain the reliability of scientific beliefs and also to account for the accuracy of the naturalist's epistemic norms and standards[4] (Roland 2007, p. 431).

With regard to Maddy's mathematical naturalism, Roland points out that, according to Maddy, mathematics should be regarded as being detached from natural sciences (and from philosophy), which fails to meet his requirement for disciplinary holism, and that she does not think that a naturalistic investigation of mathematics necessarily leads to ontological realism, which in turn disregards his requirement for causal realism. Judging Maddy's epistemology for mathematics in light of his own conception of scientific naturalism, Roland puts forth two criticisms of Maddy.

The first is what I call "reliability criticism". Roland claims that Maddy's accordance of autonomy to mathematics is the equivalent of a rejection of disciplinary holism (Roland 2007, p. 436). But in Roland's view disciplinary holism, as we have seen, is essential to providing naturalism with an epistemology for science which is able to guarantee the accuracy of its epistemic norms and standards. The same goes for mathematics: in Roland's view, disciplinary holism is necessary in order to provide an epistemology for mathematics which is able to guarantee the accuracy of its epistemological standards. So, by rejecting disciplinary holism, Maddy's mathematical naturalism disqualifies itself from the possibility of being considered as a genuine naturalistic position:

> [...] An account of the reliability of the method of mathematical naturalism analogous to the account of the reliability of scientific practice available to the naturalist is out of reach for the mathematical naturalist. (Roland 2007, p. 437)

The second objection, strictly connected to the first, is what I call "ontological criticism". We have seen that, in Roland's view, in order to provide an adequate epistemology for both science and mathematics it is essential to rely on causal realism. But, as we have seen, Maddy's mathematical naturalism leaves the question of the existence of mathematical objects and of the truth of mathematical statements substantially open, considering it to be an extra-mathematical question, and as such without interest for her naturalist. In order for a naturalistic epistemology for

---

[4] In Roland's words, this means providing a "dissident epistemology" for science (Roland 2007, p. 432).

mathematics to be adequate, Roland requires the identification of "truth-makers" for mathematics, "in virtue of which mathematical beliefs (statements, etc.) have the truth values they do" (Roland 2009, p. 72). Given her ontological agnosticism, Maddy cannot rely on existing mathematical objects as truth-makers of this kind:

> [...] The mathematical naturalist can countenance nothing to play a role in the epistemology of mathematics analogous to that of the causal order in naturalistic epistemology. (Roland 2007, p. 439).

Therefore, in Roland's view, since she ultimately refuses to ground her position on ontological realism, Maddy should not define her second philosophy as a form of naturalism with regard to mathematics.

It is worth noting that Roland's objections to Maddy only hold if we also accept Roland's conception of naturalism for mathematics, that is, if we think about mathematical naturalism as modeled on his conception of scientific naturalism, described in Roland (2007) as a position committed to disciplinary holism and to causal realism. A way to challenge Roland's criticisms would thus be to show that there are other conceptions of naturalism available[5] and that, in particular, Maddy's view of naturalism is different from Roland's.

But even if for the sake of argument we accept Roland's epistemic requirement of reliability for mathematical beliefs, defined as the need for a link of truth-conduciveness between mathematical beliefs and some objective facts ("what is actually the case," in Roland's words (Roland 2009, p. 71)), I argue that a concept introduced by Maddy in 2011 could be used to provide an answer to Roland's criticisms: namely, the concept of "mathematical depth."

In the following I shall present this notion and try to show how, within Maddy's (Maddy 2011) framework, mathematical depth could deliver the sort of reliability of mathematical beliefs that Roland demands.

12.3 Maddy's Mathematical Depth

Maddy (2011) uses the term "mathematical depth" to refer to the capacity for fruitfulness of mathematics. Mathematical notions, theories, and statements are ultimately fruitful both internally, in mathematics itself (e.g., the foundational role of set theory), and externally in the applications of mathematical concepts to empirical sciences (e.g., Maxwell's equations which established the foundation of classical electrodynamics).[6]

Indeed, I shall distinguish the depth of mathematics from its fruitfulness *tout court*. To this purpose, it could be useful to consider Godfrey Harold Hardy's

---

[5]As an example, consider that the requirement of a causal link between the world's facts and beliefs, in the case of mathematical knowledge, is certainly not common among naturalists.

[6]Maxwell's equations are the usual example given by Maddy (1997, p. 114, 2007b, p. 332, 2011, p. 19), but we could mention any other successful case of application.

attempt to define the notion of depth. After claiming that "there are two things at any rate which seem essential [to make a mathematical idea significant], a certain generality and a certain depth; but neither quality is easy to define at all precisely" (Hardy 2005, p. 24), Hardy attempts to characterize depth as follows:

> It has *something* to do with *difficulty*; the 'deeper' ideas are usually the harder to grasp: but it is not at all the same. The ideas underlying Pythagoras's theorem and its generalization are quite deep, but no mathematicians now would find them difficult. [...] It seems that mathematical ideas are arranged somehow in strata, the ideas in each stratum being linked by a complex of relations both among themselves and with those above and below. The lower the stratum, the deeper (and in general more difficult) the idea. Thus the idea of an 'irrational' is deeper than that of an integer; and Pythagoras's theorem is, for that reason, deeper than Euclid's. (Hardy 2005, pp. 27–28)

On the other hand, Hardy clearly separates the idea of depth from that of fruitfulness, since he does not think that mathematics must be judged in terms of its utility (Hardy 2005, pp. 32–33).

Nevertheless, for the purpose of the present work, I shall focus on Maddy's proposal and take the viewpoint that despite the depth of a mathematical notion, statement or theory does not have to be identified with its fruitfulness; it could undeniably be seen as the encoding of a set of virtues (to be further specified) which constitute a fundamental part of its fruitfulness and then a condition for it.

In Maddy (2011), Maddy suggests that mathematical practice is grounded in the phenomenon of mathematical depth:

> [...] What guides our [mathematical] concepts formation, beyond the logical requirement of consistency, is the way some logically possible concepts track deep mathematical strains that the others miss. (Maddy 2011, p. 79)

Maddy continues by saying:

> [...] Judgments of mathematical depth are not subjective [...]. [...] mathematical fruitfulness isn't defined as 'that which allows us to meet our [mathematical] goals', irrespective of what these might be; rather, our mathematical goals are only proper insofar as satisfying them furthers our grasp of the underlying strains of mathematical fruitfulness. [...] there is a well-documented objective reality underlying Thin Realism [or Arealism], what I've been loosely calling the facts of mathematical depth. The fundamental nature of sets (and perhaps all mathematical objects) is to serve as means for tapping into that well. (Maddy 2011, pp. 81–83)

It is in light of this notion of mathematical depth that Thin Realism and Arealism are ultimately equivalent positions:

> [Thin Realism and Arealism] are equally well-supported by precisely the same objective reality: those facts of mathematical depth. [...] They are alternative ways of expressing the very same account of the objective facts that underlie mathematical practice. (Maddy 2011, p. 112)

Maddy provides the reader with some examples of mathematical depth. Her examples refer to concept formation in set theory and group theory and to the different formulations and applications of the axiom of choice (Maddy 2011, pp. 78–81). Unfortunately, these examples do not seem to be clear enough to shed light on the concept we are seeking. Let me briefly show why.

Let us begin with the example of the axiom of choice: the axiom states that for every set of nonempty sets there is a choice function which selects one element within each set. Maddy's description of the fruitfulness of this axiom is not as complete as one might have wished: she references a few applications of the axiom of choice but only makes explicit reference to its application in geometry, connected to the Banach-Tarski paradox (Maddy 2011, pp. 34–35). She does not cite the axiom's other applications, such as in algebra (the existence of bases in vector spaces); topology (Tychonoff's theorem about the product of compact topological spaces); or analysis (Hahn-Banach theorem and the existence of non-Lebesgue measurable sets of reals). Maddy restricts herself to reminding us, in a note, of the "internal mathematical considerations in favor of the axiom" (Maddy 2011, pp. 35–36, note 74) she described in Maddy (1997, pp. 54–57).

In her presentation, which I have sketched out above, Maddy does not provide any real insight into the fruitfulness of the axiom nor does she with regard to the connection between the axiom and the discovery of mathematical depth.

The example of group theory appears more promising. The concept of group turns out to be essential in several mathematical domains: originally used to study permutations and the solvability of algebraic quintic equations, group theory went on to be recognized as the appropriate tool to study the concept of symmetry. Today, group theory is an indispensable tool in mathematics: it essentially occurs in model constructions within different scientific contexts.

Faced with this variety of use, Maddy states that group theory's fruitfulness lies in its capacity to unify different structures which share several properties (Maddy 2011, p. 79) (e.g., a mathematical structure and a physical one), by representing them with the same model. For this reason, the example of the concept of group seems particularly well suited to demonstrating what mathematical depth is through the study of its applications.

Despite all of this, Maddy (2011) does not elaborate about group theory and its applications, leaving the reader without a clear explanation of the connection between its fruitfulness and the phenomenon of mathematical depth we are examining.

Maddy examines the case of set theory more closely. In fact, much of Maddy's work is devoted to set theory, in particular to answering the following questions (Maddy 2011, p. 37): what are the methods of set theory? And according to what criteria must we choose new axioms to adopt in order to increase the deductive and explanatory power of set theory?

Through analysis of the history of mathematics and the evolution of the connection between mathematics and the study of the empirical word (Maddy 2011, pp. 3–27), Maddy establishes that set theory is essential to the unification of mathematical structures and their languages. Indeed, according to Maddy's naturalism, it follows from the autonomy of mathematics that the unified model allowing us to study different mathematical structures and methods, if one exists, must come from mathematics itself. That unified model is now represented by set theory.

It is worth highlighting that set theory can be seen as providing us with a theoretic framework that could be used as a model in which it is possible to represent numbers and functions without being forced to make specific claims about the existence of those objects. In addition, given Gödel's incompleteness theorems, set theory does not even give what Maddy calls, quoting Saunders MacLane (1986, p. 406), a "parachute" against the risk of incompleteness (Maddy 2011, p. 133). However, in spite of these ontological remarks that do not have much importance for Maddy's naturalist, set theory has a unifying role within mathematics. Maddy then states that set theory provides us with a shared framework within which every single mathematical problem concerning consistency and proof may be treated:

> What set theory does is provide a generous, unified arena to which all local questions of coherence and proof can be referred. In this way, set theory furnishes us with a single tool that can give explicit meaning to questions of existence and coherence; make previously unclear concepts and structures precise; identify perfectly general fundamental assumptions that play out in many different guises in different fields; facilitate interconnections between disparate branches of mathematics now all uniformly presented; formulate and answer questions of provability and refutability; open the door to new strong hypotheses to settle old open questions; and so on. In this philosophically modest but mathematically rich sense, set theory can be said to found contemporary pure mathematics. (Maddy 2011, p. 34)

Maddy's explanation of the depth of the concept of set ends here. She confines herself to saying that, due to their foundational role, we are allowed to consider sets as "maximally effective trackers of mathematical depth" (Maddy 2011, p. 82). But the connection between the meaning of the concept of set and the emergence of the concept of mathematical depth is not analyzed in detail. The concept of mathematical depth is thus left rather unclear.

On one hand, and consistently with her peculiar form of naturalism, Maddy claims:

> [...] I doubt that an attempt to give a general account of what mathematical depth really is would be productive; it seems to me the phrase is best understood as a catch-all for the various kinds of special virtues we clearly perceive in our illustrative examples of concept-formation and axiom choice. (Maddy 2011, p. 81)

> This is why I spend so much time rehearsing these various cases, to give the reader a feel for what 'mathematical depth' looks like. (Maddy 2011, p. 81, note 39)

From the examples presented in Maddy (2011), the reader is therefore supposed to obtain a satisfying understanding of what mathematical depth is. Unfortunately, the examples are not discussed thoroughly enough in order to obtain the "feel for what mathematical depth looks like" that Maddy is seeking to impart.

As already highlighted by previous quotes, mathematical depth is presented by Maddy as something objective:

> [...] the topography of mathematical depth [...] stands over and above the merely logical connections between statements, and furthermore, it is entirely objective. (Maddy 2011, p. 80)

Maddy claims that the phenomenology of mathematical practice itself guarantees the objectivity of mathematical depth. In her opinion, anyone who does even a little mathematics can easily come to recognize this objectivity: in Maddy's words, the first sensation which strikes anyone who does mathematics is "the immediate recognition" that it is "an objective undertaking *par excellence*" (Maddy 2011, p. 114).

It is worth underlining that the form of the fruitfulness of mathematical depth is an extrinsic justification for mathematical theories and statements. The justification of a statement of a mathematical theory is intrinsic if the truth of the statements follows from the properties ascribed to its objects by the theory in question; on the other hand, an extrinsic justification of a statement is a justification in terms of its consequences, inside or outside the theory.[7] Maddy writes:

> We're out to explain what underlies the justificatory methods of set theory [...]. Part of the answer, for intrinsic justifications, may be that they spell out what's implicit in our "concept of set," but the bulk of the justifications that interest us are extrinsic. (Maddy 2011, pp. 78–79)

In Maddy's opinion, the use of a mathematical theory that has certain consequences on the improvement of our knowledge, inside or outside mathematics, is in itself a good justification to use the theory in question. In favor of this conception of fruitfulness as an extrinsic justification of theories, Maddy quotes a number of selected passages of Zermelo's defense of the axiom of choice (Zermelo 1967, pp. 187–189), specifically insisting on its fruitfulness:

> [This axiom] has frequently been used, and successfully at that, in the most diverse fields of mathematics, especially in set theory. (Maddy 2011, p. 46)

Moreover:

> So long as [...] the principle of choice cannot be definitely refuted, no one has the right to prevent the representatives of productive science from continuing to use this hypothesis. [...] Principles must be judged from the point of view of science, and not science from the point of view of principles fixed once and for all. (Maddy 2011, p. 47)

This is what allows Maddy to conclude in favor of her hypothesis of the importance of the capacity for fruitfulness in the evaluation of each mathematical notion, statement, or theory. Like Zermelo, Maddy's naturalist counts the fruitfulness of a mathematical statement as a point in its favor – indeed, as the most important point.

12.3.1 Problems with the Notion of Mathematical Depth

Maddy names her conception of the phenomenon of mathematical depth "postmetaphysical objectivism" (Maddy 2011, p. 116): with this term, Maddy refers to a form of objectivity which has nothing to do with the metaphysical and ontological level and which is constituted by the practice-oriented reality of the depth of certain mathematical theories.

---

[7]For references and discussion on the distinction between intrinsic and extrinsic mathematical explanation, see, for example, Mancosu (2008).

I have already quoted Maddy talking about the "facts" of mathematical depth (Maddy 2011, pp. 83, 112), stating that they represent a level of objectivity which, from an ontological point of view, could be acceptable by Thin Realism as well as by Arealism (Maddy 2011, pp. 102–112). Here one could legitimately ask what exactly these "facts of mathematical depth" are and more broadly what exactly "mathematical depth" is.

Indeed, Maddy's concept of mathematical depth is appealing to us because it provides an answer to Roland's objections, insofar as mathematical notions and the methods which are able to identify mathematical depth seem to represent in and of themselves the connection between objective facts (the facts of mathematical depth) and the corresponding mathematical beliefs. Nevertheless, we cannot ignore that Maddy's notion of mathematical depth has several problems, engendered by the lack of precision with which the notion is presented.

First of all, consider that Maddy (2011, p. 114) introduces the idea that in order to account for the phenomenon of mathematical depth one could appeal to some sort of intuition that would be shared by anyone who practices mathematics, though she does not clarify what exactly she means by this. The suggestion that the mathematical depth could best be explained in terms of the concept of intuition[8] is certainly intriguing and should be further explored. Let me stress only that we are not dealing here with a mathematical intuition conceived as a rational faculty, somehow *à la* Gödel; as mentioned above, the intuition of the depth of mathematics that Maddy is talking about is rather the psychological intuition that, when practicing mathematics, we enter a domain where our methods and conclusions are to a certain extent imposed or forced, not arbitrary.

However, accepting the favorable intuitions of mathematicians as a sufficient criterion to judge of the depth of a concept would seem to clash with the naturalistic principles sketched above, which require grounding concepts on experiment and on well-confirmed scientific theories, or, in the specific case of mathematics, on proofs and mathematical theories. Mathematicians could be wrong in their intuitions; this explains why, even if we could use the shared intuitions of mathematicians as a clue of the depth of a mathematical concept, one should hope that the depth of a mathematical notion, theory, or statement would count as an objective feature of it and not as a psychological sensation subjectively associated with it.

Moreover, Maddy frequently uses a metaphorical language, without clearly defining the words she employs; again, it would, for example, be legitimate to ask what exactly these "facts of mathematical depth" are. Furthermore, what does it mean exactly that sets and set-theoretic methods "track strains" of mathematical depth? And what is the exact definition of "post-metaphysical objectivism"?

Thirdly, and more generally, not only does Maddy not provide us with a sufficiently clear explanatory definition of what mathematical depth is, but she also presents examples which are not explored in enough detail. If it is not possible to

---

[8]The role of intuition in philosophy is a topic of debate: for references and discussion, see, for example, the essays in Gendler (2010).

formalize a strict definition of the facts of mathematical depth, consistent with the naturalistic attitude Maddy emphasizes, we should at least be given more clarifying examples.

Nevertheless, it is true that Maddy (2011) does not pretend to have provided a satisfactory definition of mathematical depth: her sole intention is to focus our attention on the challenge of understanding the phenomenon that is supposed to drive the practice of pure mathematics in her latest account.

Maddy uses the metaphor of the "black box" (Maddy 2011, p. 85) to describe the effectiveness of mathematics. This lexical choice provides an idea of something that contains all the information we need, but which we do not know how to read in order to have complete knowledge of the issue. Beyond the metaphor, at present the only thing that seems clear is that in Maddy's account (Maddy 2011) the goal of any epistemological inquiry concerning mathematics ends in those facts of mathematical depth.

In order to provide a satisfactory answer to Roland's challenge – and even in a general sense, to make Maddy's new account stronger – clarifications are needed regarding the concept of depth in mathematics.

12.3.2 A Possible Direction Toward Clarifying the Notion of Mathematical Depth

In light of the previous analysis of Maddy's notion of mathematical depth and the related problems thereof, I suggest that in order to clarify Maddy's account we should see those "facts" of mathematical depth not as mathematical theoretic facts, but rather as the historical facts of the fruitful use of particular notions, statements, and theories during the history of mathematical practice.

A simple reference to the history of mathematical practice would probably not be sufficient because not all the history of mathematics is a history of success casting light on the depth of the concepts involved. Moreover, we should distinguish fruitful developments from the unfruitful ones. Nevertheless, I suggest that we should think about the history of mathematics as a gradual process akin to a sort of natural selection that promotes the development of fruitful mathematical notions and makes the unfruitful ones short-lived. Although I will not develop this suggestion here, it is useful to bear it in mind in order to see the facts of mathematical depth as the occurrences of certain uses of mathematical notions, statements and theories which turn out to be fruitful when we survey the history of mathematical practice.[9]

Defined in this manner, the facts of mathematical depth are beyond a doubt empirical facts, being part of the history and the current practice of mathematics.

---

[9]Note that we are not denying our initial distinction between depth and fruitfulness, since we clearly stated that, despite this distinction, the depth of a mathematical notion, statement, or theory could be seen as constituting a condition for its fruitfulness.

Maddy's depth could be defined as the capacity for fruitfulness of mathematical notions, statements, and theories, inside and outside mathematics itself. Mathematical notions, statements, and theories could then be seen as "tracking strains" of mathematical depth when their use in mathematical practice produces a useful insight or progress in the practice itself.

Even if Maddy does not explicitly express things in this way, her examples, discussed above, of notions and statements that point to the phenomenon of mathematical depth do not seem to conflict with this practice-oriented direction of clarifying the concept of mathematical depth.

Since historical facts are empirical, it is no longer necessary to explain why these facts of mathematical depth are objective. Fruitful uses of certain notions, statements, and theories in the history of the practice of mathematics stand out *a posteriori* and independently from the subjective intuitions of mathematicians.[10] This constitutes another reason to not base the judgments concerning depth on intuitions: the facts of mathematical depth are best understood as what turns out to be fruitful in the history of mathematics, independently from any subjective beliefs.

The definition of Maddy's objectivism with respect to mathematical depth as "post-metaphysical" would thus become clearer: the objectivity of the facts of mathematical depth is not a theoretic objectivity, depending on the ontological existence of mathematical objects, but is grounded on the empirical reality of the practice of mathematics[11] and is objective in this empirical sense.

Essentially, what was missing in Maddy's presentation of mathematical depth was a clarification of the definition of the concept, the nature of the facts it relates to, and the reasons why we should take them as objective; my suggestion indicates a possible direction toward solving these issues.

At this point, I put forward that looking at the facts of mathematical depth in the manner I proposed, that is, as the empirical, historical facts of mathematical practice, may allow us to answer to Roland's objections concerning the reliability issue and his request to link our mathematical beliefs to objectively existing facts.

12.4 Answers to Roland's Objections

In order to accept an epistemology for mathematics that ratifies our acceptance of mathematical beliefs as justified, Roland demands the existence of a connection between justification and truth, ultimately stated as "some sort of systematic

---

[10] Maddy's description of the objectivity of the mathematical depth seems to be robustly consistent with this sense of objective: see, for example, Maddy (2011, pp. 80–81).

[11] As McLarty clearly explains: "Maddy calls the existence claim [about sets] mathematical, since mathematicians routinely affirm it. She calls claims about possible existence, which do not occur in mathematics and are prominent in metaphysical discussion, metaphysical. She never argues against pursuing metaphysics and even the metaphysics of mathematics. She argues that we can understand what mathematics is and how it is justified by looking at mathematics and other sciences which mathematicians routinely do address, and not metaphysics." (McLarty 2013, p. 386)

connection between beliefs justified [...] and what is actually the case" (Roland 2009, p. 71). Roland identifies this "what is actually the case" with "truth *simpliciter*" (Roland 2007, p. 435), endorsing ontological realism and thus denying the possibility of including Maddy's ontologically agnostic account in his definition of naturalistic epistemology.

But if, with the aim of providing an epistemology for mathematics in a naturalistic manner, we take the facts of mathematical depth as "what is actually the case," in the sense of what is empirically the case, instead of the theoretic notion of truth *simpliciter*, then I argue that we might be able to find a connection between our mathematical beliefs and an objective reality, as Roland demands, while staying within Maddy's account.

As I have proposed, the facts of mathematical depth could be seen as the empirical facts of mathematical practice, objectively existing in the history of mathematics and in its current practice. Considering these facts in this light allows us to state the existence of a connection between mathematical beliefs we take to be justified and "what is actually the case," that is, the objectively existing facts of the fruitful use of mathematical notions, statements, and theories. In Maddy's approach, the fact on the basis of which to judge the reliability of beliefs in mathematics are these facts of mathematical depth and not the alleged ontology of mathematical objects that is commonly posited as grounding the truth of mathematical statements.

In order to have an adequate explanation of the reliability of mathematical beliefs in Maddy's mathematical naturalism, we needn't "think that beliefs [...] are true in a robust sense" (Roland 2007, p. 435), as Roland believes, but only that they are linked in a robust sense to the objective, empirical facts of mathematical depth.

The "truth-makers" of mathematical statements (i.e., "that in virtue of which mathematical beliefs (statements, etc.) have the truth values they do" (Roland 2009, p. 72)), which Roland requires for Maddy's epistemology of mathematics to be considered naturalistic, could now be seen as corresponding to the empirical facts of the successful use of mathematical notions and statements in mathematical practice.

With the facts of mathematical depth in place of ontological truth, we have the "bridge" between "theory and the world" that Roland's conception of epistemology calls for (Roland 2007, p. 433) without being compelled to endorse a causal form of realism.[12]

We are now able to answer Roland's question about the reliability of mathematical beliefs without being forced to adopt a form of ontological realism, which means we can offer an answer to the two objections he raised against Maddy's epistemological account for mathematics. We find this answer to Roland's criticisms within Maddy's account itself, thanks to the introduction of the concept of

---

[12]Even the projectibility of predicates Roland (2009, p. 431) applied to terms denoting objects in mathematical statements could still be there, because in Maddy's account the successful use of a mathematical notion, statement, or theory may be taken as a guide to future uses of the same notion, statement, or theory in mathematical practice. In Maddy's view, successful mathematical practice relies on the knowledge of the history of mathematics and of the patterns of mathematical depth that we discover studying and practicing mathematics.

mathematical depth and through seeing its "facts" as the empirical, historical facts of mathematical practice. This is why Maddy's account can continue to be seen as guaranteeing the reliability of mathematical beliefs in a deeply naturalistic way.

Maddy's work shows that, in regard to mathematics, it is indeed possible to be a naturalist without being a realist. For in Maddy (1997) and completely in Maddy (2007b, 2011), Maddy applies to mathematics the radical naturalistic approach that Quine applied to science. This is why I think Roland's ontological criticism is misplaced. I agree with Rosen (1999, p. 407) that Maddy's naturalism rectifies Quinean asymmetry: while Quine expects science to be completely autonomous from any philosophical considerations, he still views mathematics as dependent on empirical sciences, considering that mathematical statements need empirical support to be proven. Maddy on the other hand extends Quinean naturalism to mathematics, bestowing upon it methodological autonomy and independence from any extra-mathematical considerations, be they philosophical or scientific.

I do not agree with Roland that a naturalistic approach, adopted within Quinean tradition, forces us toward ontological realism with respect to mathematics. With regard to the philosophy of mathematics, a naturalistic account surely commits us to certain methodological Quinean (Quine 1969) standards (e.g., rejection of *a priori* philosophical inquiries, a claim of continuity between philosophy and sciences, employment of proper methods of inquiry for different scientific subjects), but it does not seem to force us to choose a realist ontological position.

Maddy's account does not in fact consider the ontological issue as being essential to her approach to the philosophy of mathematics; what really matters is the methodological statement of inquiry. This chapter's attempt to clarify the concept of mathematical depth moves in this practice-oriented direction, consistent with Maddy's naturalistic approach.

With regard to the disciplinary holism that Roland demands for any form of naturalism, I should emphasize that the autonomy accorded to mathematics by Maddy, criticized as not naturalistic, does not prevent her from establishing a fruitful connection between mathematical work and the results of other scientific subjects. Her frequent references to studies in psychology and cognitive sciences[13] to support her theory of mathematical reasoning demonstrate this. This is why I think that Roland's concern about the separation between mathematics and other sciences in Maddy's view is simply not grounded.

Recall, moreover, that in Roland's argument disciplinary holism is essential to naturalism in order to justify the possibility of espousing a causal form of realism in any scientific domain. But now that we have argued for the possibility of a link between mathematical beliefs and an empirically objective reality which assures their reliability without being grounded on causal realism, disciplinary holism no longer seems to be essential.

---

[13]Maddy (1997) referred to cognitive studies made by Hebb, Piaget, Phillips, and Gelman (Maddy 1997, pp. 58–67). On the other hand, Maddy (2007b) refers to more recent neuroscientific works of Dehaene, Spelke, Wynn, and others (Maddy 2007b, pp. 264–269, 319–328).

12.5 Conclusions

I briefly recalled the main features of Maddy's mathematical naturalism in order to present Roland's reliability and ontological criticisms of her account.

Thanks to the introduction of the concept of mathematical depth and seeing its "facts" as the empirical, historical facts of mathematical practice, I proposed an answer to Roland's objections that does not force us to abandon naturalism, as Roland stressed. In light of this answer, Maddy's account can continue to be seen as guaranteeing the reliability of mathematical beliefs in a naturalistic way.

However, I submitted that the concept of mathematical depth needs some important clarifications. In this respect, I suggested a possible manner in which the notion could be developed further, also through future investigations.

Acknowledgements I am indebted to Gabriele Lolli, Marco Panza, and Andrea Sereni for their valuable suggestions and comments. I would also like to thank the audience of the conferences *The Answers of Philosophy – 20thSIFA Anniversary Conference* (Alghero 2012) and *Philosophy of Mathematics: from Logic to Practice* (Pisa 2012), where former versions of this chapter were presented, and two anonymous referees for helpful remarks.

References

Gendler, Tamar Szabó. 2010. *Intuition, imagination, and philosophical methodology*. New York: Oxford University Press.
Hardy, Godfrey Harold. 1940. *A mathematician's apology*. Cambridge: Cambridge University Press. (Electronic edition published by the University of Alberta Mathematical Sciences Society, 2005).
Linnebo, Øystein. 2012. Book review: defending the axioms: on the philosophical foundations of set theory. By Penelope Maddy. *Philosophy* 87(1): 133–137.
MacLane, Saunders. 1986. *Mathematics: form and function*. New York: Springer.
Maddy, Penelope. 1997. *Naturalism in mathematics*. New York: Oxford University Press.
Maddy, Penelope. 2007a. Three forms of naturalism. In *The Oxford handbook of philosophy of mathematics and logic*, ed. Stewart Shapiro, 437–459. New York: Oxford University Press.
Maddy, Penelope. 2007b. *Second philosophy. A naturalistic method*. New York: Oxford University Press.
Maddy, Penelope. 2011. *Defending the axioms: on the philosophical foundation of set theory*. New York: Oxford University Press.
Mancosu, Paolo. 2008. *The philosophy of mathematical practice*. New York: Oxford University Press.
McLarty, Colin. 2013. Book review: defending the axioms: on the philosophical foundations of set theory. By Penelope Maddy. *Philosophia Mathematica* 21(3): 385–392.
Quine, Willard van Orman. 1969. Epistemology naturalized. In *Ontological relativity and other essays*, ed. W.V.O. Quine, 69–90. New York: Columbia University Press.
Quine, Willard van Orman. 1981. Things and their place in theories. In *Theories and things*, ed. W.V.O. Quine, 1–23. Cambridge: Harvard University Press.
Roland, Jeffrey W. 2007. Maddy and mathematics: naturalism or not. *The British Journal for the Philosophy of Science* 58: 423–450.

Roland, Jeffrey W. 2009. On naturalizing the epistemology of mathematics. *Pacific Philosophical Quarterly* 90: 63–97.
Rosen, Gideon. 1999. Book review: naturalism in mathematics, Penelope Maddy. *British Journal for the Philosophy of Science* 50(3): 467–474.
Zermelo, Ernst. 1967. A new proof of the possibility of a well-ordering, 1908. In *From Frege to Gödel*, ed. Jean van Heijenoort. Cambridge: Hardvard University Press.

Chapter 13
On the Indispensable Premises of the Indispensability Argument

Marco Panza and Andrea Sereni

13.1 Introduction

The recent debate on the indispensability argument (henceforth, IA) in the philosophy of mathematics features an impressive number of versions of the argument, all somehow pointing back to what is referred to as 'the Quine-Putnam indispensability argument', that is to Quine's several scattered remarks on the subject and to Putnam's (1971) first proper formulation of a version of IA. It is thus legitimate to wonder whether all (or, at least, most of) the versions available on the market can be really traced back to some minimal shared structure.

After rehearsing the most common stances towards IA, the main aim of this paper is to offer four minimal versions of IA, minimal in so far as they feature, according to classifications that will be explained below, the fewest or least controversial premises needed to gain the desired conclusion(s). We will submit that different formulations of IA on the market, related to the common stances to be discussed, could be retrieved from the minimal arguments to be offered.[1]

---

[1] This paper builds on the analysis of indispensability arguments presented in Panza and Sereni (2013), Chaps. 6–7; some of its conclusions, together with the four versions of the minimal indispensability argument discussed in Sect. 13.3, have thus been anticipated in that work.

M. Panza (✉)
IHPST, CNRS, University of Paris 1 Panthéon-Sorbonne, Paris, France
e-mail: marco.panza@univ-paris1.fr

A. Sereni
Institute of Advanced Studies (IUSS), Pavia, Italy

San Raffaele University, Milan, Italy
e-mail: andrea.sereni@iusspavia.it

Despite its methodological character, the following inquiry will have as a substantive conclusion that, in order to obtain the desired conclusions, some commonly required assumptions—namely, confirmational holism and naturalism—will prove dispensable.

Since it is not our intention in this occasion to take a stance in the debate on indispensability, many of the criticisms or defences of IA will not be discussed.

The structure of the paper is as follows. Section 13.2 introduces the issue by reference to the debate by considering some representative versions of the argument. Section 13.3 offers a taxonomy of four minimal versions of IA. Section 13.4 focuses on the notion of (in)dispensability. Section 13.5 considers in details which assumptions and notions are really needed if one wishes to endorse the minimal arguments. Section 13.6 indicates how the representative versions considered in Sect. 13.2 can be retrieved from the minimal arguments. Section 13.7 offers concluding remarks on the bearing of the present inquiries for the philosophical significance of IA.

13.2 The Indispensability Argument: Four Common Stances

A few expository remarks will help us to single out four common stances concerning the structure and significance of IA that we take as representative of the ongoing debate.

Putnam's (1971) well-known passage is commonly assumed as the reference formulation:

> So far I have been developing an argument for realism roughly along the following lines: quantification over mathematical entities is indispensable for science, both formal and physical, therefore we should accept such quantification; but this commits us to accepting the existence of the mathematical entities in question. This type of argument stems, of course, from Quine... (Putnam 1971, p. 347)

Putnam wavers, at different places, between considering the conclusion of the argument to be a form of platonism and rather a conclusion in favour of mathematical realism.[2] For our present purpose, it suffices that the two theses are acknowledged as distinct and both plausible: by platonism we mean the thesis that there exist objects of a certain sort, namely, such that our current mathematical theories can be taken to be about them, in short that there exist mathematical objects[3]; by mathematical realism we mean a particular form of semantic realism,

---

[2]Cf. Liggins (2008) for a reconstruction.

[3]In fact, we believe that other possible (and possibly more plausible) forms of platonism could be fashioned, and that the thesis just mentioned should then be more correctly called 'ontological platonism'. However, insofar as it is not part of our present aims to argue for such distinction, we avoid this specification and call it 'platonism' *tout court*. Nothing in this thesis mandates that

i.e. the thesis that the statements encompassed by our current mathematical theories, or better its theorems or consequences, are true (without specific commitment to what makes them true).[4]

Putnam's own views apart, his quotation above is commonly seen as a paradigmatic example of an argument for platonism. *Prima facie*, the Putnam's version of the argument appeals only to two notions, indispensability and quantification. However, many believe that beside these notions, IA relies on some additional theses of Quinean provenance: confirmational holism and naturalism.[5] The most debated formulation of IA that is faithful to this conception has been advanced by Mark Colyvan[6]:

> *i)* We ought to have ontological commitment to all and only those entities that are indispensable to our best scientific theories;
> *ii)* Mathematical entities are indispensable to our best scientific theories;
> ------------------------------
> *iii)* We ought to have ontological commitment to mathematical entities

According to Colyvan, 'the crucial first premise follows from the doctrines of *naturalism* and *holism*'.[7] Naturalism would be required in order to justify the only-direction of the implication, which, as Colyvan himself acknowledges,[8] is, in fact, redundant for drawing the conclusion, and confirmational holism to justify the all-direction. Whether this is so hinges on how the notions involved are defined.

So far, we are confronted with representatives of two different stances on IA. On the one hand, Putnam is concerned with logico-syntactical features of scientific theories and their expressive power: what is at stake is whether some particular vocabulary is necessary in order to *state* some given scientific laws.[9] On the other hand, Colyvan's formulation widens the scope by taking into account general concerns in the philosophy of science, especially as regards the relation between philosophy and science and the way in which empirical evidence is meant to confirm scientific theories.

---

mathematical objects are abstract, and indeed, though this is generally (or at least, often) admitted, IA can support the thesis without going into details about the nature of mathematical objects.

[4]Semantic realism as conceived here is distinct from what Michael Dummett called 'realism' (see, e.g. Dummett 1978): the latter is a thesis about statements possessing an objective, mind-independent truth *value*, whereas the former is the claim that the relevant statements are *true* (possibly in a mind-independent way, but not necessarily).

[5]For Quine's endorsement of these theses, cf. for instance Quine (1951), and Quine (1975), respectively.

[6]Colyvan (2001, p. 11). Cf. also Resnik (1995, p. 430).

[7]Colyvan (2001, p. 12).

[8]Cf. Colyvan (2001, p. 12): '[...] I should point out that the first premise, as I've stated it, is a little stronger than required. In order to gain the given conclusion all that is really required in the first premise is the 'all,' not the 'all and only, ' I include the 'all and only, ' however, for the sake of completeness and also to help highlight the important role naturalism plays in questions about ontology, since it is naturalism that counsels us to look to science and *nowhere else* for answers to ontological questions'.

[9]Cf. §§ V to VIII of Putnam (1971).

There is yet a third stance on IA. Hartry Field first suggested that a particular version of IA involves indispensability 'for explanations' (1989, p. 14): if we have a theory that we take to be our best explanation of a given set of (arguably empirical) phenomena, and this theory includes some statements to the extent that certain (sorts of) objects exist, then we 'have a strong reason to believe' (*ibid.*, p. 15) in the truth of the statements in question and consequently (pending a clarification of the ontological import of existential statements) in the existence of the relevant (sorts of) objects. This requires that inference to the best explanation (IBE) is considered a reliable principle. This third stance on IA thus equates (faithful to some of Quine's remarks, as well as to Putnam's overall picture) arguments for mathematical (ontological) realism and arguments for scientific realism about unobservable entities, where IBE is often appealed to. As a result of the recent vast debate on explanation, and mathematical explanation in particular,[10] Alan Baker (2009, p. 613) has claimed that 'for the purposes of establishing platonism [...] it needs to be shown that reference to mathematical objects sometimes plays an explanatory role in science'. He has thus offered the following 'Enhanced Indispensability Argument' (*ibid.*):

i) We ought rationally to believe in the existence of any entity which plays an indispensable explanatory role in our best scientific theories;

ii) Mathematical objects play an indispensable explanatory role in science;

----

iii) Hence, we ought rationally to believe in the existence of to mathematical objects.

Finally, we have a fourth stance, represented by arguments that build on pragmatic considerations, or generally considerations concerned with scientific practice and its needs. The most representative argument in this case has been presented by Resnik (1995, pp. 169–171, 1997, pp. 46–47)[11]:

i) In stating its laws and conducting its derivations science assumes the existence of many mathematical objects and the truth of much mathematics.

ii) These assumptions are indispensable to the pursuit of science; moreover, many of the important conclusions drawn from and within science could not be drawn without taking mathematical claims to be true.

----

iii) So we are justified in drawing conclusions from and within science only if we are justified in taking the mathematics used in science to be true.

iv) We are justified in doing science.

v) The only way we know of doing science involves drawing conclusions from and within it.

----

vi) So, we are justified in taking that mathematics to be true.

----

vii) So, mathematics is true.

---

[10]Cf. Hafner and Mancosu (2005), Baker (2005, 2009), Mancosu (2008)

[11]The argument is presented in slightly different terms in the two occasions. We are here using the one in Resnik (1995).

Setting apart any consideration about the validity of this version of IA, let us just notice that it differs from the previous versions to the extent that, as Resnik himself claims, it does not depend on the claim that 'the evidence for science (one body of statements) is also evidence for its mathematical components (another body of statements)', but merely requires that 'the justification for doing science (one act) also justifies our accepting as true such mathematics as science uses (another act)'.[12]

More recently, Azzouni (2009) has offered a reading of IA that is also based, though in rather different terms, on pragmatic considerations. More specifically, in a vein similar to ours (cf. footnote 17 below), Azzouni offers the following enthymematic 'blueprint' for IA:

Premise: Certain statements that quantify over mathematical entities are indispensable to science.
Conclusion: Those statements are true.

as underlying a family of arguments usually referred to as 'Quine-Putnam indispensability argument'. He then proposes to expand on this blueprint in order to offer what he labels the 'Assertoric-use QP', a version of IA based on the fact that what he calls the 'assertoric use' of mathematical statements is indispensable to science and that this use commits speakers to the truth of the statements in question. We will consider Azzouni's proposal, together with what we take to be a plausible reconstruction of his Assertoric-use QP, in more details below in Sect. 13.6.

For the time being, it is important to acknowledge that four major stances emerge when looking at the different versions of IA actually on the market, hinging, respectively, on logico-syntactic considerations related to expressive power (as in Putnam), on general views on science and confirmation (as in Colyvan), on the notion of explanation (as in Baker) and on features of scientific practice (as in Resnik or Azzouni). It is relevant to emphasize this point, since if minimal versions of IA are to be offered, they should at least be compatible with these common stances. It is not our intention to offer a minimal formulation corresponding to each argument representative of these stances. However, we will discuss to which extent and how it is possible, from the minimal versions to be offered, to retrieve something very close to them, or at least as close as to fit with the same stances.

13.3 The Minimal Indispensability Argument(s)

Since our aim is to establish, with respect to available versions of IA, what a minimal argument needs retain and what it can let go, we better start by considering the features of Colyvan's argument, which appears to be the most theoretically loaded version among the ones reviewed above. Here are some of its essential features. Firstly and obviously, it appeals to some notion of indispensability. Secondly, it is an

---

[12]Resnik (1995, p. 171).

argument for platonism, and not just for semantic realism. Thirdly, it is an argument stated in epistemic terms on two scores: on the one side, its premises and conclusions deal with what we 'ought to' believe, or what entities we 'ought to' be ontologically committed to; on the other side, it deals with the notion of justification, since 'best', in 'best scientific theories', should be understood as 'best justified'.[13] Fourthly, it appeals to the notion of ontological commitment; here, as in most cases, Quine's criterion ([QC]) is the relevant one.[14] Fifthly, it is claimed to rely, for the justification of its first premise, on naturalism and, sixthly and finally, on confirmational holism.

Are all of these features essential in order to obtain a version of IA? Obviously, any such version needs to retain the first feature: it requires appeal to some notion of indispensability. But things are different with the other features.

Clearly, some criterion is needed for selecting those scientific theories to which the argument is meant to apply. One can, however, either appeal here to epistemic notions or to non-epistemic notions, such as truth. We thus get a first broad distinction between arguments stated in epistemic or in non-epistemic terms.

Further, as already remarked, IA can be an argument for mathematical realism, rather than platonism, and we need to keep the two possibilities apart.

We thus end up with four possible varieties of IA: as an epistemic argument for mathematical realism, as an epistemic argument for platonism and, respectively, as a non-epistemic variety of each.

Let us begin with epistemic versions of IA for mathematical realism. As regards the selection criterion for theories, the most natural choice is for a criterion based on justification.[15] Justification, however, comes in many forms. For the time being, our appeal to it will be independent of any particular theory of justification. We could even not assume that justification for a theory is justification in believing the theory true: having justification for a scientific theory could be understood as

---

[13] Though Colyvan does not explicitly equate 'best' with 'best justified', the list of scientific virtues he considers in Colyvan (2001, pp. 78–79) for a scientific theory to count as good—among which are empirical adequacy, consistency, simplicity and parsimony, unificatory and explanatory power, boldness and fruitfulness and formal elegance—makes clear that he (like other supporters of IA) has much more in mind than simply currently accepted theories. Notice, in passing, that the 'ought to', as opposed to the 'best justified', has both a permissive and a prescriptive component. We will not put much weight on the latter.

[14] Cf. Colyvan (2001, pp. 22–24) for some qualifications. Briefly, [QC] states that the ontological commitment of a theory T is given by the objects that must be counted in the range of the objectual quantifiers in the existential theorems of (the canonical reformulation of) T. [QC] plays in Colyvan's argument the same role that quantification plays in Putnam's argument. Cf. Quine (1948).

[15] Notice that a weaker notion, like that of acceptance of a scientific theory, modelled, for example, on the lines suggested by Van Fraassen (1980), will not be strong enough to deliver the required mathematical realist or platonist conclusion. We will consider later the possibility of appealing merely to confirmation rather than justification.

simply having reasons, even only pragmatic ones, for adopting a scientific theory in ordinary scientific practice—e.g. because it is instrumentally helpful, or predictively accurate, or the like. This would lead to an argument along the following lines: we have a justification for some scientific theories; among them, some are such that some mathematical theories are indispensable to them; we have a justification for these scientific theories only if we a have a justification for the mathematical theories that are indispensable to them; therefore, we have a justification for the mathematical theories indispensable to these scientific theories.

No mention of truth is made here. This is *prima facie* consistent with Colyvan's argument, where no mention of truth is made either. Admittedly, however, in the debate on IA—and in Colyvan's discussion too—justification is understood as justification for the truth of a theory. We can, then, specify the argument accordingly and get the following version of IA:

Realism, epistemic [RE]

i) We are justified in believing some scientific theories to be true;
 [We are justified in believing T is true]
ii) Among them, some are such that some mathematical theories are indispensable to them;
 [M is indispensable to T]
iii) We are justified in believing true these scientific theories only if we are justified in believing true the mathematical theories that are indispensable to them;
 [We are justified in believing T true only if we are justified in believing M true]
[RE]- -
iv) We are justified in believing true the mathematical theories indispensable to these scientific theories.
 [We are justified in believing M true]

In what follows, we will take [RE] as the reference formulation for a minimal epistemic version of IA for semantic realism. [RE] is nothing but a specification of the more general argument sketched above, in which justification need not be justification for truth. Should we rest content with that more general argument, however, we could hardly obtain an argument for mathematical realism, for we would lack, unless further premises are added, any link between the justification of a theory and its truth (this is why we do not consider that version of the argument as one of our minimal versions of IA).

Since no mention of mathematical objects is made in [RE], [QC]—as any alternative criterion—is needed neither in the formulation nor in the justification of its premises. [RE] is neutral as to whether it is the existence of mathematical objects that makes the relevant mathematical theories true. [RE] would be a desirable version of IA for all those who believe in the objectivity and truth of mathematics, but would not by this fact alone qualify themselves as platonists.[16]

---

[16]Putnam's (1967) equivalent descriptions and Hellman's (1989) modal structuralism are two well-known candidates. [RE] seems also to respect the basic ideas underlying the criticisms that Pincock (2004), Azzouni (2004), and Paseau (2007) move against the standard platonist versions of IA.

Despite this, from [RE], an epistemic version of IA for platonism can be easily obtained by adding a premise introducing some appropriate criterion of ontological commitment:

Platonism, epistemic [PE]

i) We are justified in believing some scientific theories to be true;
   [We are justified in believing T is true]
ii) Among them, some are such that some mathematical theories are indispensable to them;
   [M is indispensable to T]
iii) We are justified in believing true these scientific theories only if we are justified in believing true the mathematical theories that are indispensable to them;
   [We are justified in believing T true only if we are justified in believing M true]
iv) We are justified in believing true a mathematical theory only if we are justified in believing the objects it is about to exist;
   [We are justified in believing M true only if we are justified in believing the objects it is about to exist]
[PE]----------------------------
v) We are justified in believing the objects which the indispensable mathematical theories are about to exist
   [We are justified in believing the objects M is abOut to exist]

Indispensability arguments are most of the time cast in epistemic terms. As Colyvan himself stresses, this epistemic character is due to a conception of ontology as a prescriptive and normative discipline: it tells us what we ought to believe to exist, or, in other terms, what we are justified in believing to exist.[17] However, one might rather conceive of ontology as a descriptive discipline and ask for arguments whose conclusions tell us that certain (sorts of) objects do exist, not just that we are justified in believing them to exist. Let us then formulate the arguments in a non-epistemic fashion:

Realism, non-epistemic [RnE]

i) There are true scientific theories;
   [T is true]
i) Among them, some are such that some mathematical theories are indispensable to them;
   [M is indispensable to T]
iii) These scientific theories are true only if their indispensable mathematical theories are themselves true;
   [T is true only if M is true]
[RnE]----------------------------
iv) The mathematical theories indispensable to these scientific theories are true[18]
   [M is true]

---

[17]Cf. Colyvan (2001, p. 11). Cf. also footnote 8 above.

[18]Those who believe the first premise to be too harsh can still accept a weaker formulation in which that premise is discharged and the conclusion is conditional in form, i.e. 'If there are true theories, then ...'.

Platonism, non-epistemic [PnE]

i) There are true scientific theories;
  [T is true]
ii) Among them, some are such that some mathematical theories are indispensable to them;
  [M is indispensable to T]
iii) These scientific theories are true only if their indispensable mathematical theories are themselves true;
  [T is true only if M is true]
iv) A mathematical theory is true only if the objects it is about exist;
  [M is true only if the objects M is about exist]
[PnE]- -
v) The objects which the indispensable mathematical theories are about exist
  [The objects M is about exist]

All these four minimal versions of IA are schematic, in more than one sense. First of all, 'T' and 'M', in the bracketed version of each premise, can be substituted, respectively, with names of particular scientific and mathematical theories. Furthermore, the meaning of 'indispensable', 'justification' and 'true' in all four arguments can be specified in different ways, so as to get strictly different arguments according to which specification is chosen. What is relevant is that the notions of justification and truth, however specified, must be such that justification in believing a theory true and truth itself are preserved under indispensability, however the latter is specified on its turn. In arguments for platonism, moreover, premise (*iv*) can be further qualified according to any preferred specification of the intuitive notion of aboutness (in Quinean terms, that premise would be specified, according to [QC], by reference to quantifiers and their domain).[19]

Prima facie, no such theses as naturalism and confirmational holism (or other) seem to be explicitly involved in the minimal arguments. But it remains open whether these (or other) theses are required, even as background assumptions, for the soundness of these arguments; this must be discussed in more details, especially if the relations between the minimal versions and the four arguments representative of the stances discussed in Sect. 13.2 must be spelled out. We will first pause to discuss the notion of (in)dispensability in the next section and then consider which assumptions and notions are really involved in the minimal arguments. The following discussion concerning both (in)dispensability and other relevant notions will also help clarifying how several versions of IA can be retrieved from our suggested minimal versions (cf. Sect. 13.6 below).

---

[19] Also Azzouni's blueprint reported in Sect. 13.2 above could be thought to be schematic. However, this is so in a different sense. Whereas our arguments are schematic in that they can be turned into strictly different versions of IA by further specifying some of the notions involved, Azzouni's blueprint is rather a matrix from which explicit and logically valid versions of IA can be obtained through the addition of other assumptions.

13.4 The Relational Character of (In)dispensability

Despite its obvious relevance for IA, the notion of (in)dispensability has undergone little specific analysis in the debate. What is exactly taken to be indispensable? And to what? And what does it mean to be indispensable?

As regards the first question, different aspects or ingredients of mathematics can and are taken into account in formulations of IA: the quantification over mathematical 'entities' or (putative) 'objects' (like in Putnam (1971)); mathematical entities or objects themselves, like in Colyvan (2001, p. 11) or Baker (2005) or many others; the apparent reference to such entities or objects, like in Colyvan (2001, p. 7); the assumption of the truth of some mathematical statements (viz., statements involving mathematical vocabulary), like in Resnik (1995, pp. 169–171, 1997, pp. 46–48); mathematical vocabulary (which we take to be what is often implicitly intended when authors use 'apparent reference'), like in Colyvan (2001, p. 16); some appropriate use of mathematical statements, like in Azzouni (2009); or finally, mathematical theories.

It is possible to maintain that there are significant differences stemming from these allegedly alternative choices. Those who believe so should also consider minimal schematic versions of IA in which what is taken to be indispensable is left unspecified. We will not dwell here on this matter, as well as on other possible parameters that fully schematic version of IA can involve (for instance, parameters specifying whether indispensability is constrained by certain goals we want our scientific theories to achieve), since we consider this at length elsewhere (cf. Panza and Sereni forthcoming). Here, we rather assume that there is a common idea underlying all the mentioned options regarding what is to be taken as indispensable, namely, that the relevant scientific theories have an essential recourse to a vocabulary fixed by some mathematical theories and then to the notions that this vocabulary is supposed to convey. As far as what is taken to be indispensable is some appropriate use of mathematical statements, we take it that what is relevant is that of the statements of a mathematical theory, or statements involving the mathematical vocabulary fixed by this theory, it is possible to make this appropriate use—while (appropriately) using (the statements of) a scientific theory. So also this option seems to reduce, with this proviso, to the option according to which it is the use of theories to be indispensable (to other theories; see below the answer to the second question).[20] Moreover, it seems to us that by literally accepting the option that what is indispensable are mathematical entities or objects themselves, one is open to an obvious risk of circularity (unless one is ready to concede that these entities or objects could be indispensable as such without existing). Here, thus, we will rest content with taking mathematical theories as what is said to be indispensable in IA, though leaving open the possibility of understanding this, when more details are given, in different ways.

---

[20] A different issue is whether an argument based on the indispensability of the use of theories can be retrieved from arguments based on the indispensability of theories *tout court*: we will come back to this in Sect. 13.6.

Parallel considerations also apply to the second question. We will thus rest content with taking other theories, typically scientific ones, to be that for which mathematical theories are said to be indispensable in IA, although we leave open, again, the possibility of understanding this, when more details are given, in different ways.[21]

The third question seems to us much more relevant. Let us focus on that, then. As will be clear, this will also help explain how some common versions of IA can be retrieved from the minimal versions through appropriate specifications of this notion.

What does it really mean that a theory is indispensable to another? Let T be a scientific theory and M a mathematical theory employed in the formulation of T. Let us call, for short, a statement employing the vocabulary of a theory M an 'M-loaded statement'. It seems to us obvious to take M to be indispensable to T if it is not possible to obtain from T a theory T′, equivalent to T according to some specified equivalence relation, in which M-loaded statements do not occur. But, then, given any version of IA, in order to specify the indispensability condition involved in it, one has to choose the appropriate equivalence relation according to which T′ is to be taken as equivalent to T. In most cases, it seems to be tacitly admitted that the relevant equivalence relation is such that if T′ is equivalent to T according to it, then T′ preserves the descriptive and predictive power of T. This suggests taking this relation to be something like the relation of having the same empirical adequacy or that of having the same observational consequences. Still, other choices could be pertinent.

Making this choice will not be enough, however, since it is also important to make sure that the equivalence between T and T′ is not obtained by merely formal gerrymandering. This requirement captures Field's suggestion that T′ has to be 'reasonably attractive' (1980, p. 8) or Colyvan's suggestion that it must be 'preferable' to T (2001, p. 77). Craig's Theorem is, for example, a well-known example of a purely formal method for obtaining from any (recursively enumerable theory) theory T involving M-loaded statements a (recursively axiomatizable theory) theory T′ that involves no such statement and has the same observational content.[22]

Attractiveness and preferability are aim-specific notions, to be decided on broadly scientific criteria case by case. We can express this point in full generality

---

[21] Whether we should consider a mathematical theory M indispensable to a theory T when only some parts of M are as a matter of fact used for the formulation of T depends, among other things, on whether the employed part of M is such that it can be considered an independent (sub-)theory of M. This is what Peressini (1997) labels 'the problem of the unit of indispensability'.

[22] According to Craig's Theorem (1956), given a recursively enumerable theory T, and a partition of its vocabulary into an observational one, o, and a theoretical one, t, then there exists a recursively axiomatizable theory T′, whose only non-logical vocabulary is o, comprising all and only the consequences of T expressible in o. Craig himself warned against the philosophical import of his result, claiming that the theorems of T′ obtained by his re-axiomatization method are not more 'psychologically or mathematically [...] perspicuous' than those of T, this being 'basically due to the mechanical and artificial way in which they are produced' (p. 49).

by saying that T′ has to be equally or even more scientifically virtuous than T, where the appropriate criterion of virtuosity will be fixed considering common scientific virtues, according to our specific purpose.

We can thus offer the following general clarification of the notion of (in)dispensability:

[IND] (In)dispensability

A theory M is dispensable from a given scientific theory T if and only if there is a scientific theory T′ that does not include M-loaded statements and that:

    a) is ε-equivalent to T, where ε is an appropriate equivalence relation;
    b) is equally or more virtuous than T according to an appropriate criterion of virtuosity α.

If T includes M-loaded statements, and there is no scientific theory T′ satisfying the above conditions, then M is indispensable to T.

A noteworthy consequence of this definition is that common talk of (in)dispensability is partly inaccurate. No theory is (in)dispensable *tout court* to another theory, but only relative to a certain equivalence relation and a criterion of virtuosity. We should better speak of ε-α-(in)dispensability, rather than (in)dispensability *simpliciter*. (In)dispensability is an essentially relational notion.[23]

According to which equivalence relation is selected, IA can have different philosophical significance. Some minimal notion of indispensability can be thought of, if the equivalence relation ε is chosen on logico-syntactical grounds (e.g. if it is taken to be the relation of having the same expressive power, i.e. of including either the same theorems or definitional paraphrases of them). But more demanding notions can be thought of. For instance, one could suggest using an equivalence relation such as that of having the same explanatory power, or cognate ones. Should theorists such as Field and Baker, building on the third of the four stances mentioned in Sect. 13.2, be willing to endorse any of the minimal IA suggested above, they could easily obtain a specification of them based on the notion of explanatory power: once ε is appropriately specified in this way, it will straightforwardly follow that M is indispensable to T only if it plays an indispensable explanatory role in T. We will come back to this below.[24]

---

[23]Colyvan's discussion of 'the role of confirmation theory' in his (2001, pp. 78–81) hints to the relational character of the notion of preferability. We take our clarification of (in)dispensability to improve on that suggestion.

[24]As pointed out to us by an anonymous referee, our schematic definition of (in)dispensability assigns no special role to the notion of applicability of a mathematical theory. It goes without saying that we acknowledge the greatest importance to the problem of the applicability of mathematics and to its role within the debate concerning IA, although it is impossible to discuss these issues here. We do believe, however, that, although the two notions will be certainly connected eventually, they can be beneficially treated separately at a general level of analysis as ours. Whether and how a particular conception of applicability affects a given version of IA—either

13.5 What Does It Takes to Be an Indispensability Argument?

In this section we explore which assumptions or theoretical ingredients generally, beyond the notion of (in)dispensability, are required in order either to formulate or to endorse one of the minimal versions of IA presented above. We begin by considering the role of doctrines such as naturalism and confirmational holism. On a fairly common understanding—and in accordance with working definitions to be given below—these doctrines are only relevant for epistemic argument, being concerned as they are with the justification of scientific theories. Other assumptions will turn out as involved in non-epistemic arguments too.

13.5.1 *Confirmational Holism and Naturalism*

Nothing—and a fortiori naturalism and holism—is required, in [RE] or [PE], to justify an all-and-only clause like that in Colyvan's argument, simply because there is no such clause to be justified at all. However, these doctrines may still be thought to be necessary conditions for justifying some of the premises of those arguments. A related concern is whether either doctrines might represent sufficient conditions.

Apart from some aspects to be considered shortly, we will take for granted a general understanding of the notions involved (and of the vast debate concerning their proper characterization) and will merely state them in a convenient form for future reference, taking it that these formulations are those that philosophers concerned with IA have most commonly in mind:

> [CH *Confirmational Holism*:][Since the appreciation of empirical evidence is in no way a matter of comparing a single fact with a single hypothesis,] the confirmation of a single hypothesis or of a system of hypotheses comes together with (or entails) the confirmation of a larger net of hypotheses (possibly of the whole net of hypotheses that our knowledge consists in).
>
> [NAT *Naturalism*:][Since scientific theories are the only source of genuine knowledge,] we are justified in believing to be true only scientific theories, or other theories (or statements) whose truth follows from the truth of some scientific theories.

In what follows, we assume that it is legitimate to talk about the justification of mathematical and scientific theories independently of whether [CH] or [NAT] turns out to be necessary or unnecessary assumptions for the justification of any premise in the minimal arguments: [NAT], by itself, tells us only which sort of theories can be true, but is not taken as constitutive of the notion of justification; and whereas

---

by facilitating its conclusion or by preventing it—is, indeed, something that we believe will have to be considered case by case, according to versions of the argument appropriately specified so to involve, for instance, one's preferred notion of applicability in the specification of either the equivalence relation ε or the criterion of virtuosity α.

confirmation could be seen as constitutive of the justification of empirical theories, the way in which confirmation is accrued need not be such: even if confirmation is holistic, the claim that a scientific theory is justified does not presuppose by itself [CH].

If one wishes, [CH] and [NAT] could be specified further in order to have a distinctive focus on ontology and would thus state, respectively, that the confirmation of a single hypothesis or of a system of hypotheses comes together with (or entails) the confirmation of the existence of all the entities that are quantified over in a larger net of hypotheses (possibly in the whole net of hypotheses that our knowledge consists in) and that we are justified in acknowledging the existence only of those entities that are quantified over in our scientific theories. In these particular formulations, naturalism and holism could be taken to be explicitly stated—not just posited as background assumptions—in the first premise of Colyvan's argument.

As stated, [NAT] entails the thesis that our scientific theories are our only source of genuine knowledge about the world. Many versions of naturalism are available, but in order to make it both plausible and relevant to IA, one should steer clear from at least two readings.

On the one reading, naturalism is the too strong thesis, almost indistinguishable from nominalism, that only non-abstract entities can be acknowledged to exist. Both what Colyvan (2001, Chap. 3) calls 'Eleatic Principle' and the version of naturalism endorsed by, for example, Weir (2005) and Armstrong (1997, p. 5) are cases in point. We agree with Colyvan that if naturalism is involved in IA at all, it cannot be of this kind, since it stands in clear contradiction with the latter's conclusion.

On another reading, naturalism can be too weak to be relevant to IA. In discussing Quinean naturalism, Colyvan (2001, pp. 23–24) distinguishes two strands: the 'No First Philosophy Thesis', this being the (normative) thesis that in approaching 'certain fundamental questions about our knowledge of the world' we should 'look to science (and nowhere else) for the answers', and the 'Continuity Thesis', this being the (descriptive) thesis that 'philosophy is continuous with science and that together they aim to investigate and explain the world around us'. Colyvan himself argues that these two theses are intimately related in a complex way. One thing, however, should be clear, beyond Quinean exegesis. If the Continuity Thesis is meant to claim that both science and philosophy, where this involves *a priori* methods of inquiry, are our only legitimate sources of knowledge about the world, the ensuing version of naturalism will not be suitable to the formulation of any empiricist argument. It can be suitable only in so far as one also adds that philosophy should abandon its traditional *a priori* methods and become a genuinely scientific (in the sense of empirical sciences) enterprise. But once this is added, empirical science and philosophy become utterly indistinguishable, and we come back to [NAT], with the only difference that in this new framework there is no philosophy left to be opposed to empirical sciences. The claim expressed in [NAT] seems to us to represent the most plausible reading of naturalism adequate for the formulation of an empiricist argument, as IA has been traditionally understood by Quine and his heirs. Notice, in passing, that we do believe that the 'only' in the formulation of [NAT] is crucial in order to distinguish naturalism from scientific realism (more on this below).

It now remains to be seen whether [CH] or [NAT] is either necessary or sufficient in order to justify any of the premises in [RE] and [PE]. Premises (*ii*) of each argument, as well as [PE]'s premise (*iv*), are beyond any suspect of guilty here.[25] So we better concentrate on premises (*i*) and (*iii*), common to both arguments. Premise (*i*) is liable to the charge of surreptitiously appealing to naturalism, whereas premise (*iii*) is liable to the charge of surreptitiously appealing to holism. Let us consider the latter first.

13.5.2   Can We Dispense with Confirmational Holism?

One could maybe advance a simple way to settle the question whether endorsing holism is necessary for endorsing premise (*iii*) of [RE] or [PE]. It would consist in suggesting that the relevant form of holism we are to consider in connection with [RE] or [PE] is not, properly speaking, confirmational holism, but some sort of holism concerning justification, and that this form of holism just consists in the claim that justification of a theory T whatsoever transmits to any other theory indispensable to T. In this case, the very claim that the justification of a scientific theory S transmits to a mathematical theory M indispensable to S would reduce to a mere instance of such a version of holism, so that the latter would be sufficient, yet not necessary, for endorsing this claim and, then, premise (*iii*) of [RE] or [PE].

Moreover, such a form of holism would not only be different from that which is usually at stake when the relation between holism and IA is discussed, which is confirmational holism proper, that is, a form of holism specifically concerned with confirmation, rather than with justification in general; it would also be a quite Pickwickian form of holism, since it would merely require that justification transmits from a theory T to another theory so intimately connected to it as the indispensability of the latter for the former implies. Of course, in order to admit that this is so, one should maintain that indispensability is so specified as to warrant this transmission. Still, this is not the point here. What is relevant is rather that such a form of holism would restrict the transmission of justification from theory to theory to a case in which the relevant theories are related by a certain sort of intimate connection. But, if a proper form of holism of justification should be defended, it should rather consist in claiming that justification transmits much more widely, along a larger net of theories or hypotheses (possibly along the whole web of hypotheses that our knowledge consists in), without constraints such as indispensability.

The situation would be even more outlandish if one argued that the relevant form of holism consists in the claim that justification of a scientific theory S transmits to a mathematical theory M indispensable to S (since it would be odd to maintain that a sort of holism concerned with justification, in general, is restricted

---

[25]But cf. footnote 30, below.

to the consideration of scientific and mathematical theories), or that confirmational holism, as such, consists in the claim that justification of a scientific theory S transmits to a mathematical theory M indispensable to S (since it would be odd to maintain that confirmational holism generally concerns justification, rather than confirmation). In this latter case, premise (*iii*) of [RE] or [PE] would just be the same as confirmational holism, and it would be beyond doubt that endorsing the latter is necessary for endorsing the former. It seems however plain to us that confirmational holism is a different and wider thesis: not only it is concerned with confirmation rather than with justification, but also it is not merely limited to the case in which confirmation is transferred from a scientific theory S to a mathematical theory M that is indispensable to S.

All these considerations lead us to discard from the very beginning the simple possibilities just evoked and to focus on confirmational holism proper, conceived as the very claim [CH].

A first, preliminary, difficulty is the following. Apparently, there is a striking asymmetry between the condition expressed in premise (*iii*) and that expressed in [CH]: the latter seems to express an inference from a part to the whole, whereas premise (*iii*) seems to express an inference from the whole to a part. We need therefore to understand how the two might be related in any way relevant (either sufficiently or necessarily) for lending support to premise (*iii*).

Under a quite weak reading of it, [CH] states that when a single hypothesis h of a theory S is confirmed, the whole S is confirmed, which we express in symbols by '$C(h) \to C(S)$', where '$C(x)$' stands for 'x is confirmed'. If we admit that a mathematical theory M involved in a scientific theory S counts as a cluster of hypotheses of S, one can replace here '$C(h)$' with '$C(M)$', so as to get the new implication '$C(M) \to C(S)$'. It is however clear that this implication (be it admissible or not) is hardly useful in an argument whose purpose is that of building on considerations about some scientific theories in order to draw conclusions about some mathematical theories appropriately connected to the former. At most, the reciprocal implication '$C(S) \to C(M)$' could be relevant. But if a mathematical theory M involved in a scientific theory S counts as a cluster of hypotheses of S, and we take confirmation to be cumulative, i.e. to be such that a conjunction of hypothesis (or of other items susceptible of confirmation) can only be confirmed by confirming all its conjuncts (which entails, of course, that confirmation is *d'emblée* also distributive: if a conjunction of hypothesis is confirmed, its conjuncts are so), this implication is trivial, since, whatever confirmation might come to in details, it is immediate to see that under this conception a theory cannot count as confirmed as a whole if some of its hypothesis are not so. Hence, in this case, arguing for this implication requires no appeal to any strong and/or controversial thesis, as confirmational holism appears to be.

Things change, however, if a mathematical theory M involved in a scientific theory S is rather taken to count as an auxiliary theory that S appeals to, without encompassing it, or if confirmation is not taken to be cumulative (so that it cannot be granted that it is distributive, *d'emblée*), namely, if it is admitted that a conjunction of hypothesis (or of other items susceptible of confirmation) can be directly confirmed as a whole (i.e. without going through a confirmation of all its conjuncts). Under both scenarios the implication 'C(S) → C(M)' becomes far from trivial: it asserts that confirming (in one way or another) a scientific theory S goes with (or is sufficient for) confirming either an auxiliary mathematical theory M, which S appeals to, but that is not included in S (under the former scenario), or a mathematical theory M that S encompasses, even if the confirmation of M is not, as such, involved in that of S (under the latter scenario). It seems to us that it is only under one of these scenarios that there is room for plausibly considering the possibility that [RE] and [PE] be somehow related with [CH], provided, of course, that a reading of [CH] be adopted, according to which this thesis entails that confirming a scientific theory S goes with (or is sufficient for) confirming an auxiliary mathematical theory M, which S appeals to, but that is not included in S, or a mathematical theory M that S encompasses, within a framework in which confirmation is not taken to be cumulative.

It is not our purpose here to argue in favour either of one of these two possible scenarios or of this reading of [CH]. We merely suppose, for the sake of the argument, that both one of the former and the latter are admitted, while contending that if this is not so (i.e. either both of the former or the latter are rejected), there is no plausible reason for connecting [RE] and [PE] with [CH]. In other terms, we admit, for the sake of the argument, that [CH] entails 'C(S) → C(M)', provided that S and M are, respectively, a scientific and a mathematical theory and that either S appeals to M, without encompassing it, or S encompasses M but confirmation is not taken to be cumulative. For short, in what follow we shall call 'weak condition' (for reasons that will become clear below) the condition involved in this supposition, namely, that either S appeals to M, without encompassing it, or S encompasses M, provided that confirmation is not taken to be cumulative.

Now, also admitting, under this same condition, that 'C(S) → C(M)' entails [CH] would be quite implausible, since, however it might be conceived, confirmational holism can certainly not be reduced to a thesis about the confirmational relation between scientific and mathematical theories under some condition whatsoever. The possibility still remains open, however, of admitting—for the sake of the argument, again, but without falling into an evident oddity—that, when confronted with a scientific theory S and a mathematical theory M, under the weak condition, one has no other ground than [CH] for arguing that C(S) → C(M). This is just what we admit. For the purpose of our following discussion, we can then suppose that, under the weak condition, endorsing [CH] is necessary for endorsing premise (*iii*) of [RE] and [PE] if endorsing the implication 'C(S) → C(M)' is necessary for this. As regards the issue whether [CH] is sufficient for endorsing this same premise, things are much simpler. Since, if [CH] entails 'C(S) → C(M)' and 'C(S) → C(M)' entails this premises, then [CH] entails this premise. We can then limit our enquiry to this question: is endorsing the implication 'C(S) → C(M)', under the weak condition, necessary or sufficient for endorsing premise (*iii*) of [RE] and [PE]?

Let us come back, then, to this premise. It states that justification to believe a mathematical theory M, indispensable to a scientific theory S, to be true is a necessary condition for having justification to believe that S is true. Let us call this the 'S-M justificatory connection under indispensability'. We can express it in symbols by: 'IND(M, S) → [J(S) → J(M)]', where 'J(x)' stands for 'we are justified in believing x true'. What we have to investigate is, then, whether the implication 'C(S) → C(M)' (which, for the sake of the argument, we take as a consequence of [CH]) is either necessary or sufficient to motivate the S-M justificatory connection under indispensability, under the weak condition.

13.5.2.1 Is Confirmational Holism Necessary for Premise (iii)?

That confirmational holism might be unnecessary for IA has been already suggested, on different grounds, by several authors (e.g. Resnik 1995; Dieveney 2007; Azzouni 2009), including Colyvan himself.[26] Taking much of this discussion for granted, what we are interested here is, as we have said, simply whether the implication 'C(S) → C(M)' needs to be assumed in order to defend our minimal (epistemic) versions of IA, especially whether it is required for upholding premise (*iii*) of [RE] and [PE]. In order for this to be the case, there must be no way of supporting the S-M justificatory connection without appealing to this implication. We shall argue that it is not so.

A first problem arises straightaway. Both our admissions that [CH] entails the implication 'C(S) → C(M)' and that one has no other ground than [CH] for arguing that C(S) → C(M) have been conditioned by the weak condition. This last condition is, however, much weaker than that involved in the S-M justificatory connection under indispensability, namely, that M is indispensable to S. Hence, one could contend that, in the case where the latter condition obtains, i.e. when M is indispensable to S, one should rely on grounds other than [CH] for arguing that C(S) → C(M). Insofar as we are not at all willing to discard this possibility (that we rather consider favourably, as we shall see later), we suggest to abstract from the supposition that M is indispensable to S, by resting content with the weaker supposition that S and M are, respectively, a scientific and a mathematical theory and that the weak condition obtains. We want then, firstly, to argue against the view that, under this condition, endorsing the implication 'C(S) → C(M)'—and then [CH]— is necessary for endorsing that justification to believe M is a necessary condition for having justification to believe that S is true. Let us call this last claim the 'S-M justificatory connection', *tout court*. We can express it in symbols by: 'J(S) → J(M)'.

---

[26]Cf. Colyvan (2001, p. 37): 'As a matter of fact, I think that the argument can be made to stand without confirmational holism: it's just that it is more secure *with* holism. The problem is that naturalism is somewhat vague about ontological commitment to the entities of our best scientific theories. It quite clearly rules out entities *not* in our best scientific theories, but there seems room for dispute about commitment to some of the entities that *are* in these theories. Holism helps to block such a move since, according to holism, it is the whole theory that is granted empirical support'. For discussion of this passage and other issues connected with holism in Colyvan's framework, cf. Peressini (2003, pp. 220–222).

We shall come back later to the question whether, under the weak condition, the implication 'C(S) → C(M)' (taken as a consequence of [CH]) is necessary to motivate the S-M justificatory connection under indispensability.

Our first point is as follows. Under some conceptions of science, like falsificationism, the link between the confirmation of a scientific theory and its justification is severed. In such conceptions, empirical confirmation of the (testable) hypotheses of a scientific theory S, whatever advantages it may deliver and whatever it might be considered to consist in, will not essentially contribute to the justification of S. Still, even if one endorsed such views, one may of course still maintain, under the weak condition, the S-M justificatory connection and, a fortiori, the S-M justificatory connection under indispensability. Even if it is conceded that confirmation is holistic, i.e. that [CH] is true, the fact that C(S) → C(M) will simply play no role, in these views, to support the claim that J(S) → J(M), under whatever supplementary condition.

Another way of showing that endorsing the implication 'C(S) → C(M)' is unnecessary for endorsing that the S-M justificatory connection, under the weak condition, consists in noticing that this connection could be (vacuously) endorsed by anyone considered to have reasons for taking M to be justified independently of any consideration about its role in or with respect to S (and, then, a fortiori, of M's being indispensable to S). This would be the case, for example, for anyone that maintained to have *a priori* reasons for believing in the necessary truth of M (and would then, at most, take its indispensability to S together with S's being justified as a welcome by-product): the S-M justificatory connection (and, a fortiori, the S-M justificatory connection under indispensability) will be motivated without appealing to holism, nor to confirmation at all. This would make whatever empirical confirmation we can have for S (be its nature holistic or not) immaterial to the justification of M. Such an option will certainly not be welcomed by many supporters of IA. On the one hand, many of them also adopt a form of naturalism that ban *a priori* arguments. On the other hand, and most importantly, such an option seems to make IA pointless. Still nothing prevents one from accepting IA even under these circumstances.[27] In any case, what all this shows is that the S-M justificatory connection is, in general, conceptually independent of the notion of confirmation (and a fortiori of [CH]) and is then so also under the weak condition. More generally, one of the basic idea underlying IA (cf. e.g. Putnam 1975a, p. 74) that one cannot be a realist about science and at the same time an antirealist about mathematics needs not be supported by confirmational holism.[28]

---

[27]See Sereni (2013) for a way in which Frege may be taken to have reasons—based on considerations on applicability—for endorsing premise (*iii*), despite being alien to a holist conception of confirmation and to the idea that confirmation is relevant for the justification of mathematical theories.

[28]Putnam would clearly endorse premise (*iii*). But he has recently dispelled any doubt that his endorsing it hinges on holism: 'I have never claimed that mathematics is 'confirmed' by its applications in physics' (cf. Putnam (2012, p. 188)).

Le us concede, now, both that (against the first point) empirical confirmation is an essential ingredient to the justification of a scientific theory (i.e. that $J(S) \to C(S)$) and that (against the second point) we are after a defence of the S-M justificatory connection, under the weak condition, which is not based on having reasons independent of M's role in, or with respect to, S (possibly *a priori* ones) to believe in the necessary truth of M. Since the S-M justificatory connection only consists in the implication '$J(S) \to J(M)$', for the other implication '$C(S) \to C(M)$' to be necessary for this connection, it must be the case that $(J(S) \to J(M)) \to (C(S) \to C(M))$. But once one explores ways of defending this implication, it becomes apparent that it is unmotivated.

Insofar as we have conceded that confirmation of S is a necessary ingredient of its justification, what we have to consider is whether, under the weak condition, $[(J(S) \to C(S))\ \&\ (J(S) \to J(M))] \to (C(S) \to C(M))$. Suppose that S is justified. Then, for the premise of this implication, it follows that S is also confirmed and that also M is justified. If this were enough for concluding that M is confirmed, under the weak condition, our implication would be verified. But, why should be so? It would be so if, once S is justified and then confirmed, the only way in which its justification could be transferred to M, under the weak condition, were that S's confirmation transferred to M and that M's confirmation were sufficient for its justification. But this is clearly unmotivated. Why should the only way in which justification of S transfers to M, under the weak condition, be through the transfer of S's confirmation to M? We have assumed that confirmation of S is necessary for its justification, not that it is also sufficient; so it could well be the case that M's justification be due, under the weak condition, to ingredients of S's justification other than its confirmation. Moreover, why (both under the weak condition or not) should M's confirmation be sufficient for its justification? And why, anyway, should one be forced to admit that a mathematical theory is open to confirmation, in order to conclude that M is justified provided that S is so and the weak condition obtains?

The situation does not change essentially if the supposition that the confirmation of S is a necessary ingredient of its justification is replaced by the other supposition that the confirmation of S is sufficient of its justification. In this case, what we would have to consider is whether, under the weak condition, $[(C(S) \to J(S))\ \&\ (J(S) \to J(M))] \to (C(S) \to C(M))$. Suppose that S is confirmed; then, for the premise of this implication, it follows that M is justified. Again, if this were enough for concluding that M is confirmed, under the weak condition, our implication would be verified. But, again, this is unjustified, and equally unjustified is the claim that from these suppositions it follows that a mathematical theory is open to confirmation (both under the weak condition or not).

All in all, the only way of claiming that [CH] is necessary for the justification of the S-M justificatory connection under the weak condition is to assume that, under this condition, the confirmation of S and the justification of M are sufficient for the justification of S and for the confirmation of M, that is, that $C(S) \to J(S)$ and $J(M) \to C(M)$, since it is easy to see that '$[(C(S) \to J(S))\ \&\ (J(M) \to C(M))\ \&\ (J(S) \to J(M))] \to (C(S) \to C(M))$' is a tautology. Still, supposing that the confirmation of S is sufficient for its justification is utterly implausible, since it results in disregarding the well-known issue that empirical confirmation will not distinguish between incompatible and still empirically equivalent theories, and that in most cases what bestows justification on a theory is, besides confirmation, a combination of many other virtues (some of which being even *a priori*), such as simplicity, unificatory power, explicatory power, ontological parsimony and fruitfulness. Moreover, supposing that the justification of M is sufficient for its confirmation requires admitting that a mathematical theory is open to confirmation, which is something that does seem to be required neither by the weak condition nor by the S-M justificatory connection under this same condition. Of course, if this were admitted and were also admitted that the justification of M is sufficient for its confirmation, one could add a supplementary conclusion to [RE] to the effect that the relevant mathematical theories are not merely justified but also confirmed, but this is far from mandatory for [RE] to work.

Up to here, we have avoided considering M as indispensable to S, by merely supposing that the weak condition obtains and investigating whether endorsing [CH] could be taken as necessary for endorsing the S-M justificatory connection under this condition. Let us now suppose that M is indispensable to S, as any version of AI requires, and investigate whether endorsing [CH] could be taken as necessary for endorsing the S-M justificatory connection under indispensability, that is, the premise (*iii*) of [RE] and [PE]. It seems to us that, in this case, even if one accepts that confirmation is relevant, necessary or even sufficient for the justification of M, it becomes still more evident that [CH] is not required for the purpose. Indeed, if we take M as indispensable to S, it is reasonable to expect that the grounds for arguing that $C(S) \rightarrow C(M)$ should be sought in that very fact, i.e. in whatever intimate connection is established between S and M by the very fact that the latter is indispensable to the former. This is not to take for granted that the notion of indispensability, under whatever specification, together with the admission that M is indispensable to S (under the relevant specification), will be able by itself to deliver these grounds. As a matter of fact, arguing that it is so does should be one of the main tasks of a supporter of IA. The point is just that, once the indispensability of M to S is assumed, it is reasonable to expect that it, and it alone, would allow one to claim that $C(S) \rightarrow C(M)$, without requiring the help of [CH], which is, in fact, a much wider claim not constrained by an indispensability condition.[29]

The unpalatable consequences of the assumption of [CH] in IA when compared to actual mathematical practice—especially with the conscious use by scientists of false assumptions and idealizations—has been stressed by Maddy (1992).[30] Apart from this, the main point, for us, is that [CH] is either irrelevant for AI (in the case

---

[29]Cf. the observation made at the very beginning of the present Sect. 13.5.2.

[30]Hellman (1999) attempts at avoiding these and others unpalatable consequences by suggesting that even though confirmation is holistic and it is conceded that it is transferred from the testable hypotheses of a theory to its inner parts, it should not be taken to transfer equally, so that different parts of a theory, like those expressing idealized conditions or mathematical hypotheses, could be taken to be confirmed to different degrees. Even if Hellman is right, his suggestion does not affect our present points concerning the non-necessity of confirmational holism for endorsing IA.

in which no clear appropriate connection between justification and confirmation is admitted) or it is an overkill (in the case in which this appropriate connection is admitted), since it delivers much more than it is actually required to defend IA, thus being an unnecessarily strong thesis.[31]

Indeed, the mere fact that an appeal to indispensability is thought to be relevant—which is already emphasised by the very expression 'indispensability argument' usually chosen to denote the argument under consideration—seems to be an indicator that such an argument is essentially grounded on that notion, and not on the much wider one of confirmational holism.

True, holism is usually required to rule out the possibility that statements about theoretical or mathematical objects might be confirmation resistant, as opposed to observational statements. However, if in [PE] one endorses, following Quine, a criterion for ontological commitment that is uniform across statements—i.e. based on the logical form of statements and content neutral—one will have as a consequences that all entities whose existence is entailed by one's criterion will be said to exist, observational and theoretical alike. If one believes in the truth of premises (*i*)–(*iii*) of [PE] and has a uniform criterion of ontological commitment, confirmational holism, again, is not needed, for this purpose.[32]

13.5.2.2 Is Confirmational Holism Sufficient for Premise (iii)?

As regards the sufficiency of endorsing [CH] for endorsing premise (*iii*) of [RE] and [PE], things are by far simpler. Since, if one supposes both that M is indispensable to S and, merely, that the weak condition obtains, in order to draw the implication 'J(S) → J(M)' from the other implication 'C(S) → C(M)', it is then necessary and sufficient to admit that J(S) → C(S) and C(M) → J(M). Hence, for premise (*iii*) of [RE] and [PE] to follow from [CH] it is sufficient that confirmation for the relevant

---

[31] We shall come back in Sect. 13.5.2.2 to the sufficiency of [CH] and the conditions it requires in terms of the connections between justification and confirmation.

[32] One could claim, however, that some form of holism (presumably non-confirmational in nature) is somehow presupposed by any criterion of ontological commitment uniform across statements, namely, in order to ensure that such a criterion uniformly applies both to scientific and to mathematical theories. Against this latter supposition, one could argue, for example, that the notion of aboutness employed in premise (*iv*) cannot be given a content-neutral characterization and does not apply to mathematical objects. A case in point is Azzouni's (1998, 2004) suggested alternative to [QC]. One could think, then, that an appropriate form of holism is required for rejecting this possibility. Notice, however, that this is far from necessary: even if some understanding of [QC] or, better, some specification of the schematic notion of aboutness employed in premise (*iv*) of [PE] and [PnE] (or of other schematic notions employed in some of our versions of IA) turned out somehow to presuppose some form of holism, this would leave untouched that an epistemic version of IA and, then, *a fortiori*, IA as such, do not necessarily require any appeal to it; at most this thesis would be involved in some particular instances of such an argument (just as it happens for IBE: cf. footnote 35 below).

scientific theories is taken to be necessary for their justification, and confirmation for the relevant mathematical theories is taken to be sufficient for their justification. Moreover, if we suppose that, when confronted with the relevant theories S and M, one has no other ground than [CH] for arguing that $C(S) \rightarrow C(M)$, these two conditions (taken together) are also necessary. Under this supposition, wondering whether endorsing [CH] is sufficient for endorsing premise (*iii*) of [RE] and [PE] reduces to wondering whether confirmation for the relevant scientific theories is necessary for their justification and confirmation for the relevant mathematical theories is sufficient for their justification.

The former of these conditions could be questioned, but it seems to us to be safely admissible in general, that is, for most of our scientific theories.[33] For it is plausible to grant that, in general, any virtue a scientific theory may have will not confer justification to it in absence of confirmation.

The latter condition is more questionable, instead. One reason for this has already been mentioned: consenting to it requires admitting that mathematical theories are open to confirmation, which could be plausibly questioned (and is, in any case, nor required for endorsing IA). Another reason is that contending that confirmation of a mathematical theory, whatever it might be taken to be, is sufficient for its justification results in disregarding any *a priori* virtue that such a theory could have as a necessary ingredient of its justification. Finally, even someone, presumably endorsing a proto-empiricist view on mathematics, who is ready to admit that mathematical theories are open to confirmation and that no *a priori* virtue of a mathematical theory is required for it to be justified (when it is well supported by *a posteriori* or empirical evidences) could face that the same difficulty raised above (Sect. 13.5.2.1) for the thesis that confirmation of a scientific theory is sufficient for its justification, since this thesis also applies, *mutatis mutandis*, in the case of mathematical theories.

13.5.3 Can We Dispense with Naturalism?

What about now premise (*i*) of [RE] and [PE]? Do we need to assume naturalism, in any form, in order to claim that we are justified in believing some scientific theory

---

[33]One could, however, question this condition in some quite particular cases, as those involving highly theoretical physical theories, for example, string theory. One could indeed maintain that in cases like these, the relevant scientific theories can be justified and are actually considered to be so, independently of any empirical confirmation they may receive or have received. It is more likely, however, that in the complete absence of empirical confirmation, we would not take ourselves to be justified, however weakly, in taking a scientific theory to be true; rather, such a scientific theory will be said to enjoy a number of virtues that will merely make it acceptable in the scientific community for many practical and theoretical purposes. Nonetheless, this form of acceptance, it goes without saying, will not be strong enough to support the conclusion(s) of IA, in any of the versions we have discussed here.

to be true? What seems clear is that some form of scientific realism will have to be assumed. For a fairly standard characterization, we can describe scientific realism as the thesis that our mature scientific theories are true or, at least, approximately true descriptions of an external, mind-independent reality, that their statements should be interpreted at face value (both when they speak of observable entities and when they speak of theoretical ones) and that the objects of which they speak do inhabit the world (cf. Psillos (1999, p. xix) for a more extended definition on these lines). In order to defend premise (*i*) in [RnE] and [PnE], we need to maintain that there are true scientific theories, and this implies a form of scientific realism. Moreover, some arguably milder form of scientific realism will be needed also for defending premise (*i*) in [RE] and [PE]. This milder form of realism should at least accept that we are justified in believing that there are true scientific theories (without necessarily taking the further step of claiming that there are, or were, or will be some).[34]

Now, scientific realism, in one form or other, is likely to be a basic assumption underlying naturalism. Quine himself listed 'unregenerate realism' among the sources of naturalism (Quine 1975, p. 72). But in order to justify premise (*i*) of [RE] and [PE], only some form of scientific realism is needed, and whereas naturalism implies (or might imply, depending on the chosen formulation) scientific realism, the latter does not imply the former. Where naturalism hinges on endorsing that scientific theories are the only source of genuine knowledge, with the result that we are justified in believing to be true only these theories or those whose truth follows from the latter's truth, scientific realism only implies that scientific theories are a kind of theories in whose truth we are allowed to be justified. This is not to deny that realism can be fruitfully combined with a naturalist position. It can even be maintained that the adoption of naturalism facilitates—since it implies it—the adoption of scientific realism. However, nothing prevents someone who believes, for example, that genuinely philosophical *a priori* arguments are a reliable source of knowledge, from believing, provided that conflicting results are avoided and that (mature, predictively successful, well-confirmed, etc.) scientific theories are sufficiently reliable sources too. It is not difficult to think of scientific realists that are not naturalists. For an illustrious case, consider Frege, who had realistic views about scientific inquiry but surely was not a naturalist, as his views on mathematics show.

---

[34]Notice that scientific realism, as formulated here, entails that (we are justified to believe that) the entities (both observable and theoretical) which are spoken of in mature scientific theories exist (at least, if we admit that a statement of a scientific theory cannot be true if these entities does not exist). Some remarks are in order. First, one may adopt forms of realism—e.g. structural realism—where the existence of these individual entities is not entailed; this version of realism would still be adequate to motivate premise (*i*) in all minimal arguments. Second, it would be odd to assume that scientific realism entails either the existence of the mathematical entities mentioned in mathematical statements or that the mathematics used in science is true; assuming scientific realism does not beg the question with regards to the conclusion of neither platonist or realist versions of IA and can be safely assumed in both.

It then seems that insofar as endorsing [NAT] requires (entails) endorsing scientific realism, the former could thus well be a sufficient for endorsing premise (*i*) of [RE] and [PE], but clearly it is not necessary.[35]

This being established, one could still wonder whether endorsing [NAT] is necessary or sufficient for endorsing any other premise of [RE] and [PE]. The only plausible candidate for this is premise (*iii*), which plays in these arguments the role that premise (*i*) of Colyvan's argument plays in this latter argument. As we have mentioned in Sect. 13.2, Colyvan indeed argues that naturalism is required in order to justify the only-direction of this latter premise. Still, we have also already observed that this direction is, in fact, useless for drawing the argument's conclusion and has, accordingly, no correlate in premise (*iii*) of [RE] and [PE].

At most, then, one can argue that the endorsement of [NAT] is necessary for endorsing a strengthened variant of this premise that would state that we are justified in believing true all and only the mathematical theories that are indispensable to scientific theories that we are justified in believing to be true. This variant would still be involved only in variants of [RE] and [PE] whose conclusions are, in turn, strengthened variants of the conclusions of these arguments, giving conditions which are also necessary and not merely sufficient for our being justified, respectively, in believing true the relevant mathematical theories and in believing the objects which these are about to exist.

13.6 Retrieving Common Arguments from the Minimal Arguments

It is easy to see how the suggested minimal versions of IA could serve as a basis for the stances on IA represented by Putnam's, Colyvan's and Baker's arguments, respectively (cf. Sect. 13.2). Putnam's argument is nothing but a variant of [PnE] (or, depending on the reading of Putnam's text, [PE]). Colyvan's argument is a variant of the epistemic version of IA that results from adding to premise (*iii*) of [PE] the redundant assumption that we are justified in believing true only the mathematical theories that are indispensable to scientific theories that we are justified in believing to be true.[36] Finally, as anticipated at the end of Sect. 13.4, Baker's Enhanced Indispensability Argument, can be retrieved from [PE] once an

---

[35]To our knowledge, the only other version of IA which is explicitly claimed by its proponent to dispense with naturalism is the one offered by Azzouni (2009). We'll discuss this below.

[36]Together with what we have said in Sect. 13.5.2.1, this entails that Colyvan's argument is, in fact, independent of confirmational holism, despite his own initial claim in Colyvan (2001, p. 12) (cf. the quote relative to footnote 5; but cf. also footnote 24).

appropriate equivalence relation ε is selected in the specification of the notion of (in)dispensability, e.g. the relation of having the same explanatory power.[37]

Things are more complex with arguments in the pragmatic stance.

Let us begin with Resnik's. This is very different in form from all other representative versions of IA that we have reviewed. It seems clear that it dispenses with confirmational holism.[38] It is instead not clear whether it also dispenses with naturalism. According to Resnik himself the passage from its premise (*vi*) to (*vii*) is justified by naturalism, which he briefly defines as the thesis that 'natural science is our ultimate arbiter of truth and existence',[39] and considers as suitable for ruling out scepticism about the justification and truth of our scientific theories. Still, it is not clear how naturalism can warrant the passage from the justification of some theories, and particularly of mathematical ones, to their truth, a passage which is surely hard to secure. More generally, it is far from clear that Resnik's argument is valid (or admit, at least, a valid rephrasing). It is thus not surprising that it is hard to retrieve it from our minimal versions, which are clearly valid. This remains, however, a unicum with respect to the ongoing debate.

Notice, on the other hand, that it is not clear whether the role played by the notion of justification in Resnik's argument is in any sense cognate to that played by it in our minimal arguments. We take justification as applying to statements or bodies of statements and as being justification for their truth, and not just for their adoption for whatever practical reasons. Resnik does firstly conceive justification as applying (if not uniquely, also) to acts, rather than bodies of statements; and secondly he bases our 'accepting as true such mathematics as science uses' on the pragmatic ground that 'we are justified in doing science' and that science, when considered in its practice, proceeds through various mathematical assumptions—and not rather on the fact that any scientific theory is even approximately true. If Resnik's argument were (valid and) sound, this could be seen as an advantage, since it would make the argument independent of the supposition that some scientific theories are true. If (valid and) sound, this argument would thus dispense with scientific realism, as Resnik (1997, pp. 46–47) suggests (and this should be so despite Resnik's quite doubtful claim that the passage from conclusion (*vi*) to conclusion (*vii*) depends on naturalism, understood as the thesis that 'natural science is our ultimate arbiter of

---

[37]Discussion of the Enhanced Indispensability Argument often pertains to the alleged role that inference to the best explanation (IBE) may have in IA. Still, insofar as the minimal version of this argument, provided by [PE] under the mentioned specifications, is concerned, no appeal to IBE is required, and its validity need then not be presupposed. This does not mean that IBE cannot be involved in any specification of the minimal versions of IA. Indeed, it could be involved in some such specifications in two ways: the equivalence relation ε in the schematic definition of the notion of (in)dispensability could be specified through the notion of explanation in a way that presupposes the validity of IBE or the criterion of virtuosity α in that same definition could itself presuppose the validity of IBE.

[38]This is what Resnik himself seems to imply in the quote relative to footnote 11.

[39]Resnik (1995, p. 166, 1997, p. 45)

truth and existence', a thesis which could hardly be taken as independent from a form of scientific realism).

Let us consider now Azzouni's argument, namely, his Assertoric-use QP.[40] This deserves more careful consideration, since it is the one that comes closest to resembling, at least in spirit, some of our suggested minimal versions, at least insofar as it is, as we said in Sect. 13.2, explicitly meant to avoid presupposition of either holism or naturalism.[41] Also the distinction between arguments for mathematical realism and for platonism is suggested by Azzouni (2004), who questions [QC] and argues at length that IA can at most be an argument for the truth of mathematical theories, but that it falls short of supporting platonism.[42]

The first thing to be noticed is that Azzouni's Assertoric-use QP stems from an interpretation of the 'enthymematic blueprint' quoted above in Sect. 13.2, but is not regimented in the form of a codified non-enthymematic (valid) argument with enumerated and explicitly stated premises and conclusion(s). Enquiring whether it can be retrieved from one of our minimal versions of IA requires, then, some reinterpretation of Azzouni's proposals.

Prima facie, given that there is no mention of epistemic notions like justification in Azzouni's blueprint, the relevant minimal version of IA with which his argument should be compared seems to be [RnE]. However, the way in which Azzouni motivates the passage from the blueprint to the Assertoric-use QP suggests another option.

Azzouni's strategy for obtaining the Assertoric-use QP depends, firstly, on the specification of the single premise of the former, i.e. 'certain statements that quantify over mathematical entities are indispensable to science', as the claim that the 'assertoric use' of the relevant statements, which quantify over mathematical entities, is indispensable to science. According to Azzouni,[43] it is an empirical fact that people use certain statements assertorically while doing science, for the two aims of presenting statements as following from other statements previously made ('deductive use') and of describing state of affairs ('representational use'). A second empirical fact is said to be[44] that given our ordinary understanding of the word 'true', the assertoric use of a statement p entails, via Tarksi's biconditionals, the commitment of those who assertorically use p to the truth of p. From this second empirical fact, it follows that the conclusion of the blueprint should be intended as the claim that we are committed to the truth of those statements whose assertoric use is indispensable to science. Although different readings can be given to the expression 'we are committed to the truth of p', we take it that the more plausible one, consistent with Azzouni's discussion, is as 'we are justified in believing

---

[40]Cf. § Sect. 13.2, above.

[41]Azzouni (2009), especially p. 147, footnote 11

[42]Cf also Azzouni (2009), p. 140, note 2; p. 147, note 11.

[43]Cf. Azzouni (2009, pp. 140–141).

[44]*Ibidem*, p. 141

p true'.[45] Assuming that this is so, it follows that the most appropriate among our minimal versions of IA to be compared with Azzouni's Assertoric-use QP is [RE].

Two questions arise here: whether Azzouni's version of IA is eventually to be considered more minimal than [RE] and whether his version can indeed be retrieved from [RE].

As to the first question, a clue for a positive answer could come from the fact that [RE] appeals to the notion of truth (thought of as a schematic notion, variously specifiable), whereas Azzouni's argument doesn't explicitly appear to do so. Still, the latter relies on the fact that assertoric use of mathematical statements is indispensable to scientific practice, and this is taken to entail commitment to the truth of these statements just in virtue of the Tarskian biconditionals which 'transform assertoric uses into truth commitments'.[46] Moreover, Azzouni argues[47] that 'if I *assertorically use* a sentence,[48] I recognize myself as bound by implication to the original sentence prefixed by 'It's true that … ''', which results in maintaining that there is a rather intimate connection between assertion and truth. But if this is so, then truth, if not explicitly used in the premises of the argument, is presupposed by the very notions of assertion and assertoric use by means of which the argument is built up. It may even well be explicitly mentioned in a premise of the argument once all its underlying assumptions are brought to the fore.[49] Hence, if the issue is the appeal to some notion of truth, then, it is far from clear that Azzouni's argument is in any sense more minimal than [RE].

A second related clue for a positive answer to the first question is that the conclusion of Azzouni's Assertoric-use QP appears to be derived without explicitly assuming that we are justified in believing some scientific theories to be true.

---

[45]The sense in which we are justified in believing a mathematical statement true is meant, however, to be in some sense 'stronger' (cf. Azzouni 2009, p. 147) than that licensed by Resnik's argument on pragmatic grounds: as Azzouni claims, 'it isn't that we're 'justified' in describing an assertorically used sentence as true; Tarski biconditionals make the use of the truth predicate *nonnegotiable*'. Whatever this distinction comes to in details, it does not seem that from the assertoric use of a statement p, the truth itself of p can follow, over and beyond our commitment to take p as true. Even if this entails that the conclusion of the Assertoric-use QP will be, as a matter of fact, a different, epistemic version of the conclusion of Azzouni's proposed blueprint (i.e. 'Those statements are true'), we still see this as the most reasonable outcome of Azzouni's discussion; we acknowledge, however, that this reading can be subject to controversy depending on how our 'commitment' to the truth of a statement is understood.

[46]Azzouni (2009, p. 142)

[47]Azzouni (2009, p. 141)

[48]Azzouni seems to indifferently use in his paper the terms 'sentence' and 'statement'. While maintaining the term 'sentence' in all our quotations from Azzouni's paper where it occurs, we shall, instead, invariably use the terms 'statement', as we do throughout our paper.

[49]Notice that no particular conception of truth is presupposed in our minimal versions, so that one is at liberty to use whatever notion one prefers in the specification of the schematic arguments (included a disquotational one). Hence, the question here is not whether we, as opposed to Azzouni, make use of some particular conception of truth but whether any notion of truth is involved at all in the relevant versions of IA.

However, as Azzouni himself acknowledges,[50] his argument has as accompanying premise: 'the *au*-indispensability of the scientific sentences themselves (given a commitment to the scientific project)', where the term '*au*-indispensability' is a shortened form of 'indispensability of the assertoric use'. It seems, moreover, plausible, in the light of our previous discussion, to understand the commitment to the scientific project as a justification in the truth of these scientific statements and then, presumably, in some scientific theories.

Be that as it may, these considerations suggest that Azzouni's argument is possibly not to be considered more minimal than [RE], but rather very close to it, at least in spirit. We should then move to our second question.

What we have said so far suggests that both the single premise and the conclusion of Azzouni's blueprint should be specified by making explicit the assumption of our commitment to the scientific project intended as a commitment to (the truth) of those statements whose assortoric use is indispensable to this project, and by stating the conclusion in an epistemic form. The ensuing formulation will thus be this:

Assertoric-use QP (I)

i) We are committed to (the truth of) those statements that are *au*-indispensable to the scientific project;

ii) Some statements that quantify over mathematical entities are *au*-indispensable to the scientific project;

----

iii) We are committed to (the truth of) these statements.

Above we have mentioned the accompanying premise of Azzouni's argument concerning the '*au*-indispensability of the scientific sentences themselves'. The whole passage where this premises is put forward seems to suggest that the indispensability of the assertoric use of the relevant statements that quantify over mathematical entities is to be, as it were, split into the *au*-indispensability of both some scientific and some mathematical statements. It goes as follows[51]: 'I [...] read the premise of the ethymemic blueprint of the QP as stating that many sentences of mathematics are *au*-indispensable to science. Accompanying this premise is the assumption of the *au*-indispensability of the scientific sentences themselves (given a committed to the scientific project)'. The statements that quantify over mathematical entities and are *au*-indispensable to the scientific project mentioned in Assertoric-use QP (I) seem here to be those that in this passage are referred to as 'sentences of mathematics [that] are *au*-indispensable to science'.[52] This suggests that the statements here referred to as 'the scientific sentences themselves' do not quantify over mathematical entities, but are such that they can be assertorically used only if some mathematical statements (which rather do) are so used (e.g. insofar as the former involve individual or predicate constants defined through the latter). It would

---

[50]Azzouni (2009, p. 144)

[51]The reference is, of course, the same as in footnote 50.

[52]Cf. footnote 46 above.

follow that the *au*-indispensability to the scientific project of the mathematical statements mentioned in Assertoric-use QP (I) just depends on this, namely, on their being such that those scientific statements whose assertoric use is indispensable to the scientific project are such that they can be assertorically used only if these mathematical statements are so used. If we express this condition by saying that the latter statements are *au*-indispensable to the former, Assertoric-use QP (I) can then be restated and further expanded as follows:

Assertoric-use QP (II)

i) We are committed to (the truth of) those statements that are *au*-indispensable to the scientific project;
ii) Some mathematical statements are *au*-indispensable to some statements that are *au*-indispensable to the scientific project;
iii) We are committed to the truth of those statements that are *au*-indispensable to the scientific project only if we are committed to the truth of those mathematical statements that are *au*-indispensable to them;
----
iv) We are committed to (the truth of) these mathematical statements.

This latter version of IA is clearly similar, in structure and content, to [RE]. Once it is conceded that 'being committed to the scientific project' can be interpreted as 'being justified in believing some scientific theories to be true', premise (*i*) of Assertoric-use QP (II) can, indeed, be conceived as a specification of premise (*i*) of [RE]. Following Azzouni, one may say that the former is justified by the empirical fact that when we assertorically use a statement p, we are committed (via Tarski biconditionals) to the truth of it. If premise (*ii*) of [RE] is then specified by employing the notion of assertoric use in the specification of the ε equivalence relation involved in the schematic notion of indispensability—so that the assertoric use of certain mathematical statements turns out to be indispensable to the assertoric use of those scientific statements that we must assertorically use if we are committed to the scientific project—what one gets is just premise (*ii*) of Assertoric-use QP (II). As regards premise (*iii*), it is easy to see that the S-M justificatory connection can be specified by considering that the commitment to the truth of the statements of S, following via Tarski biconditionals from their assertoric use, can be granted only in presence of a similar commitment, via the same route, for the mathematical statements of M that are *au*-indispensable to S: this does not seem to add nothing that Azzouni would not consider as implicit in his own argument.[53] Premise (*iii*) of

---

[53] Notice also that Azzouni explicitly objects to forms of fictionalism that constitute the most obvious strategies for rejecting premise (*iii*). In the following passage (Azzouni 2009, p. 143), it is easy so read something very close to the suggested specification of premise (*iii*) of [RE]:

> One issue to be explored in this paper is whether the assertoric use of many statements of ordinary science is compatible with one or another construal of the mathematical statements utilized in science as not assertorically used (and therefore, as either not true-apt or as false). I'll show that a position that takes us as truth-committed to statements in any area where mathematics is applied, while assuming that we aren't simultaneously truth-committed to that mathematics, is unstable.

Assertoric-use QP (II) just results from this specification of premise (*iii*) of [RE], with the implicit assumption that assertoric use of *p* entails commitment to the truth of *p*. Once [RE] is appropriately specified, then, the conclusion of Assertoric-use QP (II) is nothing by an alternative though equivalent formulation of the conclusion of [RE], as one may expect by alternative versions of an argument that are meant to support a common conclusion.

If we are correct, and Assertoric-use QP (II) is a plausible reconstruction, in explicit non-enthymematic form, of the version of IA that Azzouni has in mind, then it seems susceptible of being retrieved from [RE] through appropriate specifications of the notions involved.[54]

13.7 The Philosophical Significance of the Indispensability Argument

Much of the recent discussion on IA has focused on holism and naturalism. Many authors have either criticized IA, taking Colyvan's as the most relevant formulation, or offered alternative versions of it. Both supporters and critics, with few exceptions we have mentioned above, take naturalism and holism to be essential to this argument or, more generally, to what is usually referred to as 'the Quine-Putnam Indispensability Argument' and discuss the alleged dependence of the argument on these doctrines. For example, Maddy (1992, 2007) claims that IA fails because of inescapable clashes between the notions of holism and naturalism (as Quine conceived of it) and essential features of mathematical and scientific practice and methodology; relying on a non-holistic notion of confirmation, Sober (1993) argues that empirical evidence cannot even indirectly justify mathematical theories.

As it turns out, minimal versions of IA can be devised that are far less demanding than the so-called Quine-Putnam Indispensability Argument. Only scientific realism (beyond, obviously, a proper characterization of (in)dispensability) will be an essential ingredient in justifying premise (*i*) in [RE] and [RnE]. In order to obtain a platonist conclusion, thus to support [PE] and [PnE], only an appropriate criterion

---

This, if needed, seems to be another piece of evidence that premise (*iii*) can be upheld without appealing to confirmational holism.

[54]'This is, of course, not intended to suggest that Azzouni's Assertoric-use QP is already included, *in nuce*, in [RE]. What we argue is rather that [RE] is schematically general enough in order to provide an argument form that Azzouni's Assertoric-use QP (which, as a matter of fact, has been offered beforehand and independently of [RE]) can be taken to instantiate via appropriate specifications.

for ontological commitment—arguably, Quine's[55]—needs to be further assumed (together with what is needed to justify the remaining premises).

It is often thought that IA is especially suited for those mathematical (semantical or ontological) realists working in a broadly empiricist framework. As Shapiro claims:

> Indispensability arguments are anathema to those, like the logicists, logical positivists, and neologicists, who maintain the traditional views that mathematics is absolutely necessary and/or analytic and/or knowable *a priori*.[56]

This is an obvious philosophical outcome for versions of IA that proceed under the assumption of naturalism and deliver necessary conditions for their conclusions (cf. the end of Sect. 13.5.3). But this need not be so: when naturalism is left out of the picture, the argument only gives sufficient conditions for either semantic realism or platonism.

Clearly, if we espouse a naturalist ideology, we will better make our argument for mathematical (semantic or ontological) realism rely, more or less explicitly, on naturalism and thus secure (semantic or ontological) realism in a way that is consistent with our naturalist viewpoint. But it is not, as it were, in the very nature of IA to give sufficient and necessary conditions for its conclusions. It might just be our particular interest to have a version of it giving both.

IA is not a naturalist argument per se. We see no ban in principle, for those who believe in the *a priori* character of mathematical truths, against the acceptance of indispensability arguments.[57] Modest antinaturalists of this sort[58] will claim at most that IA is superfluous, or ancillary, since they can offer reasons for the same conclusion(s) that are by far more certain than the contingent grounds on which IA hinges. But this is definitely different from rejecting the argument.

The point is that the real anathema for all those philosophers listed by Shapiro in his quotation is not IA itself: it is naturalism. Any argument relying on naturalism will be anathema for them: IA can, but need not, be a good candidate.

---

[55][PE] or [PnE] can be seen as instances of a general way in which Quine would draw ontological conclusions. One might object that the minimal formulation would make nothing of the special subject matter of mathematics (thanks to Matti Eklund for raising this). However, we don't find anything, in Quine's reluctant acceptance of platonism, like assuming something special about mathematics and building a form of IA on this (contrary to what is suggested by Steiner's (1978, pp. 19–20) 'transcendental' interpretation of IA). The special character of mathematics seems rather to be proved by the very fact that we cannot dispense with it in science. All posits are ontologically on a par until we are faced, as Quine would call it, with an unabridged language of science. Not all posits will come out as indispensable. Propositions and meanings don't. Mathematics does.

[56]Shapiro (2005, pp. 13–14). Shapiro remarks is only cursorily made, and nothing special hinges on it in his discussion; we just take it as an indication of a widespread feeling.

[57]Some of Frege's remarks (1893–1903, §91) have sometimes be taken as a statement of a form of IA *in nuce* (but see Garavaso (2005) and Sereni (2013)). But it would be utterly implausible to claim that anything like IA was Frege's main argument for believing in the existence of mathematical objects.

[58]Radical antinaturalists, like sceptics, would deny that science is any source of knowledge at all.

On this respect, it is remarkable that, after long time, Putnam himself recently felt the need of making his voice heard again in this debate. Commenting on Colyvan's argument, Putnam clearly remarks that the 'only' direction of premise (*i*) of this argument—the one committing a supporter of the argument to naturalism—expresses a thesis he '*never*' subscribed to in [his] life'.[59] He stresses that his adoption of IA was meant to show—as already suggested in Putnam (1971, 1975b)—that it is incoherent to adopt scientific realism and at the same time reject (semantic) realism about mathematics. Moreover, he is explicit in claiming that:

> nevertheless, there was a common *premise* in my argument and Quine's [...] That premise was "scientific realism", by which I meant the rejection of operationalism and kindred forms of "instrumentalism". I believed (and *in a sense* Quine also believed) that fundamental physical theories are intended to tell the truth about physical reality, and not merely to imply true observation sentences.[60]

At the very least, our conclusions can be seen as a way of setting the debate straight to its origin and showing that the minimal versions of IA are more closely related than others to Putnam's argument.

If IA is meant as delivering only sufficient conditions for its conclusions, there is a clear sense in which it is being revisionary with respect to Quine's original views. In a number of different passages Quine makes controversial claims about parts of mathematics (e.g. higher set theory) that have no applications in empirical sciences and are *a fortiori* not indispensable to scientific theories. Quine denies that the mathematical objects to which those parts of mathematics are committed to deserve any ontological rights (he famously spoke of 'mathematical recreation'[61]). This attitude has engendered quite a wide debate.[62] If indispensability (in versions of IA for platonism) is, as it were, the mark of existence for mathematical objects, then the objects of unapplied mathematics are banned from our ontology.

It is indeed possible to maintain a version of IA for which naturalism is necessary. This argument delivers the sort of platonism that Quine endorsed. But this argument is in tension with many forms of platonism, which would not distinguish among the ontological rights of different parts of mathematics, not at least on grounds of applicability and indispensability (Maddy has long insisted on this; see e.g.

---

[59]Cf. Putnam (2012, p. 183).

[60]*Ibid.* The 'in a sense' qualification concerns Quinean themes (indeterminacy of translation, differences with a standard realist view of language) discussed in Putnam (1988). They do not affect our present point.

[61]Quine (1986, p. 400). In later writings, Quine admitted that this would create an unjustifiable asymmetry between different parts of mathematics; hence, he resorted to the idea that we cannot completely deny meaningfulness to unapplied parts of mathematics, but that we can arbitrarily decide whether to call those parts true or false (cf. Quine 1995, pp. 56–57).

[62]Cf. Parsons (1983), Maddy (1992), Leng (2002, 2010), and Colyvan (2007). Putnam (1971, pp. 346–347) suggests a view similar to Quine's on unapplied mathematics. His is however a milder position (unapplied mathematics 'should today be investigated in an 'if-then' spirit'), and he is wary of restricting his claims to 'the case for 'realism' developed in the present section'.

Maddy (2005)). It could even be argued that such an argument is not an argument for platonism (as standardly conceived), but rather for the proto-empirical (or 'quasi-empirical', to borrow from Putnam (1975b, p. 62)[63]) character of mathematics.

Versions of IA giving only sufficient conditions leave open the possibility that we are justified in believing that unapplied mathematical theories are true or that the objects they are about exist, wholly independently of IA: IA is understood as an argument for mathematical (semantic or ontological) realism among others, not as the only argument.

If Quine suggests that a proper indispensability argument hinges on naturalism, then, it is only because he was a naturalist on independent grounds in the first place. Nothing in the argument mandates that this is so. That is just an example of the philosophical use of the argument (in a non-minimal version) that can be made in an empiricist framework. It is not a philosophical outcome that the argument can secure by itself.

Acknowledgements Earlier versions of this paper have been presented on several occasions in seminars and conferences. The authors wish to thank all the audiences for their helpful comments. Special thanks go to: Andrea Bianchi, Francesca Boccuni, Jacob Busch, Annalisa Coliva, Matti Eklund, Mario De Caro, Jan Lacki, David Liggins, Paolo Mancosu, Sebastiano Moruzzi and Eva Picardi. Many thanks also to two anonymous referees for this volume, who provided precious suggestions.

References

Armstrong, D. 1997. *A world of states of affairs*. Cambridge: Cambridge University Press.
Azzouni, J. 1998. On 'On what there is'. *Pacific Philosophical Quarterly* 79: 1–18.
Azzouni, J. 2004. *Deflating existential consequence. A case for nominalism*. Oxford/New York: Oxford University Press.
Azzouni, J. 2009. Evading truth commitments: The problem reanalyzed. *Logique & Analyse* 206: 139–176.
Baker, A. 2005. Are there genuine mathematical explanations of physical phenomena? *Mind* 114: 223–238.
Baker, A. 2009. Mathematical explanation in science. *British Journal for the Philosophy of Science* 60: 611–633.
Colyvan, M. 2001. *The indispensability of mathematics*. Oxford/New York: Oxford University Press.
Colyvan, M. 2007. Mathematical recreation versus mathematical knowledge. In *Mathematical knowledge*, ed. M. Leng, A. Paseau, and M. Potter, 109–122. Oxford/New York: Oxford University Press.
Craig, W. 1956. Replacement of auxiliary expressions. *Philosophical Review* 65: 38–55.
Dieveney, P.S. 2007. Dispensability in the indispensability argument. *Synthese* 157: 105–128.

---

[63] According to Putnam (1975b), mathematics could count as quasi-empirical in that we can account for it in terms of quasi-empirical methods of inquiries (other than deductive proof from axioms) based on successful applications. This is for Putnam consistent with a non-platonist interpretation of mathematics.

Dummett, M. 1978. Realism. In *Truth and other enigmas*, ed. M. Dummett. London: Duckworth.
Field, H. 1980. *Science without numbers*. Oxford: Blackwell.
Field, H. 1989. *Realism, mathematics and modality*. Oxford: Blackwell.
Frege, G. 1893–1903. *Grundgesetze der Arithmetik*, 2 vol. Jena: H. Pohle.
Garavaso, P. 2005. On Frege's alleged indispensability argument. *Philosophia Mathematica (III)* 13: 160–173.
Hafner, J., and P. Mancosu. 2005. The varieties of mathematical explanation. In *Visualization, explanation and reasoning styles in mathematics*, ed. P. Mancosu et al., 215–250. Dordrecht/Norwell: Springer.
Hellman, G. 1989. *Mathematics without numbers*. Oxford/New York: Oxford University Press.
Hellman, G. 1999. Some ins and outs of indispensability: A modal-structural perspective. In *Logic in Florence*, ed. A. Cantini, E. Casari, and P. Minari. Dordrecht: Kluwer.
Leng, M. 2002. What's wrong with indispensability? (Or, the case for recreational mathematics). *Synthese* 131: 395–417.
Leng, M. 2010. *Mathematics and reality*. Oxford/New York: Oxford University Press.
Liggins, D. 2008. Quine, Putnam, and the 'Quine-Putnam' indispensability argument. *Erkenntnis* 68: 113–127.
Maddy, P. 1992. Indispensability and practice. *Journal of Philosophy* 89(1992): 275–289.
Maddy, P. 2005. Three forms of naturalism. In *Oxford handbook of philosophy of mathematics and logic*, ed. S. Shapiro, 437–460. Oxford/New York: Oxford University Press.
Maddy, P. 2007. *Second philosophy*. Oxford/New York: Oxford University Press.
Mancosu, P. 2008. Mathematical explanation: Why it matters. In *The philosophy of mathematical practice*, ed. P. Mancosu. Oxford: Oxford University Press.
Panza, M., and A. Sereni. 2013. *Plato's problem. An historical introduction to the philosophy of mathematics*. Houndmills: Palgrave Macmillan.
Panza, M., and A. Sereni. Forthcoming. The varieties of the indispensability argument.
Parsons, C. 1983. Quine on the philosophy of mathematics. In *Mathematics in philosophy*, ed. C. Parsons. Ithaca: Cornell University Press. Also in Hahn, L., and P. Schilpp (eds.). 1986. *The philosophy of W.V. Quine*, 369–395. La Salle: Open Court.
Paseau, A. 2007. Scientific realism. In *Mathematical knowledge*, ed. M. Leng, A. Paseu, and M. Potter, 123–149. Oxford/New York: Oxford University Press.
Peressini, A. 1997. Troubles with indispensability. Applying pure mathematics in physical theory. *Philosophia Mathematica (III)* 5: 210–227.
Peressini, A. 2003. Critical study of Mark Colyvan's 'The indispensability of mathematics'. *Philosophia Mathematica* 3: 208–223.
Pincock, J. 2004. A revealing flaw in Colyvan's indispensability argument. *Philosophy of Science* 71: 61–79.
Psillos, S. 1999. *Scientific realism: How science tracks truth*. Oxford: Routledge.
Putnam, H. 1967. Mathematics without foundations. *The Journal of Philosophy* 64: 5–22. Reprinted in Putnam (1975a), pp. 43–59.
Putnam, H. 1971. *Philosophy of logic*. New York: Harper & Row. Reprinted in Putnam (1975a), Chap. 20.
Putnam, H. 1975a. *Mathematics, matter and method, philosophical papers*, vol. 1. Cambridge: Cambridge University Press (2nd ed. 1985).
Putnam, H. 1975b. What is mathematical truth? *Historia Mathematica* 2: 529–543. Reprinted in Putnam (1975a), Chap. 6.
Putnam, H. 1988. The greatest logical positivist. *London Review of Books* 10.8: 11–13. Reprinted in Putnam, H. 1990. *Realism with a human face*, ed. J. Conant. Cambridge: Harvard University Press.
Putnam, H. 2012. Indispensability arguments in the philosophy of mathematics. In *Philosophy in an age of science*, ed. M. De Caro and D. Macarthur, 181–201. Cambridge: Harvard University Press.
Quine, W.V. 1948. On what there is. *Review of metaphysics* 2: 21–38. Reprinted in Quine (1953), Chap. 1.

Quine, W.V. 1951. Two dogmas of empiricism. *Philosophical Review* 60: 20–43. Reprinted in Quine (1953), Chap. 2.
Quine, W.V. 1953. *From a logical point of view*. New York: Harper & Row (2nd ed. 1961).
Quine, W.V. 1975. *Five milestones of empiricism*. In Quine (1981), pp. 67–72.
Quine, W.V. 1981. *Theories and things*. Cambridge: Harvard University Press.
Quine, W.V. 1986. Reply to Charles Parsons. In *The philosophy of W.V. Quine*, ed. L. Hahn and P. Schlipp. La Salle: Open Court.
Quine, W.V. 1995. *From stimulus to science*. Cambridge: Cambridge University Press.
Resnik, M.D. 1995. Scientific vs mathematical realism: The indispensability argument. *Philosophia Mathematica (III)* 3: 166–174.
Resnik, M.D. 1997. *Mathematics as a science of patterns*. Oxford: Clarendon.
Shapiro, S. (ed.). 2005. *The Oxford handbook for the philosophy of mathematics and logic*. Oxford/New York: Oxford University Press.
Sereni, A. 2013. Frege, indispensability, and the compatibilist Heresy. *Philosophia Mathematica*. doi:10.1093/philmat/nkt046.
Sober, E. 1993. Mathematics and indispensability. *The Philosophical Review* 102: 35–57.
Steiner, M. 1978. Mathematics, explanation, and scientific knowledge. *Noûs* 12: 17–28.
van Fraassen, B. 1980. *The scientific image*. Oxford/New York: Oxford University Press.
Weir, A. 2005. Naturalism reconsidered. In *The Oxford handbook of the philosophy of mathematics and logic*, ed. S. Shapiro, 460–482. Oxford/New York: Oxford University Press.

Chapter 14
Naturalness in Mathematics

Luca San Mauro and Giorgio Venturi

> *Is perhaps the right way of tackling the question just this – to write down a long list of actually observed uses, taking note of the frequency of each use and distilling the whole into a statistical table? But is this the sort of a thing a philosopher wants to do? Is he interested in the random fluctuations of speech, that sea with its endless waves and ripples?*
>
> F. Waismann, Analytic-Synthetic IV

14.1 Introduction

This article consists of two parts.[1] In the first one, we aim to propose a philosophical characterization of the notion of naturalness in mathematics. First of all, we have to acknowledge that this is not an easy task for many historical, methodological, and intrinsic reasons. To start with, there is not a wide and well-structured literature on this topic,[2] thus every step in this direction will be almost like groping in the

---

[1]The paper is the product of the intellectual collaboration of the two authors and originated from discussions that took place in Gargnano del Garda, Paris, Pisa, and Rome. Both from a conceptual and a practical point of view, it is hard to attribute each of the ideas or parts of this work to one of the two authors; indeed, they are the result of objections, mediations, and syntheses. A previous and different version of this paper, except Sect. 14.4.2.3, appeared as a chapter in the PhD thesis of the second author. The two authors would like to thank Gabriele Lolli, Chris Pincock, and two referees for useful comments on early drafts of this paper.

[2]We list here all the relevant works, according to our knowledge, that address directly the problem of naturalness in mathematics: Tappenden (2005, 2008a,b), Corfield (2004), Koepke (2009), and Bagaria (2000).

L. San Mauro
Scuola Normale Superiore, Pisa, Italy
e-mail: luca.sanmauro@sns.it

G. Venturi (✉)
FAPESP and Centre for Logic, Epistemology and the History of Science,
State University of Campinas, São Paulo, Brazil
e-mail: gio.venturi@gmail.com

dark. On the contrary, there is an important philosophical tradition that is labeled as naturalism, and this latter will make our investigation even harder, because it is quite far from the position expounded here. We will mainly discuss Penelope Maddy's position, in the attempt to clarify our view of a dialectical relationship between mathematics and philosophy.

On the methodological side, our analysis pertains to the philosophy of mathematical practice. This leads us to face the difficulty that such a fairly new branch of philosophy encounters, that is to say the absence of a well-established method of inquiry. As a matter of fact, the first part of this work will be concerned with explaining our argumentative line. Once shown that the concept of naturalness deserves an analysis, we will claim that its most suitable philosophical treatment belongs exactly to this new wave of the philosophy of mathematics. While doing so, we hope at the same time to shed some light on the general methodology of the latter. Moreover, although every work that can be labeled as philosophy of mathematical practice brings within itself an inevitable attention to concrete cases, we will distinguish the relevance we give to mathematical and historical examples from the one that is normally given by naturalism.

Our analysis will start from the statistical evidence that the use of the word "natural" has noteworthy increased in the last 70 years. The presence of linguistic tools in a philosophical work may be, at first sight, surprising: we will ground them on some methodological considerations. However, a statistical overview does not exhaust our analysis, since the descriptive temptation is rather a pernicious solution that has to be avoided at both the linguistic and the conceptual level. On the other hand, we will also reject a philosophy-first approach, looking instead for a sufficiently neutral starting point for our study. Hence, we will outline a general method, in a semiformal shape, for dealing in a philosophical way with vague concepts.

In the second part, we will take advantage of the literature on mathematical explanation, that is already enough structured to help us in the search for the proper philosophical context where to place our initial comprehension of the notion of naturalness. Following our method, we will test some working hypotheses thanks to chosen case studies. Through the analysis of two historical examples, we will conclude that the object of our study manifests both a dynamic component, as opposed to a static one,[3] and a prescriptive component, as opposed to a descriptive one.

In the end we will take a stand with respect to the role of common sense meaning in the philosophical inquiry on the semantics of words. When calling something natural, we rhetorically evoke the idea of "pertaining to nature," even if such causal context may not apply to mathematics. The semantic ambiguity of the notion of naturalness leads to a third difficulty, that is to say its intrinsic tension toward different poles. Indeed, the fast development of a more and more abstract and artificial mathematics, in the last century, seems to diverge from a

---

[3] We will clarify the meaning we assign to the dynamic-static dichotomy in a moment.

natural perspective. Then how to fit this historical phenomenon with the increasing appeal of natural components in the mathematical discourse? Also, if – as we claim – dynamical, referential, and contextual aspects are fundamental in forming judgments of naturalness, how is it possible to harmonize them to something apparently as static and objective as nature?

Just to counter some trivial criticism that can possibly arise, we acknowledge that there are pieces of mathematics so clearly stable in their naturalness that any attempt to give a philosophical account of their natural character would necessarily deal with some sociological, or cognitive, or transcendental aspect of our ways of doing mathematics – as, for instance, the natural numbers. On the same par, we maintain that there are phenomena so unnatural that they would always exceed any reasonable account of naturalness in mathematics. Instead, our main interest is not related to the stable components we can find at the extremities of the dichotomy natural-unnatural, but we would rather focus on the different nuances of gray that we can find in between. We believe that it is exactly here that the reference of naturalness diverges more substantially from the common use of this notion, thus deserving a specific philosophical investigation. To these concerns, we sympathize with Friedrich Waismann:

> In all these cases you notice that expressions which have a trivial use in everyday life, when made part of a certain trains of thought, lose their triviality, become, as it were trascendentalized, and acquire metaphysical status.[4]

Far from arguing for a truly metaphysical component of the mathematical work, we will try to make sense of an inner tension within the concept of naturalness, showing that the increasing reference to natural components in mathematics is not philosophically innocent. At this level, we will find one of the most relevant points of distance between our position and Maddy's. To refrain a famous Quinean slogan, the notion of naturalness is philosophy in mathematical clothing.

The paper is organized as follows: it is a joint product of both authors, but Sects. 14.4.2.1 and 14.4.2.2 have been written by the second author, and Sects. 14.4.2.3 and 14.4.2.4 have been written by the first author.

14.2 A Methodology

14.2.1 How to Deal with This Concept?

When facing a new conceptual problem, as this is the case, where to begin if not from the literal evidences that such concepts produce in their use? We shall not argue for a holistic point of view toward mathematics and natural language, but it

---

[4] Weismann (1951), p. 56.

seems just appropriate to start from what is more certain and secure, in order to explore what is less known and more obscure.

Let us begin with the primary definition of the word "natural"[5] from the Oxford Dictionary: "Existing in or derived from nature; not made or caused by humankind." By hinting at this possible answer to the question "what is naturalness?" we just want to stress that aside from what we will find in our analysis, there is also a commonsense meaning of this word that needs to be considered. But is this reference to the dictionary's meaning enough to grasp the whole semantic of naturalness in mathematics? We doubt it. Consider, for instance, these few occurrences of the term[6]:

> The proof presented in Section 4.1 is similar to Kruska's original proof in (5). However, we add more clarification to it in order to show that the proof is natural and intuitive.[7]

> Since the operations TC are commutative, associative, monotonic, continuous in the topology of weak convergence, etc., this shows that there are very natural operations on distribution functions that do not correspond in any simple fashion to operations on random variables.[8]

> Semirings, in the general setting as described above or with more restrictive assumptions, arise naturally in such diverse areas of mathematics as combinatorics, functional analysis, topology, graph theory, Euclidean geometry, ring theory including partially ordered rings, optimization theory, automata theory...[9]

To claim that the use of naturalness in these examples is fully captured by our dictionary entry is by no means conceptually easy. If that were the case, what does it mean for semirings to exist in nature? What kind of nature are we talking about? Also, in which sense a class of operations on a topology is not made or caused by humankind? Any of these questions implies a certain number of well-known and challenging philosophical issues; thus, it is simply unrealistic to consider them to be solved by every person who uses the term "natural" in her mathematical practice. On the contrary, one may simply acknowledge that, although its use is well understood, every such appeal to naturalness in mathematics contains a semantic that is not philosophically trivial to clarify. In this inquiry, our feeling is quite similar to that of Augustine about time: "Quid est ergo tempus? Si nemo ex me quaerat scio;

---

[5] We agree with Bertrand Russell that "The study of grammar, in my opinion, is capable of throwing far more light on philosophical questions than is commonly supposed by philosophers. Although a grammatical distinction cannot be uncritically assumed to correspond to a genuine philosophical difference, yet the one is *prima facie* evidence of the other and may often be most usefully employed as a source of discovery" (in Russell 1903, p. 42).

[6] We deliberately chose these examples randomly from the mathematical literature.

[7] Stegeman and Sidiropoulos (2007), p. 542.

[8] Sklar (1973), p. 457.

[9] Weinert (2004), p. 314.

si quaerenti explicare velim, nescio.[10]" Indeed any appeal to naturalness is perfectly understood by the mathematical community, with almost no attempt to define and formalize this concept.

But up to now, we have neglected the basic question "why dealing with naturalness in mathematics?". A first clear answer comes from a random inspection of any contemporary mathematical journal: the presence of the words "naturalness" and "natural" is ubiquitous in mathematical literature. In order to support this evidence, we shall follow Wittgenstein's remark: "Don't think, but look!" (*Philosophical investigations* 66). Thus, we propose an informed statistics of the phenomenon, thanks to the *American Mathematical Society* database (MathSciNet).[11] Our work does not pertain to sociology: we employ this evidence in order to draw attention on a wide phenomenon that we aim to analyze with philosophical tools.

The following table provides data about the frequency of the use of "natural" and "naturalness" between 1940 and 2009:

Decade	Total articles (T)	Occurrences (N)	Rate ($\frac{N}{T}$)
1940–1949	40,538	602	0.014
1950–1959	89,158	1,935	0.021
1960–1969	168,567	4,802	0.028
1970–1979	327,427	11,500	0.035
1980–1989	483,143	21,026	0.043
1990–1999	617,522	34,032	0.055
2000–2009	841,470	47,056	0.056

The table shows that during the last decades, there has been an increasing appeal to this concept. However, these data need to be handled with particular care. The term "natural" includes various and heterogeneous meanings that might be divided in two (absolutely non-exhaustive) classes:

---

[10]"What then is time? If no one asks me, I know what it is. If I wish to explain it to him who asks, I do not know.", Augustine, *Confessions*, XI, 14.

[11]Few words concerning our corpus: the MathSciNet database entirely consists of mathematical reviews. However, we believe that there is not much difference between the typical prose of a review and that of an article. Moreover, the wideness of the phenomenon we will describe is such that a possible small distortion of the data cannot hide the manifest emergence of the use of the notion of naturalness of the mathematical literature. Finally, we would like to stress the absence of a wide corpus of mathematical texts ready for a corpus linguistics analysis. Indeed, the Corpus of Contemporary American English (COCA) counts more than 450 millions words, whereas MathSciNet consists of 2,949,420 reviews. The other attempt to perform a similar linguistic analysis in a mathematical context known to the author has been presented by Lorenz Demey at the ILLC's Logic Tea, on April 21, 2009, and it makes use of a small corpus of less than three millions words. In conclusion, we believe that our starting point, even if partial, is representative enough for the described phenomenon, although a more detailed analysis would need a much larger corpus. Nonetheless, as it will be evident later, we do not feel that the absence of such a linguistic tool should be a limitation for our work.

1. Natural number, natural deduction, natural proof, natural transformation, natural isomorphism, natural topology...
2. Natural method, natural way, natural solution, natural explanation, natural argument, natural example...

Items from the first list are formal definitions in which "natural" gains some technical meaning. On the contrary, the second list reflects an informal use of the term, showing a strong inclination to assign an appearance of naturalness to mathematical practice. The term occurs on both formal and informal sides, spanning a wide semantic domain from rigid contexts (list 1: once a formal definition is given, the whole meaning of the expression is fixed) to some other relaxed and variable uses (list 2). We argue that these two kinds of occurrences are not equivalent and that they exhibit a twofold aspect of the concept. In formal definitions, naturalness is involved only in some initial stages: e.g., there are relevant reasons why Gentzen called the system of natural deduction "natural," but almost every further use of the term is actually unrelated to any consideration about naturalness.[12] Moreover, if we want to understand the meaning of these terms, there is not much more to do than looking for some equivalence results, since their semantics is fixed by deduction rules. Attempts in this direction have been made, for example, by Shelah and Hodges, who describe their intentions as follows:

> Eilenberg and Mac Lane [...] explained the notion of a "natural" embedding by giving a categorical definition. Starting from their examples, we argue that one could equally well explain natural as meaning "uniformly definable in set theory." But do the categorically natural embeddings coincide with the uniformly definable ones?[13]

On the contrary, the picture of informal occurrences is messy and dynamic. To call a portion of a mathematical work "natural" is a rather meaningful operation that consists in assigning to a definition (an axiom, a proof, a construction, etc.) an informal feature – we may already hint that such operation is somehow metalinguistic, because it expresses a sort of comment on the margin of formalization. But there is something more. Statistically, the use of the items from the first list remains almost stable during the decades – or sometimes it even decreases. For example, this is the table concerning the string "natural number."[14]

---

[12] The same can be said for the use of the term "natural" in Category theory.

[13] Hodges and Shelah (1986), p. 1.

[14] In the case of the other formal uses, the situation is even less significant toward a general picture. "Natural deduction": (decade) 40–49, (occurrences) 0; 50–59, 15; 60–69, 37; 70–79, 150; 80–89, 148; 90–99, 295; 00–09, 254. "Natural transformation": (decade) 40–49, (occurrences) 0; 50–59, 3; 60–69, 112; 70–79, 231; 80–89, 171; 90–99, 241; 00–09, 283. "Natural isomorphism": 40–49, (occurrences) 4; 50–59, 32; 60–69, 49; 70–79, 111; 80–89, 113; 90–99, 180; 00–09, 177. "Natural topology": (decade) 40–49, (occurrences) 11; 50–59, 27; 60–69, 73; 70–79, 113; 80–89, 143; 90–99, 147; 00–09, 195.

Decade	Total articles (T)	Occurrences (N)	Rate ($\frac{N}{T}$)
1940–1949	40,538	92	0.0023
1950–1959	89,158	401	0.0045
1960–1969	168,567	1,182	0.0070
1970–1979	327,427	2,456	0.0075
1980–1989	483,143	2,565	0.0053
1990–1999	617,522	2,740	0.0044
2000–2009	841,470	3,269	0.0039

This is the one for the term "naturally":

Decade	Total articles (T)	Occurrences (N)	Rate ($\frac{N}{T}$)
1940–1949	40,538	106	0.0026
1950–1959	89,158	305	0.0034
1960–1969	168,567	702	0.0041
1970–1979	327,427	1,768	0.0053
1980–1989	483,143	3,172	0.0065
1990–1999	617,522	5,187	0.0083
2000–2009	841,470	7,670	0.0091

Therefore, if one wants to consider the growth highlighted by the tables, to focus on the informal side is quite an Hobson's choice. Moreover, it seems reasonable that a good theory of the informal uses of the notion of naturalness (where "good theory" means something fairly different from a conclusive answer to the question "what is naturalness in mathematics?") would also shed light on the formal uses. However, speaking of naturalness remains somehow obscure, because this concept lies in a wobbly geography of informal notions. Indeed, the dichotomy natural/unnatural overlaps and maybe gathers a collection of a series of classical oppositions: pure/artificial, simple/complex, primitive/derivative, general/particular, direct/indirect, easy/difficult, essential/contingent, and intrinsic/extrinsic. The boundaries are rough: in most cases, there is no particular reason for choosing one of these notions over the others, and the preference is often settled by habit. However, there is something peculiar in the case of naturalness, since the statistical weight of the other notions and their rates provide a fairly different picture. Two interesting examples are that of "simple":

Decade	Total articles (T)	Occurrences (N)	Rate ($\frac{N}{T}$)
1940–1949	40,538	2,688	0.066
1950–1959	89,158	6,128	0.068
1960–1969	168,567	10,379	0.061
1970–1979	327,427	19,380	0.059
1980–1989	483,143	28,408	0.058
1990–1999	617,522	35,032	0.056
2000–2009	841,470	44,648	0.053

and that of "essential":

Decade	Total articles (T)	Occurrences (N)	Rate ($\frac{N}{T}$)
1940–1949	40,538	439	0.0108
1950–1959	89,158	935	0.0104
1960–1969	168,567	1,702	0.0100
1970–1979	327,427	3,459	0.0105
1980–1989	483,143	5,258	0.0108
1990–1999	617,522	6,464	0.0104
2000–2009	841,470	8,340	0.0099

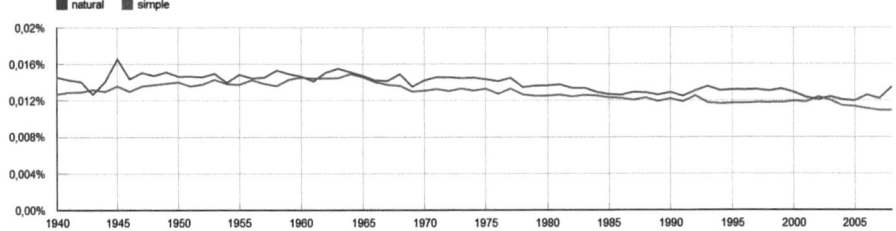

Fig. 14.1 Graphic Google ngram

In the latter two tables, we can see a constant decrease of the use of "simple" and an alternating trend in the case of "essential" that ends up with a substantial drop of its use. Our guess is that in both cases, the reason is a partial semantic erosion accomplished by the term "natural."

We may wonder if this trend is limited to mathematical language or if it is rather a general tendency of natural language. Figure 14.1 supports the former option.

In principle, this kind of analysis might proceed with more and more refined tools – indeed researchers did so for the expression "it is easy to see that".[15] But this approach brings within itself a radical form of naturalism, one that we may call "linguist naturalism," as suggested by Maddy:

> Mathematics is a form of human activity, a distinctive linguistic practice, and as such it can be studied like any other such practice [e.g.] by linguistics [...]. Here the naturalist will face questions about the similarities and dissimilarities between mathematical and natural scientific language.[16]

As far as we are concerned, we do not pursue a purely syntactic analysis; thus, we shall conclude here our statistical overview. We developed it only in order to

---

[15]This analysis has been pursued by Lorenz Demey at the ILLC's Logic Tea, on April 21, 2009. However, his starting point is quite different from ours, because it follows the same path as Corfield's approach, that is to say trying to avoid the "foundational filter."

[16]Maddy (2005), p. 453. Of course this is not the ultimate result of Maddy's inquiry. Indeed her naturalism also focuses on mathematical practice, but it sympathizes with every descriptive philosophical enterprise.

sketch the problem, in fact we believe that such a methodology is not sufficient for the philosophical solution we are looking for, and this for three distinct reasons:

1. First of all, a merely linguistic description of the occurrences of the term "natural" is too inclusive, because it groups together different motivations that might have led an author to refer to the notion, including stylistic or idiosyncratic ones, rather than conceptual. But once engaged in a linguistic analysis, how can we discern between the relevant and proper use of a term and the inappropriate ones? This problem is what we may call "the problem of relevance." It is not a merely methodological difficulty, because it regards a crucial conceptual constraint for every inquiry that aims to have its feet grounded on a philosophically neutral position – as many form of naturalism aim.

   In brief, the problem of relevance consists in discerning what is relevant and what is not in the choice of the data under examination, without imposing on the bare facts the structure we would like to find within the data. Indeed, the process of identifying some uses as canonical and others as deviant presupposes a framework in which concepts have already been embedded. Therefore, this very framework comes to shape the results of our analysis. In other words, the result of an inquiry is somehow determined by its own setting. Thus, to what extent can we recognize a statistical discovery as authentic within a philosophical study? This is a meta-theoretical problem when engaged in the analysis of the notion of naturalness, because this is exactly the task accomplished by naturalness judgments in a mathematical work.

2. On the other hand, a purely linguistic analysis can also be regarded as too limited, because it fails to acknowledge the implicit uses of naturalness in mathematics. This aspect has been suggested by Harvey Friedman in 2006, within a discussion about this topic on the Foundation of Mathematics (FOM) list:

   > One can attempt to formally justify the constant and pervasive use [of naturalness] by taking some major Journals and textbooks, and counting up the number of uses, or counting up the number of *implied uses*.

   But while trying to do so, we end up in a loop, and we have to deal with the problem of relevance again. Indeed, in order to take into consideration the implied uses of the term, one would need some general criteria apt to establish which kind of uses do refer to naturalness and which do not. In other words, it is necessary to have a conceptual frame in which the notion of naturalness can be embedded, and this is necessary prior to the analysis itself that should have defined the very same notion.

   Once formulated this frame, one could theoretically use it as the answer to the starting problem. We may call this a philosophy-first approach that accepts a sort of unsolvability for the problem of relevance. Thus, a supporter of this kind of approach, instead of attempting to shape a philosophical account on the base of the concrete uses of the term, would claim that to describe a preliminary version of this very same account would be somewhat inescapable. In other words, she would argue in favor of a substantial priority of philosophical attitudes over concrete case studies.

We reject such a solution, because in our opinion it is exactly the vagueness of the concept and the irregular geography of its semantics that call for an analysis.

3. Finally, we believe that a purely linguistic analysis does not explain the historical increase in the use of the term "natural." As a matter of fact, a perfectly detailed linguistic analysis of the notion can provide, at best, a tautological description of the phenomenon itself – nothing more than pointing at what mathematicians call "natural."

Moreover, it is common practice to call something "natural" in relation to other pieces of mathematics already partaking in the notion, hence without giving any critical account of it. It seems to us that this is a pivotal indication of the fact that a purely linguistic approach is too local, for it frequently misses the role played by global relations when defining the notion of naturalness. This recursive feature is well expressed by Chow, in the same thread from where the above quotation by Harvey Friedman was taken:

> I would incline towards modeling the space of mathematical statements as something like a graph, with vertices being known theorems and conjectures, and edges representing "similarity" or "relatedness" or some such. Then a statement would be natural if it has high degree and is near the center of a giant connected component, or something like that.
>
> In other words, a statement is likely to be natural if it is similar to many other statements that have been considered before, and/or if it is conceptually linked with many other natural statements. In contrast, a statement that is easily stated but has a strange form and is not related to other known statements is probably unnatural.[17]

Thus, a context-dependent and dynamic character of naturalness needs to be taken into account. Furthermore, it seems to us that this dynamicity addresses toward some qualitative and global philosophical considerations for a good description of the phenomenon. However, a mere quantitative proposal – such as counting the number of edges connecting two different natural objects – can hardly shed any light on what naturalness is. Indeed, if the predicate of naturalness is gained in virtue of some connection between different portions of mathematics, then it follows that this connection somehow rests on some peculiar quality of the mathematical objects. In other words, we would like to give an answer not only to the question "what is naturalness in mathematics?" but also "why mathematicians call a piece of mathematics natural?"

Therefore, if we reject this form of linguistic naturalism, we are left with the following problem. On the one hand, we wish not to decide from the outset which cases are relevant; on the other hand, we necessarily need to recognize some specific charter of naturalness in concrete cases.

---

[17]Chow, FOM-list on Jan 28, 2006.

In order to solve the impasse, a philosophical step is needed. This is what we shall try to do in the next section. Since the beginning, it is important to acknowledge that there is no painless way out, if not a slight shift of the problem.

14.2.2 A Tentative Methodology

So far, we have taken a stance against a linguistic naturalistic approach, while maintaining that a weak form of naturalism is implicit in the methodology of philosophy of mathematical practice. However, we have not excluded some refined form of naturalism, such as the one of Penelope Maddy. It shall then be fruitful to explain our methodology by means of confronting it with – and taking distance from – Maddy's position. This juxtaposition will allow us to define our proposal about the role played by philosophy in this kind of analysis.

The first attempt to find a solution to the relevance dilemma could consist in looking for concrete examples and in extracting a moral out of them, without imposing a philosophical prejudice on our scrutiny but instead relying only on the methodology established by the practice of the mathematical community – this is precisely Maddy's strategy. However, even assuming a consistent and Well-organized attitude, within the mathematical community, toward the issue of naturalness, two problems arise.

Firstly, Maddy's position forces us to accept a methodological naturalism that, because of its programmatic absence of any philosophical posture, depends on others' prejudice. Moreover, it runs into the trouble of deriving a philosophical lesson from a finite number of cases without any hope of reaching a significant approximation to the general problem. This drawing of a philosophical lesson from partial examples is simply too ambitious. In the present case, the situation of naturalness is even worse, because a sufficiently clear conceptual analysis of the notion of naturalness is missing also from the side of the mathematical community.

Secondly, Maddy's work argues for a form of naturalistic holism that defends a strong autonomy of mathematics. This thesis claims that philosophical considerations do not find place within a mathematical enterprise. While maintaining that those latter might have a role in the inspirational moment of discovery of a theorem – as in the case of Gödel's realism in the occasion of his discovery of the coherence of the continuum hypothesis – she discards their role in the process of justification.[18] However, when dealing with naturalness, a concept that exhibits a strong informal character, are we allowed to discard so easily a philosophical component in the mathematical work? Maddy refuses any exception:

> After uncovering corresponding methodological argumentation in a range of cases, the Second Philosopher concludes that though metaphysical theories on the nature of mathematical truth and existence undeniably do turn up in such debates, they are not

---

[18]cf. Maddy (2007), p. 366.

in fact decisive, they are in fact distractions from the underlying purely mathematical considerations at work. Actual methodological decisions, she sees, are based on a perfectly rational style of means-ends reasoning: the most effective methods available toward the concrete mathematical goals in play are the ones endorsed and adopted. Acting on her assumption that the actual methods of mathematics are the ones that should be followed, she resolves to apply such typically mathematical methodological reasoning to any contemporary debates she might face.[19]

Then, should we only look for "mathematical methodological reasoning" to address the problem of naturalness, in spite of its intrinsic vague character? In the light of our position, Maddy's suggestion is difficult to maintain. Mathematicians make more and more use of a term that has a strong philosophical flavor – and clearly each of those uses, pertaining to the informal side of mathematical work, may be possibly omitted without a significant loss in terms of strictly mathematical reasoning. Moreover, naturalness can be assumed as a surface detector of some much deeper theoretical phenomenon that a Second Philosopher would always miss. In this respect, we think that our case is even more compelling than those of "simplicity" and "fruitfulness" that motivated the following defense by Tappenden of a philosophy of mathematical practice:

> The assessments of simplicity or fruitfulness we make would no doubt be different if our brains were wired differently, and this would affect the mathematics and science that we produced, but still the judgments we actually make are too systematically embedded in our actual practices to be simply shrugged off in studies of either scientific or mathematical method.[20]

Then we are left with our dilemma. If our goal is to avoid possible philosophical prejudices and at the same time we aim to call into question the appropriateness of every argumentative step, then how to proceed?

Let us step back for a moment and ask a more general question than "what is naturalness in mathematics?"; that is to say, let us address the following issue: which form are we expecting from an answer to the question "what is naturalness in mathematics?"

In order to answer this latter question, the context of the philosophy of mathematical practice seems to be too wide and too vague. We need to narrow it through an inspection of its different methodologies and then choose the most convenient ones. This strategy is also motivated by the absence of a related literature structured enough to provide any general methodological guidelines.

In doing so, two aspects deserve particular attention. First of all, for importing other methodologies, we have to justify why we consider them related to the problem of naturalness. Secondly, we need to preserve our point of view as neutral as possible, dismissing the risk of distorting the concrete uses of the term just for the sake of our argumentative line. A possible solution to this second concern could be to import into the framework not a single methodology but a debate (if possible

---

[19]Maddy (2007), p. 349.
[20]Tappenden (2005), p. 154.

already well developed) between different alternative methodologies. In this light, our proposal will be to make reference to two possible approaches to the notion of mathematical explanation.

For the moment, let use sharp this argument in outlining a methodology that can be used in order to analyze vague notions, such as that of naturalness.

14.3 A Semiformal Method (For a Philosophical Analysis of a Mathematical Term)

1. Look for empirical evidence of concrete uses of the term in the mathematical literature. If its frequency is marginal, then stop. If it calls for an explanation, then proceed. If the term has a common sense, look for its definition in the dictionary.
2. (If necessary) Enforce your analysis with tools from corpus linguistics.
3. Find the right philosophical context where to place your analysis and give convincing philosophical reasons to support why it is relevant for a proper understanding of the term.
4. Inspect the possible methodologies of the context you found in (2) and look for the philosophical ideas that motivate them.
5. Formulate a (possibly binary) dichotomy, in accordance with the philosophical ideas you found in (3).
6. List historical examples in which the term is involved in some explicit form, or add contexts clearly connected to the dichotomy, where the role of the term is relevant.
7. Test items from (4) using items from (5).
8. Verify plausibility with previous outputs of the method.
9. Connect the horn of the dichotomy, to which the examples point, with the philosophical idea that motivated its proposal.
10. If the term has a common sense, compare your results with the common sense and see if it informs the philosophical ideas you found in (3).
11. Go to (1).

Remark 14.3.1. Point 7 is needed in order to allow the possibility that the various appeals to naturalness do not refer to any precise notion. Indeed, even if our goal is to show that there are philosophical ideas shaping mathematical practice, we do not argue that every informal notion that is extensively used in mathematics is philosophically meaningful or that it deserves a particular explanation.

Remark 14.3.2. Point 10 is just a device that goes back to the statistical evidence whenever it is needed. For instance, if the outcome of this method points to some similitude between the notion of naturalness and a different one, a statistical analysis can be used to support this fact.

We believe that this method proposes a third way between a naturalistic approach and a philosophy-first approach.

Against naturalism, while providing reasons for the use of a term, the method is not limited to intra-mathematical ones. Rather, since the beginning (point 2, 3), it asks for a philosophical frame in which the analysis has to be embedded in order to be pursued. Thus, according to the method, the looking at mathematical practice prescribed is by no means uncritical, since the very setting of the problem, within any conceptual analysis, cannot be completely neutral.

Against a philosophy-first approach, the method requires (point 4) to formulate a choice between opposite methodologies, in order to keep a balanced point of view through a dichotomy. In this way, it will be the analysis of mathematical practice that will solve this dichotomy, by means of tipping the balance in favor of one side or the other.

In sum, in the continuation of this work, we will make use of the proposed method of inquiry.

14.4 On the Static vs. Dynamic Opposition

14.4.1 Applying the Method

In the first part of this work, we have developed a method to study some vague notions, such as naturalness, widely used in mathematical practice. Our aim was to establish a method safe from both naturalism and philosophy-first approaches. According to the proposed method, what we need now is an appropriate context, conceptually close to that of naturalness, from which we can import a dichotomy between two different features that a naturalness account might exhibit.

This strategy is known. Indeed, Tappenden suggests to look at how the notion of naturalness has been used in the contemporary metaphysical debate:

> It's unlikely that mathematical and non-mathematical reasoning are so disjoint as to exclude interesting points of overlap. In recent decades there has been a revival of old-fashioned metaphysical debates about the reality of universals, the artificial/natural distinction, and cognate topics. It might seem initially promising to draw on these debates to illuminate the questions appearing in the survey essay.[21]

Nonetheless, he proceeds with showing through convincing arguments that the debate on metaphysical natural properties is not the right context. While considering Sider's paper *Naturalness and Arbitrariness* (Sider 1996), Tappenden argues that often mathematical debates cannot be settled by means of appealing only to metaphysical intuitions, (e.g., Benacerraf problem of *What numbers could not be*); on the contrary, they can find a solution thanks to intra-mathematical, pragmatical reasons (e.g., Von Neumann's identification between sets and numbers show[22]). Tappenden then hints to a partial coincidence between the notion of naturalness and

---

[21]Tappenden (2008b), p. 3.
[22]See Steinhart (2002) in this respect.

that of fruitfulness. Thus, he seems to agree with Maddy's suggestion to consider only intra-mathematical reasons without any independent philosophical analysis of the problem.

We cannot consider this analysis to be satisfactory. The main reason for this is that arguments proposed in Tappenden (2008a,b) do not address the problem of "why mathematicians call something natural?" But this latter question become absolutely relevant once a study of this notion is conducted within the context of the philosophy of mathematical practice – as Tappenden himself seems to admit.

Considering that metaphysics is not the proper context of analysis, Tappenden indicates an alternative route:

> It will help us sharpen the issues[23] to look for a philosophical niche served up by treatments of explanation and understanding in the natural sciences, since these have been extensively addressed.[24]

We propose to develop this hint. There are significant similarities between the notion of naturalness and that of mathematical explanation, and these call for a similar treatment. First of all, they both belong to the informal side of a mathematical work. Indeed, one of the main contributions provided by these two notions is to witness the fact that results, in mathematics, are the outcome of a process of discovery involving many informal components. Furthermore, this process can be only partially formalized in the proof of a theorem. Thus, these two notions play a role that significantly resists to any attempt to reduce mathematical practice to a mere iteration of three basic formal components: definitions, theorems, and proofs. Of course, naturalness and explanation share such a role with various other notions, such as pureness, simplicity, generality, and so on.

However, at least one of these similarities is peculiar: the expression "it is natural" sometimes just means "it is self-explanatory." So, one might employ naturalness judgments as a means of relieving the need of explanation. This latter use of naturalness is at times referred to as some sort of end-of-the-argumentation.

Let us then consider what happens in the field of mathematical explanation. What we encounter are two different and antithetical approaches: the so-called bottom-up and top-down methodologies.[25]

> It should be obvious from the above that mathematicians seek explanations. But what form do these explanations take? It is here that two possibilities emerge. One can follow two alternative approaches: top-down or bottom-up. In the former approach one starts with a

---

[23]In this work, Tappenden is not addressing primarily the problem of naturalness, but many problem related to it. This quotation is taken after the presentation of a case study where visualization seems to be a fundamental character of the representation of the multiplication table for octonions. At this point, he is discussing the naturalness of the formulation of a problem, the essentiality of its presentation, and its fruitfulness. Then, also considering the relevance that fruitfulness plays, for him, in the context of naturalness – as one sees in Tappenden (2008b) – we believe that this passage is relevant and well placed in this discussion.

[24]Tappenden (2005), p. 158.

[25]See for example Mancosu and Hafner (2005), or Molinini (2011).

general model of explanation (perhaps because of its success in the natural sciences) and then tries to see how well it accounts for the practice. In the latter approach one begins by avoiding, as much as possible, any commitment to a particular theoretical/conceptual framework.[26]

The main methodological opposition called into question by these two alternative approaches evoke is the one between a monistic account vs. a pluralistic account. A top-down attitude, as outlined in Mancosu and Hafner (2005), begins with a general model and proceeds with attempting to conform the case studies to its standards. The outliers are just discarded as non-pertinent: there is not much room for dissimilarities. This approach forces to ignore anything that diverges from the description provided by the general model chosen as a starting point – in Hegel's words "*Desto schlimmer fur die Tatsachen.*"[27] So monistic attitudes has a strongly static character, since the burden of relevance is left on the kind of arguments that an author presents in order to defend her theoretical point of view.

A bottom-up approach, instead, does not accept to sacrifice the complexity of a phenomenon for a solid but strict frame. On the contrary, it aims to provide a detailed picture of the use of a concept, even at the risk of discovering that behind such uses do not lie a single separate idea but a cluster of different ones. So a bottom-up approach is highly context dependent: while attempting to identify a general pattern for a notion, it also attempts to register every modification that occurs in the use of a term. Thus, a pluralistic stance focuses on the dynamic side of the notion.

Of course there are differences between the concept of naturalness and that of explanation. As a matter of fact, explanation evokes the idea of a process, while – as we hinted before – naturalness points at a more non-processual phenomenon. This dissimilarity rests on theoretical grounds. Indeed, to some extent, explanation pertains more to epistemology and naturalness to ontology. In this respect, it is useful to recall that the common sense meaning of naturalness calls into play an objective – realist – character.

Shall be now clear that the theoretical dichotomy we want to test is the one between static and dynamic accounts. This division echoes the two alternative positions expressed by Friedman and Chow on the FOM list in 2006.

We call *static* any approach according to which naturalness is an inherent and stable property (or class of properties) of the "object" or "action" that we call natural – even if it is not possible to characterize it properly.

We call *dynamic* any approach according to which naturalness rests on some contextual properties (as those relations that the "object" has with similar objects in mathematical discourse or its position in the development of a mathematical theory) such that it is not possible to determine what is naturalness without appealing to these very same properties.

---

[26]Mancosu and Hafner (2005), p. 221.

[27]"So much worse for the facts [if they do not fit the theory]", attributed to Hegel as an answer to those who noticed that new observations did not fit in the theory formulated in his PhD thesis. See Lask (1914) for this anecdote.

A similar fluctuation between a dynamic and a static account of naturalness already appears, somewhat implicitly, in Corfield's book *Towards a philosophy of real mathematics*. Corfield claims that naturalness might carry various different meanings – thus promoting a bottom-up approach, as one would expect from his naturalistic attitude.

> In sum, a full analysis of the use of the term 'natural' by mathematicians through the ages would require a book-length treatment. As used today it possesses several shades of meaning, which blend into each other to some extent, relying as they do on a sense of freedom from arbitrariness and artificiality.[28]

However, Corfield proves himself to be in some ways divided between a dynamic and a static approach. On the one hand, he argues that to call a mathematical object natural is a process whose validity is grounded on the properties that the object shares with other similar ones – thus supporting a dynamic view. On the other hand, while discussing the concept of groupoid, he maintains that "another way of arguing for the naturalness of a concept is in terms of the inevitability of its discovery."[29] This latter position is clearly connected with a more static view, since the naturalness of the concept of groupoid is granted by some inherent properties that made its discovery unescapable.

These examples are meant to show that the dichotomy we are proposing fits well with the different approaches available for the notion of naturalness. Hence, we take them as a hint that we are on the right path. Our final question, then, will be the following: is naturalness a static or a dynamic notion?

14.4.2 Case Studies

In this section, we will test this static/dynamic dichotomy through the lens of three case studies taken from set theory and computability theory. In doing so, we will follow our method (point 5).

14.4.2.1 The Concept of Set

We start with the literal evidence that naturalness is normally assumed as a property of the concept of set.

> Faced with the inconsistency of naive set theory, one might come to believe that any decision to adopt a system of axioms about sets would be arbitrary in that no explanation could be given why the particular system adopted had any greater claim to describe what we conceive sets and the membership relation to be like than some other system, perhaps incompatible with the one chosen. One might think that no answer could be given to the question: why adopt this particular system rather than that or this other one? One might suppose that any

---

[28]Corfield (2004), p. 230.
[29]Corfield (2004), p. 225.

apparently consistent theory of sets would have to be unnatural in some way or fragmentary, and that, if consistent, its consistency would be due to certain provisions that were laid down for the express purpose of avoiding the paradoxes that show naive set theory inconsistent, but that lack any independent motivation. One might imagine all this; but there is another view of sets: the iterative conception of set, as it is sometimes called, which often strikes people as entirely *natural*,[30] free from artificiality, not at all ad hoc, and one they might perhaps have formulated themselves.[31]

We see, in this quotation by George Boolos, a clear reference to the so-called iterative conception, influentially discussed also by Parsons (1977) and Wang (1974). This conception stems from the idea of a cumulative hierarchy for the universe of set theory and it is here linked directly to the problem of naturalness. Then following our method, we should ask: is the naturalness of the notion of set a static or a dynamic one?

As it is clear from Boolos' quotation and well known from the history of the discipline, the notion of set changes through history and thus it is a dynamic one. Indeed its first appearance in the history of mathematics was shown to be inconsistent, and moreover, it was linked with a different idea, widespread and common in the mathematical community at that time: that of a set as determined by a law. We will now analyze these two notions of sets – the original Cantorian one and the iterative one – trying to understand if the dynamic character of this concept determines a similar aspect of the notion of naturalness.

In Cantor's work, the first definition of set is in 1882, in the third paper of the series of six from the period 1978 to 1984, bearing the title *Über unendliche, lineare Punktmannichfaltigkeiten*.

> I call a manifold (an aggregate [Inbegriff], a set) of elements, which belong to any conceptual sphere, well-defined, if on the basis of its definition and in consequence of the logical principle of excluded middle, it must be recognized that it is internally determined whether an arbitrary object of this conceptual sphere belongs to the manifold or not, and also, whether two objects in the set, in spite of formal differences in the manner in which they are given, are equal or not. In general the relevant distinctions cannot in practice be made with certainty and exactness by the capabilities or methods presently available. But that is not of any concern. The only concern is the internal determination from which in concrete cases, where it is required, an actual (external) determination is to be developed by means of a perfection of resources.[32]

However, the first relevant one for a conscious history of set theory is the one in the *Grundlagen einer allgemeinen Mannigfaltigkeitslehre. Ein mathematisch-philosophischer Versuch in der Lehre des Unendlichen*, from 1883.

> By a "manifold" or "set," I understand any multiplicity which can be thought of as one, i.e., any aggregate [Inbegriff] of determinate elements which can be united into a whole by some law.[33]

---

[30]My italics.
[31]Boolos (1971) p. 218.
[32]Zermelo (1932), p. 150.
[33]Zermelo (1932), p. 204.

The idea of a set as a mathematical object determined by a law explains the reason why set theory is commonly considered as a part of logic – where logic is intended to be the general science of the law of thought. As a matter of fact, we find here expounded the notion of set as extension of a concept that, properly formalized, will bring Frege to the failure of its logistic program. However, Cantor's idea of general law – naïve as it may be – is not limited to some repertory of tools of definition, but it seems opened to any possible, and future, means.

This same idea of sets as concept extension is also what guided Dedekind in his work on the foundation of number theory: *Was sind und was sollen die Zahlen?* – which will influence Zermelo, together with Cantor's work, in the axiomatization of set theory.

> It very frequently happens that different things a, b, c ... considered for any reason under a common point of view, are collected together in the mind, and one then says that they form a system S; one calls the things a, b, c ... the elements of the system S, they are contained in S; conversely, S consists of these elements. Such a system S (or a collection, a manifold, a totality), as an object of our thought, is likewise a thing; it is completely determined when, for every thing, it is determined whether it is an element of S or not.[34]

Then the question we should ask is: was the first conception of set thought as natural? Cantor had the idea that his notion of set was instrumental for the development of his theories of ordinal numbers and infinite cardinal numbers. Then, in trying to justify the former, he says that the extension from the finite to the infinite was natural and helped him to develop set theory.

> I am so dependent on this extension of the number concept that without it I should be unable to take the smallest step forward in the theory of sets [*Mengen*]; this circumstance is the justification (or, if need be, the apology) for the fact that I introduce seemingly exotic ideas into my work. For what is at stake is the extension or continuation of the sequence of integers into the infinite; and daring though this step may seem, I can nevertheless express, not only the hope, but the firm conviction that with time this extension will have to be regarded as thoroughly simple, proper, and *natural*.[35]

Moreover, Cantor, talking about the properties and laws of the infinite, says that they depends "on the nature of things.[36],[37]" But then, how it is possible that a natural notion was transformed in another different natural notion?

---

[34]Ewald (1996), p. 344.

[35]My emphasis. In *Grundlagen einer allgemeinen Mannigfaltigkeitslehre. Ein mathematisch-philosophischer Versuch in der Lehre des Unendlichen*, see Ewald (1996), p. 883.

[36]Zermelo (1932), pp. 371–372.

[37]By the way, this opinion may be questioned by the modern development of set theory. Indeed, it is important to stress that Cantor's theory of cardinals is not as "natural" as it could be seen; as a matter of fact, it hides an important choice behind it. There are two conflicting ideas: Cantor's Principle, two sets have the same size if there is a bijection between them, and Aristotle's Principle, if a set A is a proper subset of another set B, then the size of A is smaller than the size of B. As the development of a theory of numerosity has shown Benci and Di Nasso (2003) and Benci et al. (2006), the formalization of the infinite does not involve necessarily Cantor's theory of cardinal numbers.

The change in the notion of set comes from Zermelo's axiomatization in 1908, where the explicit attempt was to keep as ample as possible the concept of set, without running into the paradoxes.

> This discipline [set theory] seems to be threatened by certain contradictions, or "antinomies," that can be derived from its principle - principles necessarily governing our thinking, it seems - and to which no entirely satisfactory solution has yet been found. In particular, in view of the "Russell antinomy" of the set of all sets that do not contain themselves as elements, it no longer seems admissible today to assign to an arbitrary logically definable notion a set, or class, as its extension. Cantor's original definition of a set (1895) therefore certainly requires some restrictions; it has not, however, been successfully replaced by one that is just as simple and does not give rise to such reservations. Under these circumstances there is at this point nothing left for us to do but to proceed in the opposite direction and, starting from set theory as it is historically given, to seek out the principles required for establishing the foundations of this mathematical discipline. In solving the problem we must, on the other hand, restrict these principles sufficiently to exclude all contradictions and, on the other, take them sufficiently wide to retain all this valuable in this theory.[38]

Few comments are needed after this quotation. Zermelo says explicitly that he wants to axiomatize the "theory created by Cantor and Dedekind," but he likewise explicitly says that the theoretical framework that motivated the founding fathers is not tenable anymore because of the antinomies. The problem is found exactly in the main definition of set that came so naturally from Cantor's analysis of well-order sets and infinite cardinalities: a too loose use of the idea of sets as concept extension is dangerous. Then Zermelo's proposal, in the line of Hilbert's school, is to start from an historically given theory and try to arrange its main theorems in a logical order, while implicitly defining the basic notion of the theory. This style of reasoning is very far from Cantor's deduction – in a kantian sense – of the principles of set theory, as he attempted to do in a letter to Hilbert, dated October 10, 1898.

It should also be noted that Zermelo does not appeal to the naturalness of the concept of set as defined by his axioms. Indeed, his system is not justified in terms of the concepts involved – even less in terms of Cantor's notion of set – but motivated by pragmatic reasons, with the explicit goal to avoid paradoxes. Then, in 1930, Zermelo, while engaged in the search for a consistency proof for set theory, proved a quasi-categoricity theorem for second-order ZF. The context of Zermelo's work is quite far from our modern treatment of the subject,[39] but the main idea was to shape a model of ZF thanks to a cumulative hierarchy: a division in levels where the elements of a set lay in levels of the hierarchy that come before the one the set belongs to. These stages were ordered by ordinal numbers and the first level that formed a model for all ZF was indexed by a strong inaccessible

---

[38] Zermelo (1967).

[39] Zermelo's work was in the in the context of second-order logic, and moreover, he thought that the definition of a model of set theory had two degrees of freedom: height and width – with respect of the *urelemente* to be considered as primitive. While the former stems from the idea of a cumulative hierarchy – and then it is still actual – the latter is not anymore a concern for the mainstream modern research in set theory, which abandoned a theory of sets with *urelemente*.

cardinal. Subsequently, the adoption of the idea of a cumulative hierarchy by Gödel in his proof of the coherence of the Axiom of Choice – where he developed the Constructible Universe – helped in spreading the idea that "set" is an iterative notion.

> This concept of set (...) according to which a set is anything obtainable from the integers (or some other well-defined objects) by iterated application of the operation *"set of"*[40] and not something obtained by dividing the totality of all existing things into two categories, has never led to any antitomy whatsoever; that is, the perfectly "naïve" and uncritical working with this concept of set has so far proved completely self-consistent.[41]

From that moment on, we could see a progressive shift from the idea of a cumulative hierarchy, for a model of ZF, to an iterative notion for the concept of set. Indeed, this idea became so linked with the concept of set that people started to inverse the process that lead from ZF to the cumulative hierarchy, trying to justify the axioms in terms of an iterative notion: the conceptual counterpart of the structural, model theoretic conception of a cumulative hierarchy. This is exactly the case of Boolos's arguments in favor of the naturalness of the axioms of ZF. Moreover, notice that this argument can be proposed only after it was possible to give a clear and intuitive picture of the theory that formalizes the notion of set. In this way, the axioms that inspired and shaped Zermelo's model(s), in the search of their consistency, are justified in terms of the model(s) itself; but what does this mean, really? It is important to remember that Zermelo's theorem is a quasi-categoricity theorem: it says that a model of second-order ZF has just two degrees of freedom, its height and the width, for what concerns the *urelemente*. Then, since all the possible models of these axioms are built as a cumulative hierarchy, it would seem that there was no need to justify the axioms in terms of their iterative character. To makes sense of this operation, we have to accept that what needs a justification is not the fact that these are axioms for set theory but the fact that they capture the essence of the concept of set. What is at work here is a hidden thesis that fixes a concept. We could call it the Zermelo-Gödel thesis: being a set means being an object that belongs to a cumulative hierarchy – and, after Zermelo quasi-categoricity theorem: being a set means to be a set in a model of ZF. As in the case of Church-Turing thesis (CTT), what seems to be the natural choice is, in reality, the stipulation of a relevant aspect of a concept. Then arguments as Boolos' or Parsons' or Wang's are at par with the attempt to prove or justify CTT.[42]

---

[40]My italics.

[41]Gödel, CW II p.180, 1947 what is the continuum problem.

[42]In this discussion, we implicitly assumed that the cantorial notion of set, at least the one proposed in the *Grundlagen*, is different from the iterative one. For what concerns the strongest claim that it is not possible to find this notion in Cantor's work, we do not take a stand, even if we believe that even the definition presented in the *Beiträge* cannot be considered as cumulative, if not forcing it from our modern perspective. See Frapolli (1991) and Jané (2005) in this respect. However, it is fare to say that the iterative conception is not entirely incompatible with the latest reflections of Cantor, even if we believe that it had different conceptual motivations, as it is well shown in Hallet (1984). The main possibility of a specification of Cantor's notion of set in terms of an

Then we can conclude that the dynamic character of the notion of set informs the notion of naturalness, shaping the latter with a dynamic component – in the diachronic sense we proposed. Moreover, our analysis also showed that a normative component is hidden in the natural character of a mathematical concept.[43]

14.4.2.2 Natural New Axioms for ZFC

A second easier example is the discussion on the naturalness of the axioms that extend ZFC. For this case, we start from a quotation by Joan Bagaria, who gave a mathematical characterization of the bounded forcing axioms in terms of generic absoluteness (Bagaria 2000) and then tried to argue for their naturalness (Bagaria 2004).

> All together, the criteria [*Maximality, Fairness, Consistency and Success*] may be regarded as an attempt to define what being a natural axiom of set theory actually means.[44]

Let us analyze these criteria, in search for static or dynamic elements in the characterization of naturalness in the context of new axioms for ZFC.

Maximality. This principle is considered useless in the absence of further specifications.[45] Then it is exemplified with some of the criteria proposed by Goödel – Reflection, Extensionalization, and Uniformity – whose program is presented as the program "of finding new natural axioms which, added to the ZFC axioms, would settle the continuum problem." After discussing the issues related to these principles – and acknowledging that the criterion of maximality is not sufficient, alone, to settle CH – Bagaria concludes saying that "Gödel's principles of Reflection, Extensionalization, and Uniformity arise naturally from the systematic application of the criterion of maximality.[46]"

---

iterative conception does, indeed, sustain our thesis of the prescriptive character of the notion of naturalness.

[43]Notice that this opinion was proposed quite early, in the development of set theory, contrary to the general idea of a naturalness of the notion of set – as this quote from König shows clearly: "That the word 'set' is being used indiscriminately for completely different notions and that this is the source of the apparent paradoxes of this young branch of science, that, moreover, set theory itself can no more dispense with axiomatic assumptions than can any other exact science and that these assumptions, just as in other disciplines, are subject to a certain arbitrariness, even if they lie much deeper here – I do not want to represent any of this as something new." Van Heijenoort (1967) p. 147.

[44]Bagaria (2004), p. 6.

[45]Notice that also Maddy says something similar: "In both cases, the structure of the counterexamples suggests that the formal criterion will need supplementation by informal considerations of a broader character" (in Maddy (1997), p. 255). These supplementation are comments like: "This last [$AD^{L(\mathbb{R})}$] is a particularly natural hypothesis, stating that AD is true in the smallest model of ZF containing all ordinals and all reals", p. 226.

[46]Bagaria (2004), p. 9.

Fireness. This criterion is explained as advising his promoters not to discriminate between sentences of the same complexity. Then, the reasons for considering classes of sentences pertaining to sets with the same rank, or to sets with the same hereditary cardinality, are the following: "Now the complexity of a set may be defined in different ways, but the most natural measures of the complexity of a set are its rank and its hereditary cardinality.[47]"

Success. This criterion is easily explained in terms of solutions to natural problems. "A new axiom should not only be natural, but it should also be useful. Now, usefulness may be measured in different ways, but a useful new axiom must be able at least to decide some natural questions left undecided by ZFC. If, in addition, the new axiom provides a clearer picture of the set-theoretic universe, or sheds new light into obscure areas, or provides new simpler proofs of known results, then all the better.[48]"

For what concern consistency, this principle is explicitly considered as a regulative idea that acts only as a necessary condition for new axioms. As a matter of fact, once we are in the context of classical first-order logic, this principle can be subsumed under the one of success, because if an axiom is not consistent, it allows the proof of every proposition. Hence, it is not useful.

In light of these consideration, it is clear that the definition of a natural axiom is not statical but dynamical – in the synchronic sense we proposed – because it depends on the context, on other attempts to define naturalness, on natural ways to consider sentences of the theory, and on the naturalness of other pieces of mathematics.

However, how to make sense of an attempt to *define* – as Bagaria argues – naturalness in terms of naturalness in a meaningful way? The appeal to "natural questions" and "natural measures" is sustained by qualitative judgments, on the subject matter of the theory, that pertain to considerations of relevance and of importance that, far from being objective and necessary, gain strength in connection to other naturalness considerations. Here again, we find at work normative judgments that stem from subjective or intrasubjective – read scientific community – considerations that aim to shape mathematical work, pointing to what is relevant and what deserves attention and commitment.

14.4.2.3 Naturalness and c.e. Intermediate Degree

This section is devoted to the study of the role of naturalness in the solution of Post's problem, one of the main classical problems in computability theory. This is a quite technical topic. To present it exhaustively, or to any formal degree, largely exceeds our general interest. However, we chose it as a significant case study, because the

---

[47] Bagaria (2004), p. 9.
[48] Bagaria (2004), p. 10.

notion of naturalness appears here in a very explicit form and also because there is already a large (but absolutely nonconclusive) debate, among leading researchers of the field, on how such notion has to be intended. Our main attempt is essentially to reformulate this debate in the frame of our static vs. dynamic opposition. We aim to show that in order to properly capture naturalness in computability, one needs a dynamic account in which normative components play a decisive role.

We proceed as follows. Firstly, we recall a few classical definitions in order to give an informal description of the problem (for a full exposition, the reader is referred to any standard textbook on computability theory, e.g., Soare 1987). Then, we mainly focus on a couple of thesis – that we will call "lack of naturalness thesis" – whose consequences, at least in the version defended by Friedman, originated a huge and intricate thread of discussion in the FOM list, consisting in two non-consecutive waves, that of 1999 and that of 2005. Rather than giving a philological presentation of the cluster of positions expressed in the debate, which is sometimes very unstructured, we shall organize them in a conceptual laboratory in which different answers to the lack of naturalness thesis can be used to test different perspectives on naturalness.

Some Background

Let us begin rather schematically. In this section, we aim to give, in a quite informal fashion, a brief collection of some preliminary definitions.[49] All the sets we will consider are $\subseteq \omega$.

We take as known the following notions: Turing machine, computable function, computable set, computable enumerable (c.e.[50]) set, and Halting set. We denote the Halting set with K.

- An *oracle Turing machine* is a standard Turing machine equipped with an extra "read-only" tape on which is written the characteristic function of a set A (χ_A), called the *oracle*. Informally, a *Turing program* for an oracle Turing machine is a standard Turing program that may also contain instructions to read the oracle tape.
- Given an effective list P_e of all the Turing programs (encoded with natural numbers), we denote with φ_e^A the function executed by P_e with oracle A.
- For any set A, we denote with A' the following set: $\{e \mid \varphi_e^A(e) \downarrow\}$.
- Let A, B be two sets. We say that A is *Turing-reducible* to B ($A \leq_T B$), if $\chi_A = \phi_e^B$ for some positive integer e. Informally, a set A is Turing-reducible to B if there is an algorithm for deciding whether $x \in A$ given answers to any question of the form $x \in B$?

---

[49] We stress again that any of the multiple formal gaps that we leave behind can be filled reading a very initial segment of Soare (1987).

[50] We follow the current use to convert any single occurrence of the word "recursive" (or derivates) in classical recursion theory in the word computable.

- If $A \leq_T B$ and $B \leq_T A$, we say that A and B are Turing-equivalent ($A \equiv_T B$). Turing equivalence is an equivalence relation. Its classes of equivalence are called *Turing-degrees*. We denote with **0** the degree of computable sets and with **0'** the degree of K.
- We recall that **0'** is Turing-complete among c.e. set, i.e., for any c.e. set A, $A \leq_T$ **0'**.
- We call a degree **d** *intermediate* if **d** is c.e. and **0** $<_T$ **d** $<$ **0'**.

Then, we can state Post's problem as follows: are there any intermediate degrees? This is how Post, in 1944, presented for the first time his problem:

> As a result we are left completely on the fence as to whether there exists a recursively enumerable set of positive integers of absolutely lower degree of unsolvability than the complete set K, or whether, indeed, all recursively enumerable sets of positive integers with recursively unsolvable decision problems are absolutely of the same degree of unsolvability.[51]

The importance of Post's problem lies in the fact that its solution gives us information not only about the structure of c.e. Turing-degrees but also (through some proper coding) about any sort of c.e. problems that could appear in mathematics. In Rogers' words:

> Post's problem was significant in two respects: (1) it concerned the variety of structures possible among the nonrecursive, recursively enumerable sets; (2) it therefore concerned the variety possible among axiomatizable theories and among other sorts of recursively enumerable problems. [...] If all theories were $\equiv_T K$, then reducibilities from K would be a general method for demonstrating the undecidability of axiomatizable theories.[52]

Lack of Naturalness Thesis

Post's problem has positive answer: intermediate degrees exists. Moreover, far from being trivial, the degree structure generated by Turing-reducibility among c.e. sets – whose standard notation is $\mathcal{R}$ – is a very complex and rich one. For decades, studying its properties, both on logical and algebraic side, has been one of the main research projects in computability theory. For the record, it has been proved that every finite distributive lattice can be embedded into $\mathcal{R}$ and that its first-order theory is equivalent to the theory of true first-order arithmetic. So, where does naturalness appear?

Typically, speaking of naturalness about Post's problem and its solution concerns two different claims (frequently overlapped in the FOM debate, as we will see): (1) that there is no natural example of intermediate degrees outside computability theory and (2) that almost every known solution to the problem significantly diverges from Post's original approach, in most cases, making use of a class of

---

[51] Post (1944), p. 314.
[52] Rogers (1967), p. 144.

methods (i.e., priority arguments, spread everywhere in computability theory) that are to some extent unnatural, or artificial. We call these claims "lack of naturalness thesis" (from now on:, LoN_1 and LoN_2).

This is the presentation that Harvey Friedman made of LoN_1:

> Usually, when mathematicians undertake an intensive investigation of some specific structure or class of structures, the need for such an investigation has already been motivated by a set of specific, natural examples showing the richness and interest of the subject. For instance, group theory was motivated by a wealth of examples such as matrix groups, permutation groups, symmetries of geometrical figures, etc. Contrast this with the r.e. sets and degrees that are so much beloved by recursion theorists. The only natural examples known to date are the original ones, i.e. the halting problem and the complete r.e. degree. Thus there is really only one example, and that example is highly atypical of the way the subject has developed. It is reasonable to wonder whether this lack of examples may indicate some sort of defect or imbalance in the subject.[53]

Thus, the lack of naturalness is motivated by a very contextual consideration: being natural (in this case, for a degree) is a property that can be evaluated only in relation with some other mathematical theories. In this view, naturalness consists in a sort of familiarity: you can call a degree natural only if you have already met it before (no matter where). It is important to notice that in order to satisfy this condition, it is not enough to just prove some general existence result, but one has rather to show precisely the intermediate degree or possibly a set that can be naturally coded in it. Otherwise, LoN_1 may be immediately discarded as false. Indeed, Cooper noticed:

> It is true that all the known canonical c.e. sets (e.g. those associated with standard first-order axiomatic theories, the halting problem, etc) turn out to be computable or of complete c.e. degree, and that for a pure mathematician that is very significant. [...] However, there are mathematical criteria according to which *all* c.e. sets and Turing-degrees potentially contain "natural" information content which may be encountered in specific contexts - just to mention two well-known examples: 1) (Feferman, Hanf) All c.e. degrees contain (finitely) axiomatisable theories, and 2) (Matiasevich) All c.e. sets are diophantine[54]

To our concern, the emphasis should be on the following words: "which may be encountered in specific contexts." Firstly, let us spend a few words on this kind of images. Such an expression is not isolated in the FOM debate. Indeed, its language is frequently shattered by spatial metaphors. Taken together, they may define a global image: that of different mathematical contexts forming a well-defined geography, in which naturalness (at least in the form expressed by LoN_1) consists in a quite stable presence of a concept in more than one single region. There can be something profound: if the common meaning of natural overlaps some form of spatiality, than its use in mathematics may suggest a hidden realistic stance, that of mathematical theories located somewhere (we will focus on this point in our conclusions).

In any case, there are many ways in which the familiarity condition of LoN_1 (i.e., asking for the presence of the degree in other mathematical contexts) can be

---

[53] H. Friedman, FOM list: Jul 28, 1999.

[54] Cooper, FOM list: Jul 29, 1999.

intended; Cooper's objection lies on a very relaxed one: it is sufficient to prove the existence of the degree outside the theory of Turing-degrees. To prevent these objections, it may be reasonable to ask for a stronger version of LoN_1, such as the one entailed in this response by Shipman:

> For example, consider the result of Hanf (cited by Cooper): for every RE degree there is a finitely axiomatizable theory of that degree. This is one of my favorite papers; I recommend it to all of you. But it does not really tell us anything about degrees; it tells us about the nature of finitely axiomatizable theories. This nature is still somewhat of a mystery, despite interesting results by Kleene and model theorists such as Zilber. But I do not see how Hanf's finitely axiomatizable theory (which, roughly speaking, describes a non-standard Turing machine) can be considered as natural.[55]

This refinement of LoN_1 is less innocent than it may sound: how can we possibly distinguish between natural and unnatural finitely axiomatizable theories? The only available answer here is a sort of generalization of LoN_1: a finitely axiomatizable theory is natural if it appears outside its own theory. Moreover, given a class of objects that ground naturalness on finitely axiomatizable theories, we can ask what grounds their own naturalness – and so on. In sum, LoN_1 seems to fit perfectly in our dynamic view: there is no single property shared by a class of mathematical objects (in this case, intermediate degrees) that implies naturalness, but this notion can rather be considered as part of a dynamic and cumulative process, transmitted by a relation of familiarity. For the moment, we can stop here for LoN_1.

In order to understand the role of the term "natural" in LoN_2, let us begin with a brief exposition of Post's original approach. In 1944, in the same article in which the problem has been formulated for the first time, Post was able to prove the existence of an intermediate degree for m-reducibility (a notion stronger than Turing's one). In doing so, Post introduced a new property for c.e. sets, whose definition is quite easy: a c.e. set S is simple if (1) S is co-infinite; (2) for every infinite c.e. set W_e, $W_e \cap S \neq \emptyset$ (informally, the idea lying behind this definition is that S is large enough to meet every infinite c.e. set – and so that its complement is somehow thin, with respect to c.e. sets). Post devoted considerable effort in studying various refinements of the notion of simplicity, strengthening his original intuition of making the complement of the set as thin as possible. In this way, he hoped that he could have found a good candidate for solving the problem in the context of Turing-reducibility (for instance: introducing hypersimplicity, he was able to find intermediate degrees for tt-reducibility, a reducibility lying strictly between m-reducibility and Turing one). Thus, in general, he aimed to formulate a property for a set that would have implied both noncomputability and noncompleteness of its Turing-degree; these kinds of properties are indeed called Post's properties. Unfortunately, all these attempts failed: the problem was eventually solved with a completely different approach.

---

[55] Shipman, FOM list: Aug 12, 1999.

In 1956–1957, Friedberg and Muchnick (independently from each other, see[56]) solved Post's problem by means of using similar strategies, namely, adapting Kleene-Post theorem to the c.e. context. In 1954, Kleene and Post had proved the existence of a pair of incomparable degrees (not c.e.) below $\mathbf{0}'$ (see Kleene and Post (1954)). They built two sets $A, B \leq_T K$, such that $A \not\leq_T B$ and $B \not\leq_T A$. Their innovative strategy was to split these conditions between two infinite lists of easier *requirements* of the form:

$$R_e : \chi_A \neq \varphi_e^B; \quad S_e : \chi_B \neq \varphi_e^A,$$

so that every requirement prevents some function e from being a Turing-reduction of one of the two sets into the other. Kleene and Post satisfied all these requirements by making reference, infinitely many times, to K. Thus, their construction was not absolutely computable, but only relatively to K, while A and B were not c.e. but just $\leq_T \mathbf{0}'$. Now, Friedberg-Muchnick's proof is a reformulation of Kleene-Post construction, in which – in order to produce two c.e. sets – no noncomputable step is admitted. Of course, in lack of the information given by K, to satisfy the requirements is more complex. The key idea is to associate any requirement to multiple strategies ordered to accomplish it, without knowing *a priori* which one will eventually succeed. But the situation is delicate, because, in pursuing a strategy, a requirement may interfere with some other requirements.

Friedberg-Muchnick's solution to face these odds is based on the idea of giving to requirements a *priority* ordering. So, any requirement with higher priority may interfere with one with a lower priority, forcing it to replace its current strategy with a new one (in this case, the requirement with lower priority is said to be *injured*). However, every requirement is injured only finitely many times, and then all the requirements are eventually met.

Nonetheless, such solution does not describe a Post's property. Furthermore, while establishing the existence of intermediate degrees (no one contests its formal validity), it requests to any possible Post's property to show a further feature, that of naturalness; otherwise, one may just trivially solve Post's problem by means of taking as a property the following: A is c.e. and $\mathbf{0} <_T A <_T \mathbf{0}'$. In this light, Cooper recently presented naturalness as fundamental component of Post's problem: "Find some *natural property*[57] $\mathcal{P}$ of computable enumerable sets such that if A satisfies $\mathcal{P}$, then $\mathbf{0} <_T A <_T \mathbf{0}'$.[58]"

To sum it up, asking for a natural solution of Post's problem – in view of LoN$_2$ – essentially consists in looking for a natural Post's property, whose proof of actually being a Post's property is, to some extent, natural too. Everything is very circular, of course; we will try to gain some meaningful insights, thanks to a recasting of

---

[56] See Friedberg (1957) and Muchnik (1956).
[57] Our italics.
[58] Cooper (2004), p. 226.

our static vs. dynamic opposition. In particular, our belief is that almost every real approach to LoN_2 (i.e., every attempt to solve it, thanks to the formulation of a Post's property with alleged reasons to count it as natural) embeds a static notion of naturalness. This is not a forced path: in theory, one could think that the natural feature of a Post property may depend on contextual aspects (such as the web of familiarities implied by LoN_1), but practically this is not the case. Most of the replies to LoN_2 incorporate (in one case, at least, in a very explicit form) a general criterion for separating natural and unnatural properties. Clearly, the existence of a convincing criterion of this kind would also partially dismiss LoN_1, because in order to judge naturalness of a degree, it would be sufficient to see if its Post's property would fall among the natural ones or not. However, we aim to show that none of these proposals – even though mathematically relevant – do philosophically grasp the way in which the term natural is commonly used and intended in computability theory or in mathematics. In doing so, we proceed giving a short catalogue of these proposals, mainly discussing the principal limitations of a static account. Then, we will take LoN_1 seriously.

Staticity

In standard textbooks on computability theory, the first solution usually presented of Post's problem is that of the existence of a low simple set. Apparently, it contains a Post property, namely, being simple and low. We already know what simplicity is, so the only missing ingredient here is lowness: a set A is *low*, if $A' < \mathbf{0}'$ (intuitively A is, in some sense, computationally weak; indeed, since for every set A, $A < A'$ holds, then among c.e. sets, lowness implies noncompleteness). Unfortunately, this is not very informative. It just says where the set is in the hierarchy of Turing-degrees. Thus, considering it as a natural Post's property would be an analogue of our previous example of a trivial case: $\mathbf{0} <_T A <_T \mathbf{0}'$. Furthermore, the usual construction of a low simple set contains injuries, and its lowness requirements have a rather artificial shape. To use Nies's words:

> Post may have hoped for a different kind of a solution to the problem he posed, one that is more natural. [...] In mathematics, to be natural an object must be more than a mere artifact of arbitrary human-made definitions (for instance, the particular way we defined a universal Turing program). Natural properties should be conceptually easy. Being a simple set is such a property, satisfying the requirements in the proof of Theorem 1.6.4[59] is not.[60]

Similar considerations also appear in the FOM debate, but clearly they are not very illuminating: what does it mean to be conceptually simple? We recall that in LoN_2, there is a double request of naturalness: one that lies on the side of proofs and the other one concerning the definition of the Post property. It seems reasonable to expect a good characterization of conceptual simplicity, in this context, to meet both.

---

[59]Theorem 1.6.4 stases: "There is a low simple set A"
[60]Nies (2009), p. 34.

With respect to the former aspect, a condition frequently taken under consideration is that of being injury-free. This is not surprising. After all, the topic of priority arguments is the general frame in which the FOM debate on naturalness emerged. Nonetheless, arguing that injury-free solutions coincide with natural ones is an hard claim.

Let us consider, for instance, the case of K-triviality. This is a quite recent notion that appears in the context of algorithmic randomness. Its formal definition requires a certain amount of preliminary technicalities that rapidly lead us far from our present scope. However, to describe its intuitive meaning is much more feasible. Firstly, recall that any c.e. set A can be viewed as the final outcome of an algorithm that runs at stages. So, at any stage s, only a finite portion of the set is given. For this reason, we can always properly refer to its first n bits of information. Now, a set A is K-trivial if, for every n, the information encoded in the first n bits of A is not bigger (up to an additive constant) than the information encoded in its length. In other words, the informational content of every initial segment of A is as low as possible. To our concerns, K-triviality has two noteworthy features: (1) its definition may claim to be conceptually simple (at least in its informal presentation) and (2) the construction of a c.e. noncomputable K-trivial set takes only few lines and is injury-free. Is it enough to consider it a natural Post's property? There are a couple of significant constraints to a positive answer.

Firstly, proving that each K-trivial set is Turing-noncomplete is a very challenging task. It was firstly demonstrated, in 2003, by Downey, Hirschfeldt, Nies, and Stephan through the introduction of a new technique, the so-called decanter method, whose explanation lasts several pages and is by no means "simple." Can we still consider it conceptually simple? Here the problem is that, even in the presence of a very relaxed semantic of conceptual simplicity, there seems to be a gap between what we are willing to consider as conceptually simple (or natural) and what is characterized as conceptually simple by a formal condition. For instance, being injury-free gives us, formally, a certain information on the complexity of a proof, but on the other hand, the same property does not always enlighten us on its conceptual simplicity.

On the side of definitions, the situation is even more desperate. It is not clear what a naturalness condition wants to discard (in the context of proofs, "being injury-free" was at least an initial hint for a possible solution), and moreover, there is apparently no way to establish – through a static account – which set of positive features transforms a definition in a natural one. Once again, the case of K-triviality can be fruitfully used. As we said, its definition may be regarded as conceptually simple (so, maybe, even as natural). Then, a pure static approach has to explain this fact singling out some feature of K-triviality responsible for conceptual simplicity and without referring to any contextual property. Yet, a deeper look shows that its form of simplicity lies in the fact that K-triviality can be easily described in the general frame of randomness (in Kolmogorov complexity, it is exactly the opposite of randomness; thus, informally, K-triviality means "being far from random," i.e., its information is very easy to compress). Such scenario is highly contextual: what grounds conceptual simplicity for K-triviality is its relation with some other notions

14 Naturalness in Mathematics

and even its position within the general frame of computability. Furthermore, most of the notions involved in this definition reinforced their meaning in the last decades. In this direction, a good thought experiment may consist in asking if notion like K-triviality would have been viewed as conceptually simple, or natural, in the same way in a period in which computational aspects were less considered in mathematics – say, anytime before 1930. According to this view, the main deficit of a static approach is that of being totally blind with respect to any of these contextual and dynamic observations and then unable to capture the phenomenon of naturalness in its fullness.

We conclude this section with a brief discussion of a radical version of a static view of naturalness, that of Friedman. Supporting the idea of a strong analogy between naturalness and simplicity, in the FOM debate, Friedman proposed to tackle the problem in the following way:

> I now wish to move to some precise conjectures.
> For some time now, I have been very interested in "simplicity." The crudest approximation to simplicity is tiny.
> Yes, I hear from the some readers that this is too crude for them, that this is beneath them. But even the first 500 obvious questions in f.o.m. in terms of "tiny" are more than enough in technical depth to occupy 1 billion people for 1 billion years.
> Now let's relate this to "naturalness" in recursion theory.
> 1. It has been a well known occupation of some to try to find the "tiniest" Turing machine whose behavior is nonrecursive, in terms of the accepted inputs. A pretty much generally accepted standard has emerged for talking about how "tiny" these TM's are. I think one measures the number of states and symbols used in the set of quadruples.
> [...] The numbers are very very small, if I recall, and the accepting strings of the TM's written down are COMPLETE RE SETS.
> CONJECTURE: There is no tiny TM whose acceptance set is an r.e. set of intermediate degree. There is no TM with at most 10 symbols and 10 states whose acceptance set is an r.e. set of intermediate degree.
> Of course, negative results are going to be excruciatingly hard in this direction. But
> CHALLENGE. Write down the smallest sized TM you can whose acceptance set is an r.e. set of intermediate degree[61]

Even if such proposal would probably be mathematically interesting, the claim for which it would offer any conclusive insight about the notion of naturalness can be strongly disputed. Indeed, which kind of relation Friedman does establish between tininess (of a TM) and naturalness (of its acceptance set)?

Firstly, saying that tininess is a sufficient condition for naturalness is in direct contrast with the fact that there exist many tiny TMs, whose behaviors are typically considered as profoundly unnatural. Consider, for instance, the so-called busy beaver. On an alphabet $\Sigma = \{0, 1\}$, a TM of n states is called a n-busy beaver if, on input 0, it produces the maximum number of 1 in its output tape, among all the other TMs with n states. The busy beaver function – namely, the function assigning to n the output of a busy beaver with n states, started on n – grows extremely fast; indeed, it grows faster than any computable function. For example, a 5-states busy beaver,

---

[61] H. Friedman, FOM list: May 24, 2003.

produces (at least) 4098 "1," after 47176870 steps.[62] It is very difficult to label any n-busy beaver as natural (even for little value of n). Furthermore, one cannot just escape by means of arguing that the result is counterintuitive, because the general aim of Friedman's proposal is exactly to clarify what naturalness in mathematics means; there is no room here for the Hegel's motto we cited before.

On the other hand, taking tininess as a necessary condition for naturalness means defending the following thesis: that any function we would count as natural may be executed by a TM with few states. As a philosophical consequence for this stance, we would have a sort of computationalist dream: a clear mirroring between parameters of naturalness for human beings and simplicity in the architecture of TMs. Once again, the practice unveils a rather different landscape. For instance, the search for optimal algorithm (in terms of the number of states) is often anything but a trivial task.

In light of these difficulties, one may ask for a weaker interpretation of Friedman's proposal. Instead of looking at tininess (or other formal features) as something that fixes our concept of naturalness in a rigid way, we can maybe just take it as indicators of a much more general and quite incompressible phenomenon. This is precisely our belief. So, in order to gain a wider point of view, it is time to go back to the dark side of naturalness: the dynamic one.

14.4.2.4 Dynamicity and the Missing Ingredient: Normativity

As we tried to show, any attempt to formulate a static account for naturalness leaves out various contextual aspects that profoundly characterize the phenomenon in its generality. Indeed, the (possible) lack of naturalness of intermediate degrees cannot be fully explained without considering contextual relations, like the intercourse with some other mathematical theory, or the construction of a notion of admissibility (in terms of naturalness) of mathematical proofs, on which we can evaluate the priority arguments. Furthermore, one has to admit that this large class of contextual relations exceeds the pure mathematical domain. For instance, in order to describe why K-triviality may be regarded as natural, it is necessary to consider the evolution and the relevance of some mathematical concepts in time (like the notion of randomness), while the simplicity of TMs does not always reflect our feelings on what makes a function's behavior natural or unnatural.

All of this may lead to the formulation of the thesis, according to which naturalness of a mathematical object depends on the sum of all the contextual relations that an object has with respect to other pieces of mathematics. This definition is not completely new. It recalls Chow's image of a graph of similarities that we quoted in our methodological section, and also, a graph structure for naturalness (formulated in order to take into account its dynamic components) is sometimes evoked in the FOM debate.

---

[62]The function values are known exactly only for $n < 5$. This lower bound is given in Marxen and Burntrock (1990).

Nonetheless, such a solution is at the same time persuasive and frustrating. Indeed, this kind of graph has to admit among its edges a gigantic class of different notions of similarity. But if mathematical ones are extremely hard to classify, most of the other ones are simply too vague to be embedded in a coherent model. Moreover, to establish how these multiple notions of similarities interact with each other is by no means clear. For example, once granted that Turing-reducibility is a natural notion, how is it possible that a notion completely based on it, that of intermediate degree, appears to be unnatural (as argued by supporters of LoN_1 and LoN_2)? To solve these complications, a possible hint may be found in this comment by Shoenfield:

> There is one common feature of the above uses of natural: the assumption that whatever natural means, what is natural is good. I see some signs of this assumption in some of the communications to FOM, which seem to argue that if there is no natural intermediate r.e. degree, then there is something wrong or deficient in the study of r.e. degrees. To justify such a claim, it is not merely necessary to explain the meaning of natural; it is necessary to explain why lack of this type of naturalness is a deficiency in a theory.[63]

We assist, here, to a shift analogue to something we already assist to. In our methodology, while debating which kind of tools could prevent preserved us from both naturalism and Philosophy-first approaches, we proposed a subtle but significant change to our main question: from "What is naturalness?" to "What is naturalness for?" Or, rather, "Why mathematicians make use of naturalness?" Shipman's formulation of this very same question can be stated as follows: "Why do we ask for naturalness in our mathematical theories?"

Brought to the extreme conclusion, Shipman's suggestion, depurated of its provocative aspects, is very poignant. In this perspective, naturalness might be considered as a request posed to our theories, rather than an attribute of the elements of the theories themselves. As we can see, even in the presence of our graph, judgments on naturalness are, most of all, requests about the ways in which we do mathematics.[64] Consequently, a graph of similarities in which the edges are stably given would be deprived of a crucial element for the comprehension of naturalness, namely, normativism.

It is now clear, after this *tour de force*, that our proposal is precisely to consider the normative aspects as decisive. If we admit that to refer to a portion of mathematics or a mathematical object as "natural" is an operation that includes an element of normativity in itself, then any general request for a criterion to distinguish, once for good, between natural and unnatural characteristics would become meaningless. Thus, even though some very local border of the geography given by our graph comes to be persuasive and accepted enough to be considered stable – recall what we said about natural numbers in the introduction – most

---

[63] Shoenfield, FOM list: November 3, 1999.

[64] This aspect of naturalness is also linked with a commonsense meaning of naturalness that points to our habits and the familiarity we have, in this context, with some pieces of mathematics. As we will see in the last section of this chapter, this temptation to reduce unfamiliar to familiar aspects of our mathematical work goes hand in hand with an even stronger attitude towar mathematics.

of the judgments of naturalness are still the outcome of a continuous process of negotiation; a negotiation that takes place fully inside the practice. Thus, this geography is, essentially, in movement.

Therefore, we might conclude that to call a piece of mathematical discourse "natural" is a normative operation that consists in connecting it with something already labeled as "natural." It is important to notice that this kind of operation is, to some extent, a sort of baptism – i.e., neither inherent properties of the object that we call "natural" (i.e., static aspects) nor the relations it establishes with similar ones (i.e., dynamic aspects) can be taken as full justifications for its naturalness. In other words, the process of labeling something as "natural" is not purely descriptive.

Whenever this process becomes critical, as in the case of intermediate degrees, we have a crisis of naturalness. So naturalness is not to be intended as a specific attribute of our mathematical objects but rather as an issue about our mathematical practice. Then these crises cannot be solved, thanks to the elaboration of a "stable" theoretical frame. If we want to fix these contradictions, then we have to go back to the practice.

In conclusion, we suggest that naturalness should be considered as a device of self-regulation within mathematical practice, a device that through a dynamic and communitarian process informs us of the ways in which we want this practice to be performed. At this point, one may believe that we are fallen back into a form of naturalism, for which a mathematical problem, such as the one of naturalness, cannot be solved in a philosophical dimension. But it would be a wrong interpretation, since what we tried to show, with the particular analysis of naturalness, is instead precisely that mathematical practice is continuously exposed to philosophical issues that address it and shape it. This latter aspect is what we want to elucidate in the conclusion of this chapter.

14.4.3 Philosophy in Mathematical Clothing

By means of testing our case studies on the static-dynamic dichotomy, we showed that the notion of naturalness has a mostly dynamic character. Moreover, as a by-product of our analysis, we highlighted a normative component in the judgments of naturalness. This is to say that a piece of mathematics is natural when it fits with a background idea that is chosen to be relevant. In the case of the concept of set, the background ideas were the logical principle of comprehension (according to Cantor and Dedekind) and, later on, Zermelo's quasi-categoricity theorem in terms of a cumulative hierarchy (according to Gödel and Boolos). In the case of Bagaria's criteria for natural axioms, the naturalness of the bounded forcing axioms is guaranteed by some background ideas that inform set theoretical practice and that, focusing on the ongoing research, determine the relevant problems. Finally, in the case of c.e. intermediate degrees, the naturalness conditions can be properly understood only through a dynamic scenario, in which the (possible) existence of these degrees is considered with respect to other mathematical theories already labeled as natural.

In conclusion, it might be interesting to reconsider the meaning of the word "natural," as defined by the Oxford dictionary, in the light of our analysis: if "natural" is something "existing in or derived from nature; not made or caused by humankind," can we fit this definition with the use of the word that emerges in mathematical practice? In other words, how to harmonize such a reference to a given and independent existence, that is something conceptually close to a realistic stance, with the dynamic and normative components that we have unveiled in the concept of naturalness as used within the mathematical discourse?

These two meanings seem hard to reconcile. This apparent incompatibility, though, can be solved if we consider the philosophical aim for defining something as natural in mathematics. We believe that this aim is orthogonal to the features we described. Indeed, the dynamic and prescriptive characters of naturalness on which we have been focusing point in the direction of a nontransparent character of the notion. In other words, when something is labeled as "natural," the reasons are to be found in rational arguments, and these latter do not gain strength by the inevitability of a definition or by the evidence of a concept but rather on the ground of the context where the appeal to naturalness is given. Of course, success and fruitfulness play a role, but to put it in Cantor's words: "every mathematical concept carries within itself the necessary corrective: if it is fruitless or unsuited to its purpose, then that appears very soon through its uselessness and it will be abandoned for lack of success.[65]" Hence, which are the rational arguments and the philosophical reasons, besides success, that might satisfactorily explain the widespread use of the notion of naturalness within the mathematical discourse?

We believe that the answer to this question might reconcile the opposition between mathematical naturalness and commonsense naturalness. Nonetheless, in order to properly tackle the issue and develop it fully, another article would be needed; thus, we chose to leave these conclusions in the form of a suggestion that may trigger further research.

In the end, it seems important to us to reassert that the philosophical ideas that we found in the analysis of the use of "naturalness," far from being metaphysical nonsense, consistently inform mathematical practice. They show that a wide conceptual framework is essential to mathematics, and they call for a global philosophical approach.

References

Bagaria, J. 2000. Bounded forcing axioms as principles of generic absoluteness. *Archive for Mathematical Logic* 39(6): 393–401.
Bagaria, J. 2004. Natural axioms on set theory and the continuum problem. *CRM Preprint* 591: 19.
Benci, V., and M. Di Nasso. 2003. Numerosities of labelled sets: A new way of counting. *Advances in Mathematics* 173: 50–67.

---

[65]Zermelo (1932), p. 206.

Benci, V., M. Forti, and M. Di Nasso. 2006. An aristotelian notion of size. *Annals of Pure and Applied Logic* 143: 43–53.
Boolos, G. 1971. The iterative conception of set. *Journal of Philosophy* 68(8): 215–231.
Cooper, B. 2004. *Computability theory*. Boca Raton: Chapman & Hall.
Corfield, D. 2004. *Towards a philosophy of real mathematics*. Cambridge: Cambridge University Press.
Ewald, W. 1996. *From Kant to Hilbert: A source book in the foundations of mathematics*, vol. II. Oxford: Clarendon Press.
Frapolli, M. 1991. Is Cantorian set theory an iterative conception of set? *Modern Logic* 1(4): 301–318.
Friedberg, R. 1957. Two recursively enumerable sets of incomparable degrees of unsolvability. *Proceeding of National Academy of Science of the United States of America* 43: 236–238.
Hallet, M. 1984. *Cantorian set theory and limitation of size*. Oxford: Clarendon Press.
Hodges, W., and S. Shelah. 1986. Naturality and definability i. *Journal of the London Mathematical Society* 33(2): 1–12.
Jané, I. 2005. The iterative conception of sets from a Cantorian perspective. In *Logic, methodology and philosophy of science. Proceedings of the twelfth international congress*, ed. P. Hájek, 373–393. London: King's College Publications.
Kleene, S., and E. Post. 1954. The upper semi-lattice of degrees of recursively unsolvability. *Annals of Mathematics* 2(59): 379–407.
Koepke, P. 2009. Naturalness in formal mathematics. Preprint
Lask, E. 1914. *Fichtes Idealismus und die Geschichte*. Mohr.
Maddy, P. 1997. *Naturalism in mathematics*. Oxford: Clarendon Press.
Maddy, P. 2005. Three forms of naturalism. In *The Oxford handbook of philosophy of mathematics and logic*, ed. S. Shapiro, 437–459. New York: Oxford University Press.
Maddy, P. 2007. *Second philosophy: A naturalistic method*. Oxford/New York: Oxford University Press.
Mancosu, P., and J. Hafner. 2005. The varieties of mathematical explanation. In *Visualization, explanation and reasoning styles in mathematics*, ed. P. Mancosu, K. Jorgensen, and S. Pedersen, 215–250. Dordrecht/Norwell: Springer.
Marxen, H., and J. Burntrock. 1990. Attacking the busy beaver 5. *Bulletin of the European Association for Theoretical Computer Science* 40: 247–251.
Molinini, D. 2011. Toward a pluralist approach to mathematical explanation of physical phenomena. PhD Thesis.
Muchnik, A. 1956. On the unsolvability of the problem of reducibility in the theory of algorithms (Russian). *Doklady Akademii Nauk SSSR* 108: 194–197.
Nies, A. 2009. *Computability and randomness*. Oxford/New York: Oxford University Press.
Parsons, C. 1977. What is the iterative conception of set? In *Logic, foundations of mathematics, and computability theory*, ed. R. Butts, and J. Hintikka, 335–367. Dordrecht: D. Reidel.
Post, E. 1944. Recursively enumerable sets of positive integers and their decision problems. *Bulletin of the American Mathematical Society* 50(5): 284–316.
Rogers, H. 1967. *Theory of effective functions and recursive computability*. New York: McGraw-Hill.
Russell, B. 1903. *The principles of mathematics*. Cambridge: Cambridge University Press.
Sider, T. 1996. Naturalness and arbitrariness. *Philosophical Sutdies* 81: 283–301.
Sklar, A. 1973. Random variables, joint distribution functions, and copulas. *Kibernetika* 9(6): 449–460.
Soare, R. 1987. *Recursively enumerable sets and degrees*. Berlin/New York: Springer.
Stegeman, A., N.D. Sidiropoulos. 2007. On Kruska's uniqueness condition for the Candecomp-Parafac decomposition. *Linear Algebra and its Applications* 420(2–3): 540–552.
Steinhart, E. 2002. Why numbers are sets. *Synthese* 133(3): 343–361.
Tappenden, J. 2005. Proof style and understanding in mathematics I: Visualization, unification and axiom choice. In *Visualization, explanation and reasoning styles in mathematics*, ed. P. Mancosu, K. Jorgensen, and S. Pedersen, 147–214. Dordrecht/Norwell: Springer.

Tappenden, J. 2008a. Mathematical concepts and definitions. In *The philosophy of mathematical practice*, ed. P. Mancosu, 256–275. Oxford/New York: Oxford University Press.

Tappenden, J. 2008b. Mathematical concepts: Fruitfulness and naturalness. In *The philosophy of mathematical practice*, ed. P. Mancosu, 276–301. Oxford/New York: Oxford University Press.

Van Heijenoort, J. 1967. *From Frege to Gödel: A source book in mathematical logic, 1879–1931*. Cambridge: Harvard University Press.

Wang, H. 1974. The concept of set. In *From mathematics to philosophy*, 181–123. New York: Humanities Press.

Weinert, H.J. 2004. Review of *The theory of semirings (with applications in mathematics and theoretical computer science)* by J.S.Golan. *Bulletin of the American Mathematical Society* 30(2): 313–315.

Weismann, F. 1951. Analytic-syntethic III. *Analysis* 11(3): 49–61.

Zermelo, E. 1932. *Georg Cantor. Gesammelte Abhandlungen mathematischen und philosophischen Inhalts*. Springer.

Zermelo, E. 1967. Investigations in the foundations of set theory I. In *From Frege to Gödel*, 199–215. Harvard: Harvard University Press.

Chapter 15
An Inquiry into the Practice of Proving in Low-Dimensional Topology

Silvia De Toffoli and Valeria Giardino

15.1 Introduction

Philosophy of mathematics has recently become more attentive to the practice of mathematics and in particular to the everyday work of mathematicians. One reason behind this "practical turn" is that mathematics should be acknowledged not only as an abstract science, but also as a human enterprise with its own dynamics that still need to be investigated in depth. Accordingly, it is common today to refer to the "philosophy of mathematical practice" (Mancosu 2008).

Setting our research in this context, our starting point is the analysis of some of the material representations used in the practice of mathematics. In Thurston's words, mathematicians

> use wide channels of communication that go far beyond formal mathematical language. They use gestures, *they draw pictures and diagrams*, and they make sound effects and use body language. (Thurston 1994, p. 166, *emphasis added*)

Among the wide variety of externalizations used by experts to convey and practice mathematics, only some of them are material and therefore easily shared, inspected and reproduced. Such material representations are introduced in a specific practice, and once they enter into the set of the available tools, they have in turn an influence on the very same practice. This process plays a significant role in

---

S. De Toffoli (✉)
Technische Universität Berlin, Institut für Mathematik, Straße des 17. Juni 136, 10623 Berlin, Germany
e-mail: toffoli@math.tu-berlin.de

V. Giardino
Département d'Etudes Cognitives, Ecole Normale Supérieure, Institut Jean Nicod (CNRS - EHESS - ENS), PSL Research University, Paris, France
e-mail: Valeria.Giardino@ens.fr

the practice of mathematics. One of the aims of this article is to investigate the conditions for its manifestation. Moreover, we will analyze what kind of cognitive abilities are triggered by the use of pictures in low-dimensional topology. In fact, we claim that a specific kind of imagination comes into play when dealing with visual representations in this field, which we label *manipulative* imagination. This notion will be used to characterize what it means to "see" in topology. We will propose that reasoning in low-dimensional topology is based on preexisting cognitive capacities—mathematicians imagine a series of possible manipulations on the representations they use—and is modulated by expertise: representations are cognitive tools whose functioning depends in part from preexisting cognitive abilities and in part from specific training.

Moreover, the actual practice of proving in low-dimensional topology involves a kind of reasoning that cannot be reduced to formal statements without loss of intuition. In this sense, visualization plays a specific epistemic role in this practice. We will show several examples of reasoning which are representationally heterogeneous, i.e. neither entirely propositional nor entirely visual.[1] This form of reasoning is shared by experts: it is the kind of reasoning that one has to master to become a practitioner. Moreover, the manipulations allowed on the representations—what we will define as *permissible* actions—as well as the representations themselves are epistemologically relevant. This is because they are integral parts both of the reasoning and of the justification provided. Inferences involving visual representations are permissible only within a specific practice and in this sense context dependent: this leads to the establishment of local criteria of validity.

In Sect. 14.2.2, we will introduce simple examples of reasoning in topology involving pictures and text in order to make the reader acquainted with various representations. In Sect. 14.4.1, we will focus on a specific proof in low-dimensional topology: Rolfsen's demonstration of the equivalence of two presentations of the Poincaré homology sphere. To do so, we will introduce the mathematical background only to the extent needed in this context. In Sect. 14.4.2, we will analyze our case study. We will discuss the form of topological arguments, in particular the role of pictures in topological reasoning. Moreover, we will consider how the notion of "seeing" in topology depends on our spatial-motoric intuitions of three-dimensional space and can be characterized using the notion of *manipulative* imagination. Lastly, we will analyze the reliability of this specific practice. In Sect. 14.4.3, we will sum up our conclusions and present possible lines of further research.

---

[1] Examples of representationally heterogeneous arguments can be found also in other areas. For example, Shin (2004) describes the project of funding a heterogeneous logic by Barwise and Etchemendy (1996) as a very fruitful one.

15 Proving in Low-Dimensional Topology

15.2 Reasoning in Topology

Topology is a branch of geometry that focuses on qualitative properties of objects, while ignoring quantitative ones. In order to represent a topological object, we have to choose one of its particular geometric shapes. Low-dimensional topology focuses on the study of objects of dimension four or less. It is particularly interesting to investigate the practice of proving in this subfield because, as we will see, it is deeply influenced by our intuition of space.

Let us start with an example in dimension two. Representations of surfaces can be "manipulated in space" by exploiting our familiarity in manipulating concrete objects, as if the objects of topology were made of modeling clay. For example, in topology the surface of a cup and that of a doughnut are equivalent: they are both homeomorphic to a *torus*. (In topology, objects are considered up to homeomorphism, i.e. continuous transformations whose inverse is also continuous.) To prove this, we have only to exhibit an appropriate deformation that takes the surface of a cup into the one of a doughnut. Students are trained to "see" transformations such as this and to move freely among different geometric shapes of the same topological object without need to justify these equivalences in other ways. More sophisticated arguments could be used by introducing the machinery of algebraic topology, but this is often not requested by the practice.

For example, the *torus* can be defined as a square with its sides identified. In order to explain how this is possible, let us first analyze a simpler example. Given a square with boundary, that is, a surface homeomorphic to a disk $\mathbb{D}^2$, we glue (i.e. identify) two of its opposite sides in order to obtain another surface. If the two sides are glued in the same direction, as indicated by the arrows in Fig. 15.1a, we will obtain a *cylinder* (Fig. 15.1b). If the two sides are glued in the opposite direction, as indicated in Fig. 15.2a, we will obtain a *Möbius band* (Fig. 15.2b).

In order to obtain the torus from a square, we identify all its four sides in pairs. First, we identify two of them in the same direction, as in the case of the cylinder

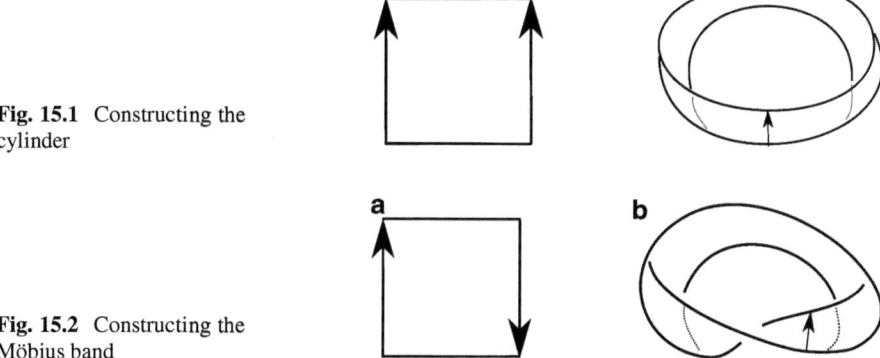

Fig. 15.1 Constructing the cylinder

Fig. 15.2 Constructing the Möbius band

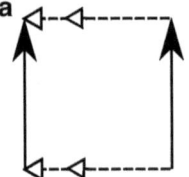 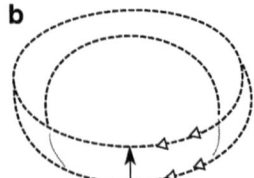 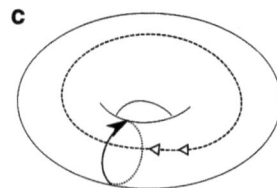

Fig. 15.3 Constructing the torus

(Fig. 15.3b), and then the other two again in the same direction (Fig. 15.3c).[2] In Fig. 15.3c, one can see the torus with two marked curves, where the gluings, i.e. the identifications, were made.

In discussing the role of notation in mathematics, Colyvan takes into consideration these square diagrams with arrows indicating the gluings:

> This algebraic topology notation is something of a halfway house between pure algebra and pure geometry. It is both notation and a kind of blueprint for construction of the objects in question. The first seems to belong to algebra, while the second is geometric. But whichever way you look at it, we have a powerful piece of notation here that does some genuine mathematical work for us. (Colyvan 2012, Ch. 8)

This visual presentation of the torus can be formalized as a quotient of the unit square $\mathbb{D}^2/\sim$. We consider the square in a coordinate system such that the edges are situated in the points $(0, 0)$, $(0, 1)$, $(1, 0)$, and $(1, 1)$ and the equivalence relation $\sim$ identifies points (x, y) and (x', y') according to the following (see Fig. 15.3a):

$$(x, y) \sim (x', y') \Leftrightarrow \begin{cases} (x, y) = (x', y') \text{ or} \\ \{x, x'\} = \{0, 1\} \text{ and } y = y' \text{ or} \\ \{y, y'\} = \{0, 1\} \text{ and } x = x' \end{cases}$$

Without visualizing the transformations from Fig. 15.3a to Fig. 15.3c, which are simplified by the particular notation, one can hardly topologically understand what the above formula defining the equivalence relation is about. In this sense, visualization is essential to the specific kind of understanding proper to topology. The same holds for more complex examples.

For instance, consider, the decomposition of the 3-sphere (the equivalent of the sphere in one more dimension) as the union of two solid tori.[3] Jones uses this

---

[2] If we identify two sides in the same direction and the other two in the opposite direction, we obtain the *Klein bottle*.

[3] A solid torus is a torus that is filled. While a torus is a surface without boundary homeomorphic to $\mathbb{S}^1 \times \mathbb{S}^1$ (i.e. the product of two circles), a solid torus is homeomorphic to $\mathbb{D}^2 \times \mathbb{S}^1$ (i.e. the product of a disk and a circle), and its boundary is a torus.

15 Proving in Low-Dimensional Topology

Fig. 15.4 A circle is homeomorphic to a line plus a point

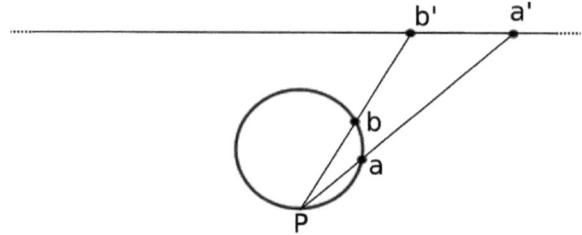

example to emphasize the importance of visualization in low-dimensional topology. He compares a visual presentation with the following formal one: $\mathbb{S}^3 = T_1 \cup T_2$, with

$$T_1 = \{(x_1, x_2, x_3, x_4) \in \mathbb{R}^4 : x_1^2 + x_2^2 + x_3^2 + x_4^2 = 1, x_1^2 + x_2^2 \leq 1/2\}$$

$$T_2 = \{(x_1, x_2, x_3, x_4) \in \mathbb{R}^4 : x_1^2 + x_2^2 + x_3^2 + x_4^2 = 1, x_3^2 + x_4^2 \leq 1/2\}$$

Let us now "see" how the 3-sphere can be filled up by two solid tori beginning with two preliminary remarks.[4]

1) The 1-sphere $\mathbb{S}^1$ can be decomposed into a line $\mathbb{R}$ and point at infinity: Fig. 15.4 illustrates that each point in a circle except one (point P) can be put in correspondence to a point in a line. Point a goes to the point a', b to b', etc. Then, point P will be sent to infinity. Through this map, we get an homeomorphism between the circle $\mathbb{S}^1$ and the line $\mathbb{R}$ with a point at infinity added to it $\mathbb{R} \cup \infty$. Analogously, a sphere $\mathbb{S}^2$ can be obtained from a plane $\mathbb{R}^2$ by adding a point at infinity; and the 3-sphere $\mathbb{S}^3$ can be obtained from $\mathbb{R}^3$ by adding a point at infinity.

2) Given a line, we can rotate it around one of its points, e.g. the origin, and obtain the plane. We can also take a half-line, the one with the origin as its endpoint, and still obtain a plane after the same rotation. Similarly, we can start with a plane or a half-plane and rotate it around a line. For example, the xy plane in the standard coordinates if rotated around the y axis gives rise to the three-dimensional space $\mathbb{R}^3$.

In Fig. 15.5, the 3-sphere is represented as $\mathbb{R}^3 \cup \infty$ and $\mathbb{R}^3$ is the result of the rotation of the plane of the paper (the xy plane) along $\mathbb{R}_y$, the y axis. Let $\mathbb{D}_1$ be a disk (with boundary) in the half-plane $x < 0$; after the rotation of the plane around $\mathbb{R}_y$, it will form a solid torus, intersecting the xy plane in another disk: $\mathbb{D}_2$. In this way we get a representation of the 3-sphere with a solid torus inside it. Now, in order to prove that the 3-sphere can be decomposed as the union of two solid tori, we want to prove that also the complementary space of this solid torus is a solid torus.

Each of the segments depicted in Fig. 15.5 connecting a point of the boundary of $\mathbb{D}_1$ to the corresponding point of the boundary in $\mathbb{D}_2$ will give rise to a disk after the rotation. These disks can be parametrized by the line $\mathbb{R}_y$ plus the point at infinity since each intersects $\mathbb{R}_y \cup \infty$ in its midpoint (we can choose all these segments to intersect the circle $\mathbb{R}_y \cup \infty$ orthogonally). Thus, the complementary space is homeomorphic to $\mathbb{D}^2 \times \mathbb{S}^1$, a disk times a circle, which is a solid torus.

---

[4] The following argument can be found in Fomenko (1997, pp. 123–124).

Fig. 15.5 Decomposition of $\mathbb{S}^3$ in two solid tori

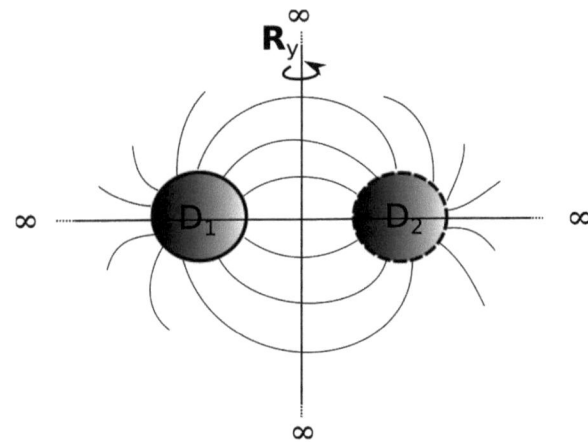

Jones, considering a formal and a visual way of presenting the 3-sphere as the union of two solid tori, claims that

> [the] formal picture [...] is complete but inadequate. If one does not 'see' the other picture [...] one is not ready to take the next step in low-dimensional topology. Of course this is just the beginning. There are more complex things to 'see' and *sequences* of such visions are compounded one upon another *in the same way as the elementary logical steps in a formal argument*. If one 'sees' the pictures, then one understands, but otherwise one cannot follow. In principle one could formalize the whole argument, but *that would add nothing*. (Jones 1998, p. 213, *emphasis added*).

In Jones' view, an argument in this domain can be broken down into units, no matter if they are elementary logical steps or pictures. As we will later see, a formalization of an argument whose units are pictures would often not be relevant for the practice. This is characteristic of topology in general: often such formalizations would hide the relevant (topological) reasoning that is externalized by the pictures.

In the following, we will illustrate a proof in topology as an example of an argument composed of a sequence of pictures plus textual instructions on how to interpret them.

15.3 Rolfsen's Proof

We will now present the core of our case study: a proof of the equivalence of two presentations of the Poincaré homology sphere, taken from a popular graduate textbook: *Knots and Links* by Rolfsen (1976). The first presentation of this 3-manifold[5] is a *surgery code*, while the second one is a *Heegaard diagram*. In order to explain this proof, we will briefly define the representations and techniques used to obtain them.

15.3.1 Dehn Surgery and Heegaard Diagrams

First, let us consider *Dehn surgery* on the 3-sphere. To do so, we introduce mathematical *knots*, which are smooth simple closed curves in the 3-sphere.[6] Knots are considered up to smooth deformations, i.e. we are not interested in their specific geometric shape, but in the way they are knotted.

In *Dehn surgery*, first we take a knot in the 3-sphere, then we thicken the knot to a tube in order to obtain a knotted solid torus. Next, we cut it out from the 3-sphere and glue it back in a different way to obtain another 3-manifold. The way in which the tube is glued back in can be coded by a rational number. Any closed compact 3-manifold can be obtained by Dehn surgery on some knot, i.e. it can be coded by a knot plus a rational number.[7] For example, a code for the Poincaré homology sphere is represented in Fig. 15.6.

To give a hint of the reasons why Dehn surgery is possible at all, consider an analogous process in lower dimension: in Fig. 15.7 we start with a circle (i.e. a 1-sphere, which corresponds to the 3-sphere in our example) and two points on it (i.e. a 0-sphere, which corresponds to the knot, i.e. a 1-sphere); then, we thicken these two points (this corresponds to thicken the knot) and get two segments (instead of a tube). After that we cut them out from the circle and glue them in a different way. We obtain two circles, which is a different topological manifold from one circle, the manifold we started with.

Let us now consider another way to present manifolds: *Heegaard diagrams*. To do so, we introduce *handlebodies of genus g*, which are balls with g handles attached. For example, a solid torus is, a handlebody of genus 1. A handlebody of genus g has g holes.

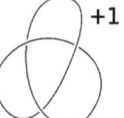

Fig. 15.6 The surgery code for the Poincaré homology sphere

---

[5]A 3-manifold is the equivalent of a surface in one more dimension. It is a three-dimensional topological space which is locally homeomorphic to $\mathbb{R}^3$, the Euclidean space.

[6]In a previous study, we considered the use of knot diagrams in relation to knot types. We claimed that the key feature of knot diagrams is their "dynamicity": experts manipulate them according to different sets of possible transformations (De Toffoli and Giardino 2014).

[7]See Fomenko (1997, Ch. 9) for details.

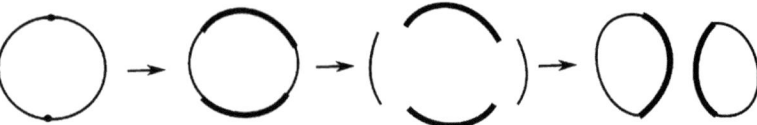

Fig. 15.7 Surgery in lower dimension

Fig. 15.8 A system of meridians of a genus 2 handlebody

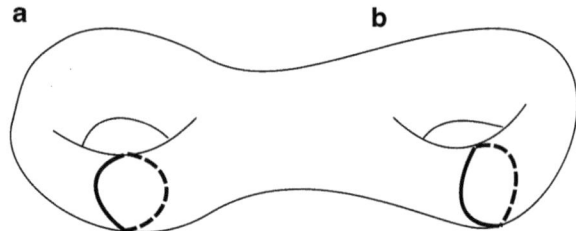

It is possible to construct closed compact 3-manifold by gluing two handlebodies H_1 and H_2 of the same genus along their boundaries ∂H_1 and ∂H_2, which are two orientable closed surfaces of the same genus.

Let us focus on the case of genus two, which is the one that interests us for the proof which we will later analyze. To describe the gluing, i.e. a homeomorphism between the two boundaries, we consider a and b, the meridian loops of the first handlebody, which are two curves in ∂H_1, and the boundary of the first handlebody, as in Fig. 15.8. Then, in order to specify the result of the gluing, it is enough to know $f(a)$ and $f(b)$, the images of these curves under the attaching homeomorphism on ∂H_2, the boundary of the second handlebody. So, all we need to know in order to construct a 3-manifold from two handlebodies of genus two is a pair of simple closed curves in the boundary of a handlebody (these curves are interpreted as coding the gluing, since they are the images of the meridian loops of the first handlebody on the surface of the second.) This information, however, cannot be so easily conveyed, since it would require drawing curves on pictures of three-dimensional objects. To overcome this problem, Heegaard diagrams are introduced, which are two-dimensional diagrams containing all the relevant information. Their two dimensionality makes the presentation easier to draw and can be effective in a sense that we will explore later.

To construct a Heegaard diagram, we imagine cutting open ∂H_2, the boundary of the second handlebody, so that it lays flat on the plane and then we trace the image under this transformation of the curves determining the gluing (these curves were on ∂H_2 and thus will still be represented on this "flat" presentation of ∂H_2). First, we consider the meridian loops of ∂H_2 (call them A and B); then, we cut the surface ∂H_2 along them in order to create a surface with four circles as boundary (for genus g, we will have $2g$ circles), as in Fig. 15.9a. This is equivalent to a sphere with four holes, as in Fig. 15.9b: we imagine "inflating" the object represented in Fig. 15.9a so that it becomes a sphere.

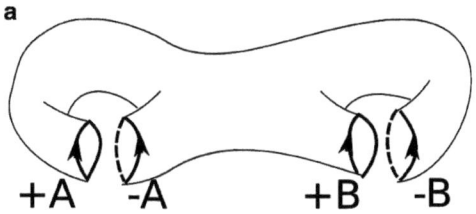 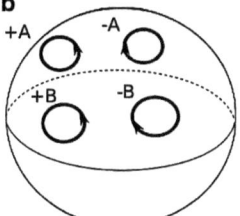

Fig. 15.9 Constructing a Heegaard diagram

Fig. 15.10 "Flat" presentation of the boundary of a genus two handlebody

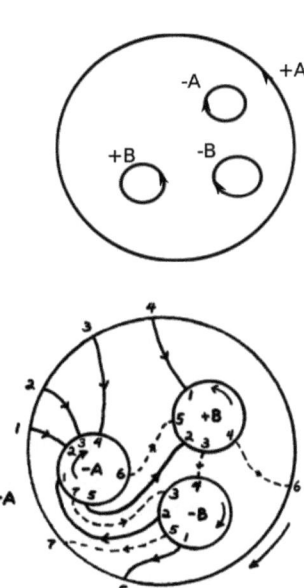

Fig. 15.11 The Heegard diagram taken from Rolfsen representing the Poincaré homology sphere

Then, from one boundary hole, we stretch the surface onto the plane, obtaining a disk with three holes. In Fig. 15.10, we stretched from the hole labeled $+A$.

While applying these transformations, we have to keep track of where the curves $f(a)$ and $f(b)$ defining the gluing are going, i.e. their image under this transformation. After the modification of the boundary of the second handlebody, these curves will be lying in the planar disk with holes. Then, a Heegaard diagram of genus two is a diagram of a disk with three holes endowed with a set of curves; see Fig. 15.11 for an example. Note that the set of lines in this diagram is actually to be interpreted as two closed curves, since the boundary circle are pairwise identified: $+A$ with $-A$ and $+B$ with $-B$, the signs denoting the opposite orientations which we need in order to glue them correctly.[8] In particular, there is a solid curve and a dotted one. Let us follow the solid one. We can start from $(1, +A)$, the point marked 1 in the external circle $+A$; then, we reach 2 in $-A$, but $-A$ is identified

---

[8] See Fomenko (1997, Ch. 5) for details.

with $+A$, so we arrive at 2 in $+A$; then for similar reasons, we reach in the order $(3, -A) = (3, +A)$, $(4, -A) = (4, +A)$, $(1, +B) = (1, -B)$, $(5, +A) = (5, -A)$, $(2, +B) = (2, -B)$, $(1, -A) = (1, +A)$. At the end we have returned at the starting point: the curve is indeed closed.

The information about these closed curves, so encoded, is enough to uniquely determine the gluing of the two handlebodies.

15.3.2 Two Presentations of the Poincaré Homology Sphere

In 1900 Poincaré, in the second supplement to his *Analysis Situs* (Poincaré 1900), had conjectured that a compact 3-manifold with the same homology groups as the 3-sphere would be homeomorphic to it. Five years later, he gave a counterexample constructing a manifold now known as the *Poincaré homology sphere*. This is a compact 3-manifold with the same homology as the 3-sphere but not homoeomorphic to it, because its fundamental group[9] is not trivial.

In Fig. 15.12, we present Rolfsen's direct proof of the equivalence of a surgery code and a Heegaard diagram of the Poincaré homology sphere by reporting the pictures and the accompanying text with the instructions on how to interpret them.[10] This proof (Rolfsen 1976, pp. 249–250) is accepted as a valid argument and, to our knowledge, is the only direct proof of the equivalence of these two presentations. We will discuss this proof in the next section.

We start with Dehn surgery, considering a trefoil knot with associated rational number $+1$; see Fig. 15.6. In the first picture of the proof, we see a thick trefoil forming a knotted tube with a curve indicating Dehn surgery to which another tube with a curve has been added.

The main idea of the proof is to continuously deform this object in order to obtain another known presentation of the same 3-manifold. The transitions between the first four pictures are topologically interpreted as homeomorphisms. While applying these transformations, we are going to keep track of the two curves. Training is normally required to follow these deformations, but we nevertheless hope to convey an idea of what a proof in topology might look like (it is not necessary to be able to follow all the transformations in order to get the gist of the proof).

We have to follow the instructions given in the text in order to identify the various transitions connecting the pictures. For example, to move from the fourth to the last picture, as the knife suggests, we cut the surface open and lay it flat on the plane. This is "almost" a Heegaard diagram: we just have to pick one of the holes and stretch it "outside" in order to obtain one. If we choose the hole labeled $+A$, we get the Heegaard diagram in Fig. 15.11.

---

[9]The fundamental group is a very important algebraic invariant used to study the shape of topological spaces. See Massey (1991, Ch. 2) for details.

[10]We sincerely thank Professor Rolfsen who gave us permission to reuse the material from his book.

The argument is contained in the series of pictures[...] Starting with Dehn's example, we drill out a tube from the trefoil's exterior and add it to the $\mathbb{S}^1 \times \mathbb{D}^2$, making the latter into a handlebody of genus two which is to be attached to the outer part of (1) by the characteristic curves shown. The isotopic deformation (1) – (4) shows that this outer part is also a handlebody. (Since the deformation is done equivariantly with respect to 180° rotation about the vertical, the back of the figure looks exactly like the front). Again the solid and the dashed curves of (4) are directions for sewing a handlebody onto the handlebody on the <u>outside</u> of the surface pictured. Cutting this surface open and flattening with the the inside toward us gives us the pictures (5) in the plane (plus a point at ∞). Moving the point at infinity to be in the hole labelled +A gives us exactly the Heegaard diagram of Poincaré's example.

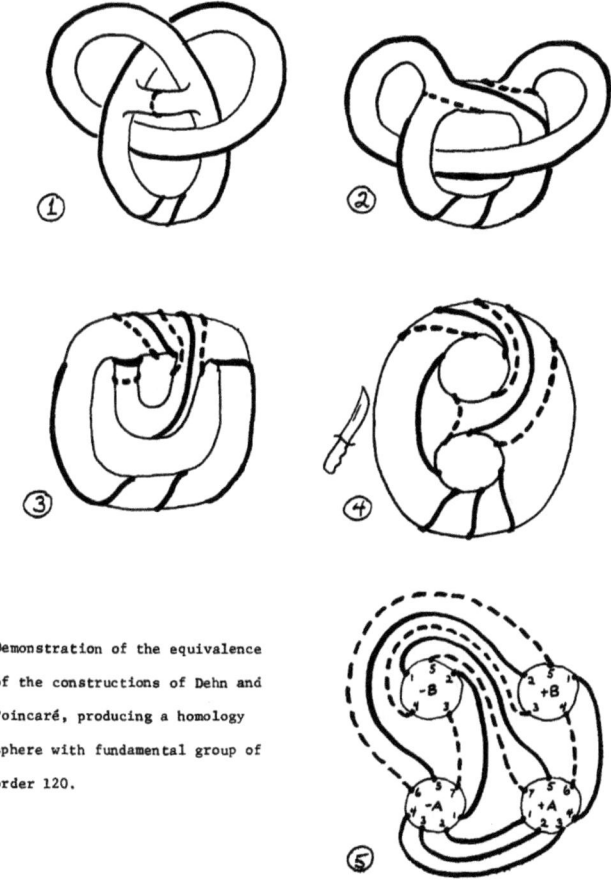

Demonstration of the equivalence of the constructions of Dehn and Poincaré, producing a homology sphere with fundamental group of order 120.

Fig. 15.12 Rolfsen's proof (Rolfsen 1976, pp. 249–250)

15.4 Discussion of the Case Study

What is striking about Rolfsen's proof is that even if it is clearly very far from a formal proof, and also far from a non-formal proof written in natural language, it is nonetheless (accepted as) valid. In this specific example, a formal description of the topological deformations would be beside the point.[11]

However, this should not be surprising, since, as Thurston writes,

> ...we should recognize that the humanly understandable and humanly checkable proofs that we actually do are what is most important to us, and that they are quite different from formal proofs. For the present, formal proofs are out of reach and mostly irrelevant: we have good human processes for checking mathematical validity. (Thurston 1994, p. 171)

The question is: which "human processes" would make mathematicians conclude that this proof is valid? This issue is also crucial to determine what makes low-dimensional topology so special as a subfield of mathematics.

In the following, we consider some of the features of this proof.

15.4.1 Heterogeneous Arguments

One evident feature of Rolfsen's proof is that it consists of pictures *and* text. The sequence of pictures is accompanied by a set of instructions given in natural language. Without the pictures, it would be impossible to understand the text; conversely, without the text, it would be very hard to correctly interpret the pictures: neither one is complete without the other. The argument requires both for its cogency.[12] Thus, the proof is not purely visual, for the same reason for which it cannot be considered as purely syntactic: it is representationally heterogeneous. In order to understand it, one has to be guided by the text on how to imagine the continuous transformations whose discrete steps are represented by the pictures.

The interdependence of pictures and text is not an exclusive feature of proofs in low-dimensional topology. Manders (2008) has discussed in depth how diagrammatic and propositional content are interacting and are both essential in the case of Euclidean geometry. Nevertheless, there are some differences between this proof (and others) in topology and most proofs in Euclidean geometry. For example, in the case of Rolfsen's proof, once the context, which is linguistically defined, is understood and the goals are set, then the text becomes superfluous

---

[11]This is not an isolated case. Similar phenomena can be observed also in research articles, for example, in the subfield connecting low-dimensional topology and hyperbolic geometry. See, for reference, Adams and Reid (1993).

[12]In our view, pictures do not always have to be physically drawn: it is sometimes sufficient to imagine them. Nonetheless, we will also argue that the fact that they can be physically drawn has cognitive advantages, since it makes inspection and reconfiguration easier and is the condition for sharing content with other practitioners.

for our understanding. That is to say that once we know what *to do* with the pictures—and this can only come along with expertise—we can concentrate on them exclusively. As Sullivan (2013) writes about this same proof: "you can't say you really understand the proof until you reach the stage where the pictures alone would suffice." The only place where we would still need instructions in words would be to go from the fourth to the fifth picture, but in this case, the instruction "cut" is substituted by the icon of a knife. This does not happen in Euclidean proofs, where the text guides us through the interpretation of the diagrams step by step and gives us necessary information that the picture alone does not provide (the text describes the articulated construction required to draw and work with Euclidean diagrams).

Let us now return to the argumentative form of Rolfsen's proof. Recall Jones' quotation in Sect. 15.2, where he claimed that "sequences of [...] visions are compounded one upon another in the same way as the elementary logical steps in a formal argument." Accordingly, we have seen that the majority of the steps into which Rolfsen's proof can be broken down are transformations that lead from one picture to another, as the text giving the instructions on how to interpret the pictures clarifies. Therefore, the representations give a material form to the transformations (and in this sense they "externalize" them) because they allow experts to perform "epistemic actions" (and in particular to draw inferences) on them. By *epistemic actions*, we mean actions that are performed with an epistemic aim.[13] Moreover, these actions are controlled by the shared practice: the set of legitimate transformations is limited and determined by the context.

As Larvor explained in a recent article by comparing different forms of mathematical arguments:

> ... if an argument includes an inferential action that manifests or manipulates the subject-matter, or a representation thereof, then formalising this argument in a general logical language must either misrepresent or fail to include this action. Moreover, we can say something in the direction of explaining how informal arguments work as arguments: they are rigorous if they conform to the controls on *permissible actions* in that domain. (Larvor 2012, p. 724, *emphasis added*)

Through an inferential action, it is possible to manipulate the representation itself for epistemic purposes. In Rolfsen's proof, the inferential actions consist in manipulating pictures; for this reason, any formalization would fail to capture the inferential actions performed. Moreover, we can proceed from one representation to another by applying actions that are permissible, i.e. allowed and controlled by the shared practice of the subfield. These epistemic actions are inferential actions: they constitute the units of a mathematical argument.

The commented pictures in Rolfsen's proof—*in sequence*—are thus the argumentative form of the proof. The representations constituting the argument are heterogeneous and yet adequate to its mathematical context: pictures thus play a

---

[13]*Epistemic actions* have been characterized by Kirsh and Maglio (1994) as "actions performed to uncover information that is hidden or hard to compute mentally" in contrast to *pragmatic* actions, which are "actions performed to bring one physically closer to a goal."

relevant epistemic role. This is in line with the actual practice of mathematics, as the first words of the proof makes clear: "The argument *is contained in* the series of pictures on the next page..."

15.4.2 Representations Externalizing Reasoning

We propose to consider the pictures in Rolfsen's proof as externalizing part of the reasoning: in order to grasp the validity of this proof, it is necessary to imagine the manipulation that leads from one to the other. Moreover, it is the very use of these external representations that triggers *manipulative* imagination, which is, as we will argue, crucial in topology.

This supports an approach to mathematics, according to which mathematical representations and symbols are intimately linked to the concepts they represent. De Cruz and De Smedt (2013, p. 4) claim that "symbols are not merely used to express mathematical concepts" but are "constitutive of the concepts themselves. Mathematical symbols enable us to perform mathematical operations that we would not be able to do in the mind alone, they are epistemic actions."

In Rolfsen's proof, the sequence of pictures externalized the reasoning, allowing us to "see" the transformations involved. These pictures follow some convention, but their pictorial features are also relevant. For this reason, they cannot be considered as purely symbolic, but can be recognized as hybrid representations presenting symbolic as well as visual elements, which are both to be taken into account by the experts. For example, in the last picture of the proof, the curves are represented as lines on the diagram that are the images of the lines of the previous pictures under the applied transformations, but at the same time, they must be interpreted as codes for gluings.

Not only in topology, but also in different mathematical fields, visual and symbolic elements come in different and often complementary degrees. A notation for which a syntax is explicitly defined can reach a higher degree of abstraction and therefore allow for wider generalization. Nonetheless, most of the times, this happens at the cost of losing a straightforward intuitive interpretation and in some cases the very possibility of exploiting intuition.[14]

A paradigmatic example of this is the use of closed curves to represent syllogistic reasoning in its development from Euler to Shin. Peirce's (as well as Shin's) introduction of new conventions has increased the diagrams' expressive power, but at the expense of the visual clarity and the intuitive interpretation of Euler's original diagrammatic system. The new conventions are more arbitrary and the new representations more confusing (Shin, Lemon and Mumma 2013).

---

[14]Of course, as Giaquinto suggests, it could be that we develop more sophisticated forms of intuition and imagination allowing us to manipulate also arrays of symbols, or syntactic expressions in general (Giaquinto 2007, Ch. 12). We tend to agree with this claim.

Let us go back to Rolfsen's proof, taking into account this balance between visual and symbolic elements. The topological pictures represent geometric objects that are straightforwardly interpreted as topological objects. On the contrary, in the case of Heegaard diagrams, the representation of a topological object is more codified, so as to allow for a stronger syntactic control. As a notation,[15] Heegaard diagrams have the potential to allow for generalizations: we can code all closed compact 3-manifolds with them. Nonetheless, the price for this is that the interpretation of the representations cannot be driven anymore mainly by intuition. We have to be aware of the conventions introduced in order for the diagram to "make sense".

Two examples will help to better clarify how these degrees of intuitive and conventional elements converge in different mathematical diagrams and how algebraic and symbolic reasoning can interact. (1) Knot diagrams present clear visual elements because they "intuitively" represent geometric objects, but at the same time, they allow for a syntactic control (through local moves specifically defined on them) (De Toffoli and Giardino 2014). (2) Commutative diagrams of homological algebra display a more evident syntactic component: these diagrams no longer describe geometric objects but abstract structures and relations. Nevertheless, their arrangement in space is essential and thus visual features also characterize them and allow us to "manipulate" them effectively (De Toffoli, Diagrams in homological algebra, manuscript).

Furthermore, it is clear that not all representations of the same topological object would have the same degree of effectiveness in giving a material form to the relevant reasoning and thus in promoting inference: specific conditions have to be met. First of all, different presentations are suitable for different purposes. Not only we do observe different degrees of symbolic and visual elements, but for a given mathematical object there can be more or less effective representations.

Let us consider a specific example to clarify: diagrams representing the torus. We can choose among various possibilities. On the one hand, Fig. 15.3c is a classical diagram of a torus (in this specific picture two curves are added to it), where just a few lines are easily interpreted as a three-dimensional object. A more detailed picture, for example, depicting thickness or shadows, would make the representation more similar to the corresponding material object, but would be less useful since it would distract the viewer from the essential topological features of the object by adding "noise."[16] The connection between topological and material objects is crucial in topology, but only in so far as it stimulates topological imagination that takes inspiration from the one used to manipulate concrete objects but afterwards develops independently. The similarity with concrete objects has its limits: we must be able to detach ourselves from this analogy and perform an abstraction in order to extract relevant topological features. On the other hand, Fig. 15.3a is definitely

---

[15]Two features of mathematical diagrams allow for an interpretation of them as a notational system: (i) they follow certain conventions and are a codified way to present different mathematical concepts and (ii) they can be used in sequences and constitute a system through which it can be possible to "calculate" effectively.

[16]See Sullivan (2013) for a survey of different mathematical, in particular topological, pictures.

more complicated for novices (e.g., one must know what the arrow signifies to correctly interpret the diagram), but nevertheless it can be useful. As Colyvan (2012, Ch. 8) claims, this notation presents both geometric and algebraic features (which can be exploited in different ways). A side remark is that the consideration of such representations leads to a rejection of a sharp distinction between algebraic and geometric reasoning in mathematics (Giaquinto 2007, Ch. 12). The abstract character of this hybrid notation opens the way to new discoveries; in fact, it leads to generalizations: not only does the same notation allow us to present the cylinder, the Möbius band, the torus, and the Klein bottle, but if we generalize the notation and instead of a square we take a polygon with n sides, then we can present every closed compact surface (see, for instance, Massey 1991, Ch. 1). Moreover, Colyvan points out at the construction of the Klein bottle. The Klein bottle is straightforwardly presented with the square diagrams (we just have to invert one arrow in the diagram of the torus) and in a sense it is the very notation that drives us toward the study of a surprising new object that is two dimensional, but cannot be embedded in three-dimensional space and does not have an outside or inside.

This example illustrates how for topological diagrams, by adding more syntactic elements we often get a more powerful notation from an expressive point of view and at the same time we lose the analogy with material objects.[17] This is not to say that some diagrams are better than others in principle: it depends on the specific context and on the particular purpose for which they are used. The possibility of choosing among a wide variety of representations enriches the set of tools available to the mathematician.

A final remark about representations concerns their materiality. To avoid confusion, we have to keep in mind the distinction between the material pictures and the imagination process, which, especially in the case of trained practitioners, tends to vanish. Actual topological pictures trigger imagination and help us see modifications on them, but experts may not find it necessary to actually draw all the pictures. The same holds for algebra, where experts skip transitions that non-trained practitioners cannot avoid writing down explicitly. This does not mean that experts do not need pictures to grasp the reasoning, but only that, thanks to training and thus to their familiarity with drawing and manipulating pictures, they are sometimes able to determine what these pictures would look like even without actually drawing them. More generally, for each subfield it would be possible to define a set of "background pictures" that are common to all practitioners. This set of pictures determines what Thurston calls a "mental model" for a group of mathematicians in a subdiscipline (Thurston 1994, pp. 174–175). For instance, any knot theorist knows without need for material pictures what a diagram of the trefoil knot or of the figure-eight knot looks like. To give a more sophisticated example, the original proof by Alexander of his famous theorem about the possibility of representing any knot as a braid is a visual argument that requires the use of this type of imagination but does not contain a single picture (Alexander 1923).

---

[17]Of course, these syntactic elements are not arbitrary introduced, but according to specific aims.

15.4.3 "Seeing" in Topology Through Manipulative Imagination

We have just claimed that in order to understand Rolfsen's proof, we need to "see" topological transformations. As the text says, "The isotopic transformation (1) – (4) *shows* that this outer part is also a handlebody." To "see" what is "shown" by the picture is crucial. Even in simpler cases, instructions are useful in making us grasp the transformations, as for example going from Fig. 15.9a to Fig. 15.9b.

Therefore, in low-dimensional topology, following an argument, let alone being able to construct one, often requires "seeing" certain transformations. This method of proving, different from the standard one in other areas of mathematics, which is entirely propositional, consists in providing pictures plus instructions for their interpretation. Yet, what does "seeing" actually mean in this context? We can connect it with intuition. Heinzmann, writing about Poincaré's *Analysis Situs*, traces the need for three distinct types of intuition in what is now topology:

> Defining it as the science of classification of closed surfaces, called later manifolds, with respect to continuous deformations, it requires *geometric intuition* concerning the qualitative property of a n-dimensional manifold, *arithmetic intuition* insofar as he introduced computing with the topological object 'manifold' and, insofar as the strongest classification-criterion is the fundamental group, one needs *algebraic intuition*, too. (Heinzmann 1999, p. 55, *emphasis added*)

Even if the term "intuition" is vague, we can specify what it means in the present context: for low-dimensional topology, it is possible to point at some of the "intuitive" capacities involved. Our proposal is that topologists use a special kind of imagination that does not only involve vision but also spatial-motoric intuition of three-dimensional space. In fact, in Rolfsen's proof, one finds no difficult calculations—at least in the standard meaning of the term—and nonetheless the argument is not easy to follow. In order to understand the proof and check its validity, practitioners have to use their ability to imagine topological transformations *correctly*. For example, they have to interpret the transitions between the pictures as homeomorphisms. As a result, "seeing" in topology means first to interpret the representations coherently with the shared practice and then performing epistemic actions on them. In the case of Rolfsen's proof, these actions take the form of continuous deformations. The interaction with the representations is thus pivotal: mathematicians have to activate this form of imagination in order to use the representations as inferential tools.

Consequently, "seeing" is here to be intended as much more than simple vision for at least two reasons. First, because it exploits some of our spontaneous cognitive abilities such as vision, but has nevertheless to be properly trained inside the practice in order to be correctly applied. Secondly, because mathematicians do not only see particular representations, but also the possible actions that could be performed on them, i.e. the possible transitions between pictures. We chose to label this capacity *manipulative* imagination[18] as a sophisticated form of imagination that derives from our preexisting manipulative capacities with concrete objects and our motor agency in three-dimensional space: it seems to have a spatial-motoric and not a specifically visual nature.[19] An important feature of this form of imagination is that it is not exclusively innate or *a priori*; on the contrary, it needs to be specifically trained by experimenting with the available representations. This is in line with Poincaré's view that "intuition" has to be trained:

> The main goal in teaching mathematics is to develop some faculties of the mind, and among these, intuition is not the least precious.[20] (Poincaré 1889, p. 160)

Moreover, it is crucial to highlight that it is the massive use of this form of imagination that is responsible for the peculiar development of low-dimensional topology, which has been so different from that of other mathematical fields. As Jones (1998, p. 212) writes, in low-dimensional topology, we do not need to formalize every argument for the very reason that we can rely on our intuition. In other fields, this intuition is unavailable, just as it would be unavailable in low-dimensional topology if we were two-dimensional creatures without the imagination of three-dimensional space. In this case, we would have to formalize each argument and low-dimensional topology would be more similar to more abstract areas of mathematics.

15.4.4 Justifications and Criteria of Validity

As we have already discussed, the practitioners of low-dimensional topology "see" the transformations and check whether they are permissible: the representations embody their reasoning and provide at the same time evidence for their conclusions.

Of course, as in any proof, not everything has to be justified. The background knowledge, amounting to the mental models shared by the members of the community to which the proof is addressed, is assumed as already established. Moreover, particular standards of justification and criteria of validity are provided: the permissible actions on the representations are already defined, and they are part of the background material.

In Rolfsen's proof, we saw that among the permissible actions on the pictures are continuous transformations. These are part of the background material in the sense that any topologist knows immediately that these transformations can be interpreted

---

[18]We already used the expression "manipulative imagination" when studying knot diagrams (De Toffoli and Giardino 2014).

[19]Think of the blind mathematician Morin who contributed to the understanding of one of the first actual sphere eversions (Morin and Petit 1980).

[20]We have translated from the original: "Le but principal de l'enseignement mathématique est de développer certaines facultés de l'esprit, et parmi elles l'intuition n'est pas la moins précieuse."

in terms of homeomorphisms. The validity is thus based on the "practice": it is the practice itself that integrates a way of controlling the actions on the representations used, which results in the establishment of local criteria for validity. The responsibility is shared among experts: since in low-dimensional topology different forms of reasoning are employed, some of which are specific to it, purely external criteria of validity cannot exhaust all the criteria actually adopted. As Brown suggests, we should acknowledge the existence of non-formal reasoning in mathematics: "first-order logic may be well understood, but what passes for acceptable proof in mathematics includes much more than that" (Brown 1999, p. 164). If this is true, then, as Larvor has exhaustively discussed, "the cost is that we have to abandon the hope of establishing a general test for validity" (Larvor 2012, p. 723).

In our view, what Thurston refers to as "good human processes for checking mathematical validity" (Thurston 1994, p. 171), are context-dependent processes that in low-dimensional topology rely on our manipulative imagination and more generally on our intuition of three-dimensional space, duly trained according to the specific practice. Furthermore, formal proofs are "out of reach," because in order to obtain reliable formalizations, mathematicians would have to spend all of their time to rewrite already known results and conform them to general standards. As Thurston notes, on a small scale, this is easy to do, but on a large scale, where results are interconnected, we would have to check for the coherence of all the arbitrary local choices of formalization. To do so would require a huge amount of time, and topologists are not willing to undertake such a project.

This does not mean that in principle any proof in topology could not be translated so as to assume a propositional and more formal form – even if to do so would be a "nightmare," in Jones' words (Jones 1998, p. 212).[21] The point is that it is not usually done. Nevertheless, as we have mentioned already, part of the confidence of the practitioners is based on the knowledge of how to convey visual modifications in more formal expressions. For example, "gluing" in topology is straightforwardly interpreted in terms of quotient spaces and "deforming continuously" in terms of homeomorphisms. In Thurston's words:

> When people are doing mathematics, the flow of ideas and the social standard of validity is much more reliable then formal documents. People are usually not very good in checking *formal correctness* of proofs, but they are quite good at detecting potential weaknesses or flaws in proofs. (Thurston 1994, p. 169)

We would also like to mention that once we accept that the argumentative form is based on such externalizations of reasoning, given by the specific representations used and the control of the practice, then the process of discovery and that of justification seem to occur at the same time.

Let us now turn to the issue we addressed at the end of the previous section, that is, to the reasons why low-dimensional topology is such a special subfield of mathematics. We have analyzed Rolfsen's proof as a paradigmatic example of

---

[21] Jones is taking the example of formalizing a proof in knot theory.

an informal argument that can be given in low-dimensional topology, where the sequence of pictures embodies and at the same time justifies the reasoning. Any version of the proof without the pictures, let alone a formal version, would not be able to externalize the reasoning and trigger manipulative imagination and thus would completely obscure the topological permissible actions. By this feature, the case of low-dimensional topology seems to be quite distant from other areas of mathematics. Moreover, this form of reasoning is epistemologically relevant, as Jones makes clear:

> One of the interesting consequences of the use of three-dimensional intuition is that the field of low-dimensional topology has advanced in a way that is significantly different from other branches of mathematics. One is expected to "see" results in this field, and once the result, or partial result, has been "seen," it requires no further discussion. (Jones 1998, p. 212)

At this point, we can interpret Jones' claim in the light of the proposed interpretation. First, manipulative imagination is the cognitive process that modulates our three-dimensional spatial-motoric intuition in relation to the particular mathematical context. Second, thanks to manipulative imagination, we have at our disposal a set of permissible actions.[22] Pictures indicating the stages of transformations, plus instructions explaining how to interpret them, can count as justifications. Exclusively linguistic proofs, and formal proofs are thus just a small portion of the proofs accepted as valid. Of course, it is still possible to translate visual arguments into formal ones. Nevertheless, as Jones and Larvor suggest, the formal version might be complete, but it remains inadequate. As a consequence, once we accept the existence of arguments structured in sequences of pictures, we realize that although there might be good reasons to reduce the reasoning to formal statements, this move would add nothing to the topological reasoning behind the argument.

A practitioner of low-dimensional topology uses material representations, which are the condition for sharing content among the experts. These representations must be adequate to externalize reasoning and to trigger manipulative imagination so as to allow performing permissible and effective inferential actions. To establish the validity of an argument, a low-dimensional topologist shares the responsibility with other practitioners. That is, the community defines criteria of validity specific to their subfield: this is part of the normative structure of this practice.

15.5 Conclusions

In this article, we aimed to show that attention to how low-dimensional topology is practiced gives new insight on the use of mathematical representations. In particular, we unveiled some of the reasons why reasoning with these representations can be seen as an essential part of doing mathematics.

---

[22]It would be misleading to conceive this set as fixed once and for all, since it varies according to the context of use.

Mathematicians rely on an astonishing variety of proving practices, beyond the one analyzed here. As Larvor suggests (Larvor 2012, p. 723), philosophers should work in the direction of completing the list of all objects involved in mathematical argumentations that are found in the practice. We have also shown that in low-dimensional topology, a proof can take the form of a sequence of pictures accompanied by instructions. The transformations of pictures are the result of permissible epistemic actions. Moreover, the choice of specific representations plays a pivotal role, because it triggers different cognitive skills and externalizes reasoning. In fact, the mathematical practice is characterized by a continuous feedback between specific forms of reasoning and particular representations.

In further research, we aim at comparing different practices from the point of view of the relation between the cognitive abilities triggered, the representations introduced and the argumentative form employed. Another development of the present project will be compiling a taxonomy of topological pictures, which would identify the specific features that are responsible for prompting manipulative imagination through different representational conventions. Moreover, we aim at developing the present framework in order to appreciate the effectiveness of hybrid notations and visual representations, with respect to their possible generalizations and their power to trigger specific kinds of imagination. As Giaquinto claims (Giaquinto 2007, p. 265): "Visual thinking in mathematics is extensive, diverse, familiar, and yet little understood. Here is abundant terrain for research." With the present study, we hope to have given an initial contribution.

Acknowledgements We presented an early version of this work at the first conference on *Mathematical Cultures*, funded by the *Arts and Humanities Research Council* and organized by Brendan Larvor, which took place in London in September 2012. We are thankful to the organizers and to the audience for giving us useful feedback. We thank Ken Manders, John Sullivan and the anonymous referees for their useful comments that helped us improve the article.

References

Adams, C.C., and A.W. Reid. 1993. Quasi-Fuchsian surfaces in hyperbolic knot complements. *Journal of the Australian Mathematical Society* 55: 116–131.
Alexander, J.W. 1923. A lemma on systems of knotted curves. *Proceedings of the National Academy of Science of the United States of America* 9(3): 93–95.
Barwise, J., and J. Etchemendy. 1996. Visual information and valid reasoning. In *Logical reasoning with diagrams*, ed. Allwein and Barwise, 3–25. New York: Oxford University Press.
Brown, J.R. 1999. *Philosophy of mathematics*. London/New York: Routledge.
Colyan, M. 2012. *An introduction to the philosophy of mathematics*. Cambridge: Cambridge University Press.
De Cruz, H., and J. De Smedt. 2013. Mathematical symbols as epistemic actions. *Synthese* 190: 3–19.
De Toffoli, S., and V. Giardino. 2014. Forms and role of diagrams in knot theory. *Erkenntnis* 79: 829–842.
Fomenko, A. 1997. *Algorithmic and computer methods for three-manifolds*. Dordrecht/Boston: Kluwer Academic.

Giaquinto, M. 2007. *Visual thinking in mathematics*. Oxford: Oxford University Press.
Heinzmann, G. 1999. Poincaré on understanding mathematics. In *Philosophia Scientiae, Travaux d'histoire et de philosophie des sciences: Actes du Colloque France-Autriche, Mai 1995: Interférences et transformations dans la philosophie française et autrichienne* (Mach, Poincaré, Duhem, Boltzmann), ed. Heinzmann, 43–60.
Jones, V.F.R. 1998. A credo of sorts. In *Truth in mathematics*, ed. Dales and Oliveri. Oxford: Oxford University Press.
Kirsh, D., and P. Maglio. 1994. On distinguishing epistemic from pragmatic action. *Cognitive Science* 18: 513–549.
Larvor, B. 2012. How to think about informal proofs. *Synthese* 187(2): 715–730.
Mancosu, P. (ed.). 2008. *The philosophy of mathematical practice*. Oxford: Oxford University Press.
Manders, K. 2008. The Euclidean diagram. In *The philosophy of mathematical practice*, ed. Mancosu. Oxford: Oxford University Press.
Massey, W.S. 1991. *A basic course in algebraic topology*. New York: Springer.
Morin, B., and J.P. Petit. 1980. Le retournement de la sphère. In *Les Progrès des Mathèmatiques*, 32–45. Paris/Berlin: Pour la Science.
Poincaré, H. 1889. La logique et l'intuition. *L'enseignement mathématique* 1:157–162, 5.
Poincaré, H. 1900. Second complément à l'Analysis Situs. *Proceedings of the London Mathematical Society* 32: 277–308.
Rolfsen, D. 1976. *Knots and links*. Berkeley: Publish or Perish.
Shin, S-J. 2004. Heterogeneous reasoning and its logic. *The Bulletin of Symbolic Logic* 10(1): 86–106.
Shin, S-J., O. Lemon, and J. Mumma. 2013. Diagrams. In *The Stanford encyclopedia of philosophy*, ed. Zalta, Fall 2013 edition. http://plato.stanford.edu/archives/fall2013/entries/diagrams/.
Sullivan, J.M. 2013. Diagrams and visualization in mathematics. In *Band 4: Sichtbarkeiten 4: Praktiken visuellen Denkens*, ed. von Dieter Mersch, Mira Fliescher, and Fabian Goppelsröder. Berlin: Diaphanes, 15–17 Nov 2012 in Berlin, 2013 forthcoming.
Thurston, W. 1994. On proof and progress in mathematics. *Bulletin of the American Mathematical Society* 30(2):161–177.

MIX
Papier aus verantwortungsvollen Quellen
Paper from responsible sources
FSC® C105338

If you have any concerns about our products,
you can contact us on
ProductSafety@springernature.com

In case Publisher is established outside the EU,
the EU authorized representative is:
**Springer Nature Customer Service Center GmbH
Europaplatz 3, 69115 Heidelberg, Germany**

Printed by Libri Plureos GmbH
in Hamburg, Germany